AF389206

District Heating and Cooling Networks

District Heating and Cooling Networks

Special Issue Editors

Antonio Colmenar Santos
David Borge Diez
Enrique Rosales Asensio

MDPI • Basel • Beijing • Wuhan • Barcelona • Belgrade • Manchester • Tokyo • Cluj • Tianjin

Special Issue Editors
Antonio Colmenar Santos
National University of Distance Education
Spain

David Borge Diez
University of Léon
Spain

Enrique Rosales Asensio
University of La Laguna
Spain

Editorial Office
MDPI
St. Alban-Anlage 66
4052 Basel, Switzerland

This is a reprint of articles from the Special Issue published online in the open access journal *Energies* (ISSN 1996-1073) (available at: https://www.mdpi.com/journal/energies/special_issues/ district_heating_cooling_networks).

For citation purposes, cite each article independently as indicated on the article page online and as indicated below:

LastName, A.A.; LastName, B.B.; LastName, C.C. Article Title. *Journal Name* **Year**, *Article Number*, Page Range.

ISBN 978-3-03928-839-7 (Pbk)
ISBN 978-3-03928-840-3 (PDF)

Contents

About the Special Issue Editors

Antonio Colmenar Santos is Senior Lecturer in the field of Electrical Engineering at the Department of Electrical, Electronic and Control Engineering at the National Distance Education University (UNED), a position he has held since his appointment in June 2014. Previously, Dr. Colmenar-Santos served as an Adjunct Lecturer at the Department of Electronic Technology at the University of Alcalá and at the Department of Electric, Electronic and Control Engineering at UNED. He has also worked as a consultant for the INTECNA project (Nicaragua). He has been part of the Spanish section of the International Solar Energy Society (ISES) and of the Association for the Advancement of Computing in Education (AACE), working in a number of projects related to renewable energies and multimedia systems applied to teaching. He was the coordinator of both the virtualisation and telematic services at ETSII-UNED, and Deputy Head Teacher and Head of the Department of Electrical, Electronics and Control Engineering at UNED. He is the author of more than 60 papers published in respected journals (http://goo.gl/YqvYLk) and has participated in more than 100 national and international conferences.

David Borge Diez holds a Ph.D. in Industrial Engineering and an M.Sc. in Industrial Engineering, both from the School of Industrial Engineering at the National Distance Education University (UNED). He is currently a Lecturer and Researcher at the Department of Electrical, Systems and Control Engineering at the University of León, Spain. He has been involved in many national and international research projects investigating energy efficiency and renewable energies. He has also worked in Spanish and international engineering companies in the field of energy efficiency and renewable energy for over eight years. He has authored more than 40 publications in international peer-reviewed research journals and participated in numerous international conferences

Enrique Rosales Asensio (Ph.D.) is an industrial engineer with postgraduate degrees in Electrical Engineering; Business Administration; and Quality, Health, Safety and Environment Management Systems. He has served as a Lecturer at the Department of Electrical, Systems and Control Engineering at the University of León, and Senior Researcher at the University of La Laguna, where he has been involved in a water desalination project in which the resulting surplus electricity and water would be sold. He has also worked as a plant engineer for a company that focuses on the design, development, and manufacture of waste heat-recovery technology for large reciprocating engines, and as a project manager in a world-leading research centre. He is currently Associate Professor at the Department of Electrical Engineering at the University of Las Palmas de Gran Canaria.

Preface to "District Heating and Cooling Networks"

Conventional thermal power generating plants reject a large amount of energy every year. If this rejected heat were to be used through district heating networks, given prior energy valorisation, there would be a noticeable decrease in the amount of fossil fuels imported for heating. As a consequence, benefits would be experienced in the form of an increase in energy efficiency, an improvement in energy security, and a minimisation of emitted greenhouse gases. Given that heat demand is not expected to decrease significantly in the medium term, district heating networks show the greatest potential for the development of cogeneration. Due to their cost competitiveness, flexibility in terms of the ability to use renewable energy resources (such as geothermal or solar thermal) and fossil fuels (more specifically the residual heat from combustion), and the fact that, in some cases, losses to a country/region's energy balance can be easily integrated into district heating networks (which would not be the case in a "fully electric" future), district heating (and cooling) networks and cogeneration could become a key element for a future with greater energy security, while being more sustainable, if appropriate measures were implemented. This book therefore seeks to propose an energy strategy for a number of cities/regions/countries by proposing appropriate measures supported by detailed case studies.

Antonio Colmenar Santos, David Borge Diez, Enrique Rosales Asensio
Special Issue Editors

Article

Energy Efficiency Analysis Carried Out by Installing District Heating on a University Campus. A Case Study in Spain

Ana M. Marina Domingo [1,*], **Javier M. Rey-Hernández** [1,2,*], **Julio F. San José Alonso** [1,*], **Raquel Mata Crespo** [3,*] and **Francisco J. Rey Martínez** [1,*]

1   Department of Energy and Fluid mechanics, School of Engineering (EII), University of Valladolid (UVa), 47011 Valladolid, Spain
2   Higher Polytechnic College, European University Miguel de Cervantes (UEMC), 47011 Valladolid, Spain
3   Department of Statistics, School of Engineering (EII), University of Valladolid (UVa), 47011 Valladolid, Spain
*   Correspondence: ana.mar.dom@hotmail.com (A.M.M.D.); javier.rey@uva.es (J.M.R.-H.); julsan@eii.uva.es (J.F.S.J.A.); raquel.mata@uva.es (R.M.C.); rey@eii.uva.cs (F.J.R.M.); Tel.: +34-983-423-685 (J.F.S.J.A.)

Received: 27 September 2018; Accepted: 16 October 2018; Published: 19 October 2018

Abstract: This article analyses the reduction of energy consumption following the installation of district heating (DH) in the Miguel Delibes campus at the University of Valladolid (Spain), in terms of historical consumption and climate variables data. In order to achieve this goal, consumption models are carried out for each building, enabling the comparison of actual data with those foreseen in the model. This paper shows the statistical method used to accept these models, selecting the most influential climate variables data obtained by the models from the consumption baselines in the buildings at the Miguel Delibes campus through to the linear regression equations with a confidence level of 95%. This study shows that the best variables correlated with consumption are the degree-days for 58% of buildings and the average temperature for the remaining 42%. The savings obtained to date with this third generation network have been significantly higher than the 21% average for 33% of the campus buildings. In the case of 17% of the buildings, there was a significant increase in consumption of 20%, and in the case of the remaining 50% of the buildings, no significant differences were found between consumption before and after installation of district heating.

Keywords: district heating; energy efficiency; baseline model; energy prediction; verification

---

1. Introduction

The building sector consumes more than a third of the world's energy and is responsible for 30% of all CO_2 emissions. These emissions were 9.0 Gt CO_2-eq in 2016 [1]. In order to reduce these emissions, the European Union (EU) has established the target for 2050 of reducing greenhouse gas emissions by 80% compared to 1990 levels [2]. The aim is to limit the increase in global temperature to 2 °C by 2050 [3]. This objective requires that emissions in 2030 compared to 2005, are limited or reduced in all developed country parties, but by different percentages, from 0% in Bulgaria to 40% in Luxembourg through to 26% in Spain [3,4].

The building sector in Spain has an approximate weight of 30% in final energy consumption, distributed at 18.5% in the residential building sector and 12.5% in the non-residential sector integrated by retail trade, services and public administration [5]. More than 65% of this consumption is used to supply heating needs, with 82% of the individual heating systems and the remaining 8% of central heating. The energy sources used mostly in heating are electricity (46%) and natural gas (32%) [6].

In front of individual and central heating systems for a single building, urban heat networks allow an easy change from fossil fuel to a renewable one, such as biomass, allowing the use of other renewable energies like: solar thermal, geothermal, urban solid waste and residual energy from other nearby processes. In addition, it can offer versatility to the energy system by cheaply storing thermal energy, for instance in hot water tanks, and reducing heating cost, especially in densely populated urban areas that have a concentrated heat demand. All of this makes it one of the best alternatives to improve the environmental behaviour of cities, as demonstrated in numerous European programs [7–9].

Therefore, it seems logical to assume that district heating (DH), since it is generated in large-scale power plants, will be more economical and efficient than heating generated in individual installations [10], as has been demonstrated in Seoul, Switzerland, Sweden, Poland, Denmark and Lucerne in Italy [11–16]. However, the energy efficiency of heating networks depends on a number of factors that can undermine the efficiency with which they are planned. Such factors include regulation, heat loss in distribution, or water leaks [17,18]. Along with these drawbacks, these systems must often compete with dominant technologies such as natural gas networks, as is the case in the United Kingdom and Latvia [19,20].

The first district energy system dates back to the 14th century [21]. Nowadays, four generations of heating networks are considered: the first generation (1880–1930) characterised by the use of steam as a thermal fluid, the second (1930–1980) in which steam was replaced by high temperature water channelled through concrete pipes, the third (1980–2020) based on the average water temperature in prefabricated pipes buried directly in the ground, and the fourth, the future generation (2020–2050), which will focus on low temperature distribution, supplying below 50 °C and return close to 20 °C or between 70 °C and 30 °C, using waste heat, municipal solid waste, renewable energies, and possibly combined with cogeneration plants and integrated into smart energy grids [21–31]. The system will be optimal for new buildings, constructed using near-Zero Energy Building (nZEB) guidelines and high energy efficiency standards [32,33].

According to ADHAC (Spanish Association of Heating and Cooling Networks), by the end of 2017 Europe accounted for 64.1% of the world's heating grids, which means more than 5000 grids, with more than 425 GW of power and more than 200,000 km of pipes laid. In the EU, district heating provides 9% of heating. The main fuel was gas (40%), followed by coal (29%) and biomass (16%) [34]. In Spain, district heating provides a non-representative percentage of the heating necessities; there are only 352 heating networks with 1280 MW of power installed, where more than 60% were concentrated between Madrid and Catalonia. In Castilla y León, a region located in the centre of the peninsula, there are 56 networks with a total installed power of 92.7 MW [35]. One of these grids, with a power of 14.1 MW, is that of the University of Valladolid (UVA). This network was built in 2015 to satisfy a heat demand of 22,000 MWh/year. It consists of two rings: one that connects the 12 buildings that make up the Miguel Delibes campus and the other that is connected by 11 buildings on the Esgueva campus together with four buildings of the regional authorities, making a total of 27 buildings, which offers the possibility of connecting more adjacent buildings.

The generation system consists of three 4.7 MW biomass boilers. The facilities with the 1800 m^3 storage silo (540 tons of wood shavings) have a constructed area of 1418 m^2. The wood chips are fed via a screw conveyor or a movable floor to the boilers.

This DH consists of 11,200 m of buried steel pipe, most of which is pre-insulated, with a diameter of between 32 and 350 mm. The system uses water as a thermal fluid at a maximum temperature of 109 °C, returning the boilers to temperatures above 60 °C. There are two 40,000 L backups. The design conditions are 90 °C/70 °C in the network primary and 80 °C/65 °C in the connected secondary building. The thermal difference considered for the calculation of the substations was 15 °C, between the exchanger of the substation and the circuit of each building, making it a third-generation network, built at a cost of five million euros. The aim is to avoid the production of 6800 tons of CO_2 per year and obtain an economic saving of 30%, together with an annual reduction in heating consumption of at least 15%. This paper focuses on the district heating part of the Miguel Delibes campus, which has 12 connected buildings. Table 1 shows the names, use of the building, installed power and heat exchange power of its installed facilities. These buildings have a capacity of 8.89 MW (Figure 1).

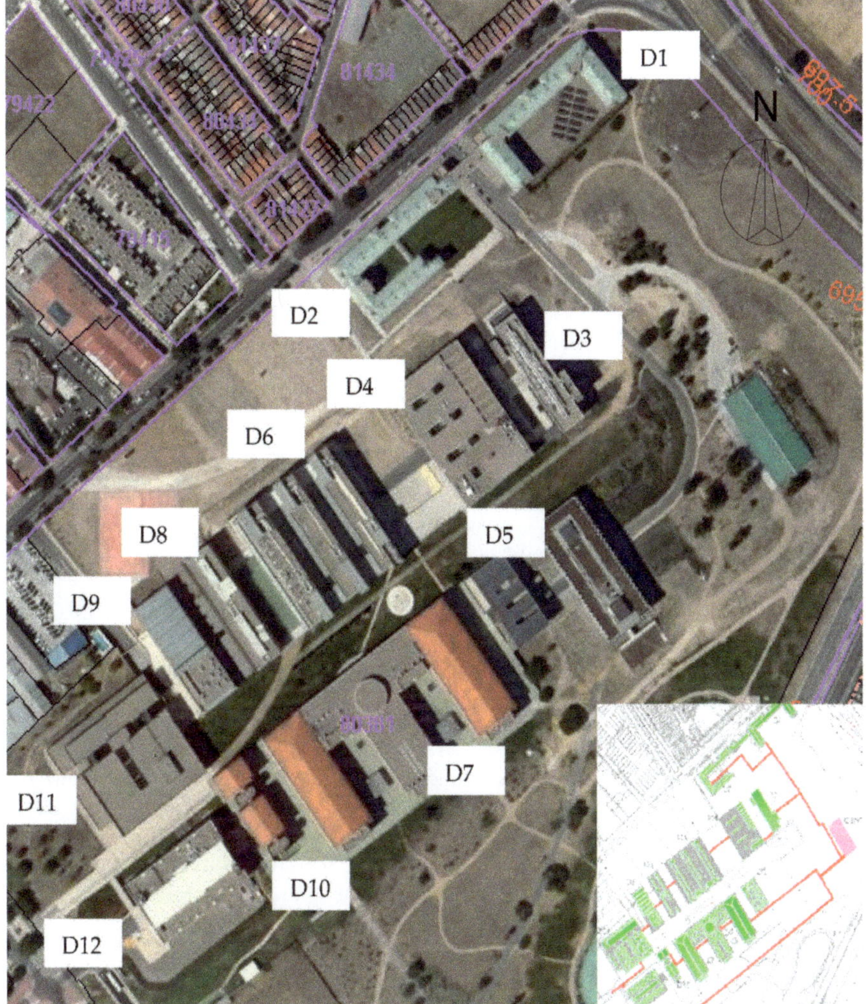

Figure 1. Miguel Delibes Campus. Buildings connected to district heating (DH).

All buildings were initially operated on natural gas. Depending on the heating program, three types of buildings are distinguished:

- Educational buildings, which use heating just weekdays from 6:00 a.m. to 10:00 p.m., and stop over the Christmas period from 24 December to 8 January.
- Residential buildings, which are working all week and use heating 24/7.
- Sports buildings, which are working the whole week with heating from 10 a.m. to 2 p.m. and from 4 p.m. to 10 p.m.

Heating is switched on for all the buildings from 15 October to 15 May every year.

The main objective is to model the energy consumption of district heating on the Miguel Delibes campus at the University of Valladolid and compare it with actual consumption to assess whether the energy savings proposed in the project had been carried out. In order to achieve this general objective, the following issues, in consecutive order, must be answered: determining the most influential variables in the heating consumption of buildings, by modeling the consumption of buildings on a baseline based on the variables indicated; and obtaining the expected consumption by modifying the value of the most influential variables.

Table 1. Buildings, installed thermal power and heat exchanger power in their facilities.

Ref.	Name of the Building	Type of Building	Thermal Installed Power (kW)	Heat Exchanger Substation Power (kW)	Total Built-Up Area (m²)	Outside Air Flow (L/s)
D3	CTTA Building (Centre for the Transfer of Applied Technologies)	Educational	348.0	342.0	5487	4600
D5	IOBA Building (University Institute of Applied Ophthalmology)	Educational	81.4	80.0	4146	3400
D10	Languages School	Educational	325.6	326.0	5636	4700
D1	Cardenal Mendoza University apartments	Residential	1554.4	1454.0	17,616	14,600
D2	Cardenal Mendoza University apartments (Library)	Educational	40.9	42.0	464	400
D4	Miguel Delibes classroom (Library)	Educational	860.0	1140.0	14,541	12,100
D6	Science School	Educational	1162.8	1120.0	19,137	15,900
D8	QUIFIMA building (Building of Fine Chemistry and Advanced Materials)	Educational	465.1	460.0	5610	4700
D9	Education School (Gymnasium)	Sports	507.0	504.0	3673	3000
D11	Education School Building	Educational	1000.0	1000.0	14,943	12,400
D12	R + D Building	Educational	802.3	802.0	7412	6200
D7	IT School	Educational	1953.5	1620.0	20,179	16,700
DELIBES	TOTAL 12 Buildings. Miguel Delibes Campus		9101.0	8890.0	118,843	98,700

2. Methodology

The method applied in this study is based on the statistical analysis of consumption before and after the installation of the network. To achieve this, the steps shown in Figure 2 were followed.

As in Sathayea, the study was based on the application of a system that examines the difference between consumption before and after the implementation of a specific project, constructing a baseline that represents the expected consumption if the project had not been carried out [36].

Below are the steps to follow in the research with the 12 buildings of the Miguel Delibes campus.

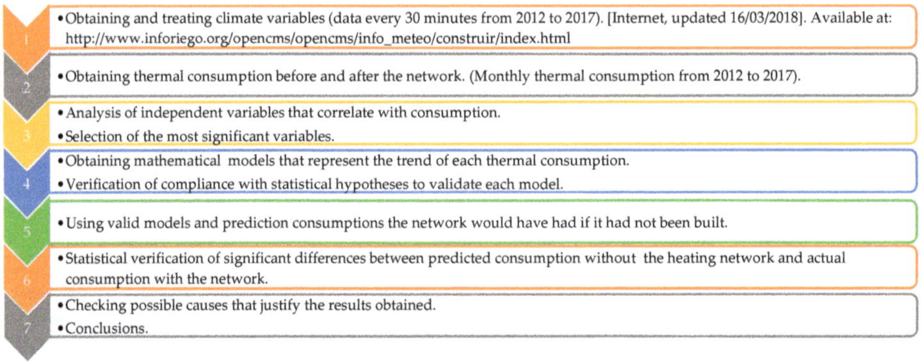

Figure 2. Methodology used in the study.

2.1. Obtaining and Processing Data on Climatic Variables

The variables are independent parameters to model the expected consumption in each building, and were obtained every 30 min over the last five years at a weather station located in Zamadueñas

(Valladolid), property of the Instituto Tecnológico Agrario de Castilla y León (Spain), and were related to the following variables: temperatures—average, average daytime, maximums and minimums.

- Degree-days: on 15 °C and 20 °C basis.
- Relative humidity: average, daily, maximums and minimums.
- Radiation: radiation intensity.
- Wind speed: average, daily, night-time and maximums.
- Wind path.
- Accumulated rainfall.
- Hours of sunlight.

The forecast of the expected demand generally depends on the outdoor temperature, when the buildings will be occupied and the indoor set point temperature. User habits and indoor temperatures were not included as independent variables in the study, since they hardly varied throughout the periods analysed.

In this paper, temperatures, humidity, velocities, wind trajectory and precipitations were processed to obtain monthly averages, maximums, minimums and accumulated. The results are shown in Figure 3.

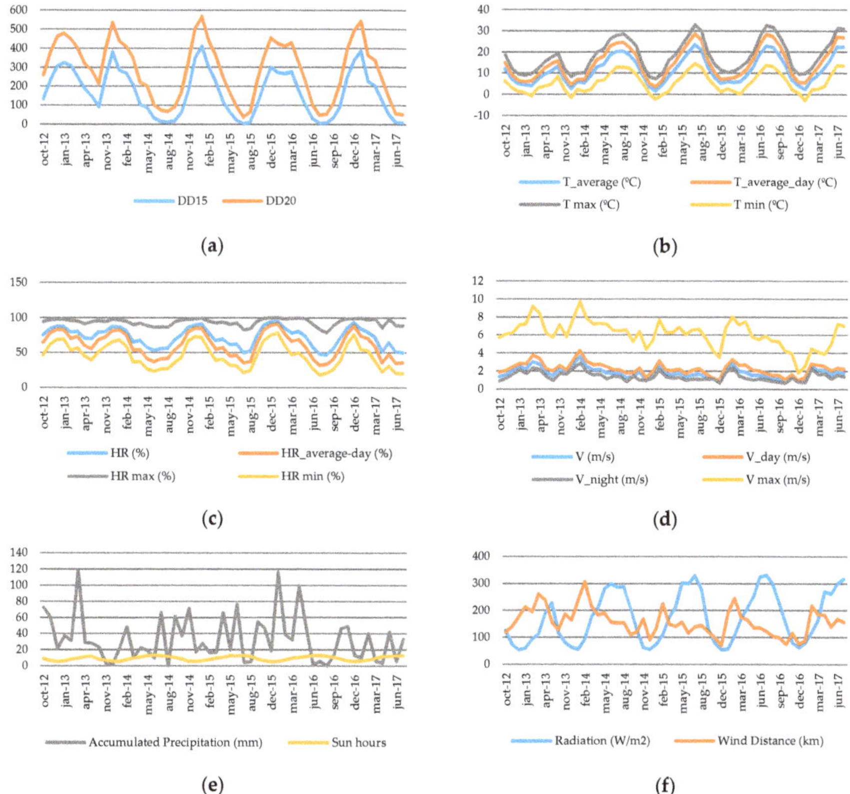

Figure 3. Climatological data used by the study variables. (**a**) DD, (**b**) Temperature, (**c**) Humidity, (**d**) Velocity, (**e**) Precipitation and sun hours, (**f**) Radiation and wind distance.

In the case of degree-days, as given by Equation (1):

$$DDBasemonth = \sum_{i=1}^{n}(Base - T_i) \tag{1}$$

where:

Base = 15 °C or 18 °C

T_i = Temperatures measured by period below 15 °C or 18 °C

n = Number of month periods

The degree-days are values that express accumulated temperature differences; they are calculated according to the UNE-EN ISO 15927-6: 2009 standard [37]. Its calculation is based on the concept of base temperature, from which the building needs to be heated. This variable has been used in numerous studies [38–42].

2.2. Obtaining the Heating Consumption before and after the District Heating Is Installed

Data on monthly heating consumption were collected between 2012 and 2017, corresponding to the 12 buildings on the Miguel Delibes campus. The district heating was built in 2015, so that the heating seasons from October 2012 to May 2013 and from October 2013 to May 2014 were considered the reference periods before the installation of the network, and the seasons 2015–2016 and 2016–2017 the periods after its installation.

Following option C of the IPMVP (International Performance Measurement and Verification Protocol) [43], corresponding to verification of saving with statistical adjustment of the entire installation, these consumption data were taken from energy invoices and from the counters available in the boiler rooms of thermal power stations. The results obtained are shown in Figure 4.

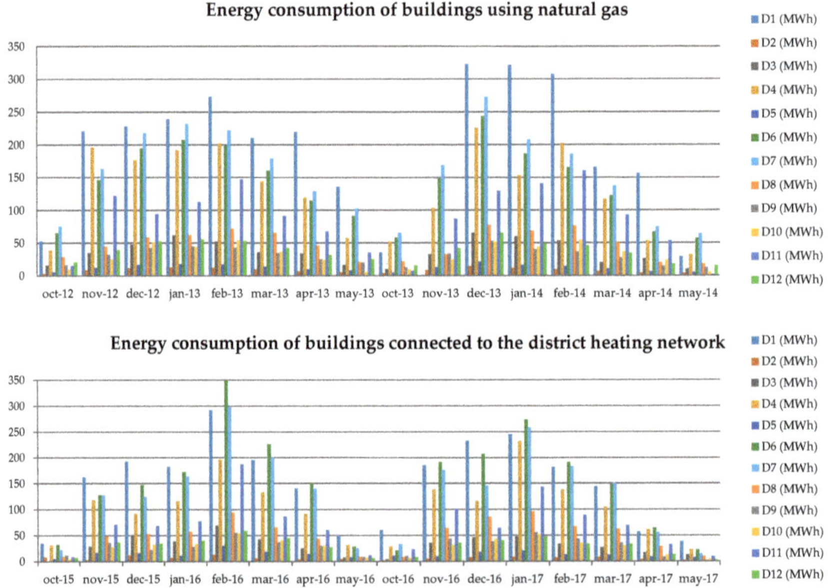

Figure 4. Heat consumption data.

The total consumption of the two campaigns prior to the start-up of the network was 14,286,109 kWh, compared to 12,558,748 kWh in the two campaigns subsequent to the installation of the heating network.

The season from October 2014 to May 2015 is considered to be the period of the first start-up of the district heating and the data has not been analysed. Figure 5 shows the total monthly consumption profile analysed. This is the usual profile of heating demand in the city of Valladolid, where the months with the highest demand are from November to March.

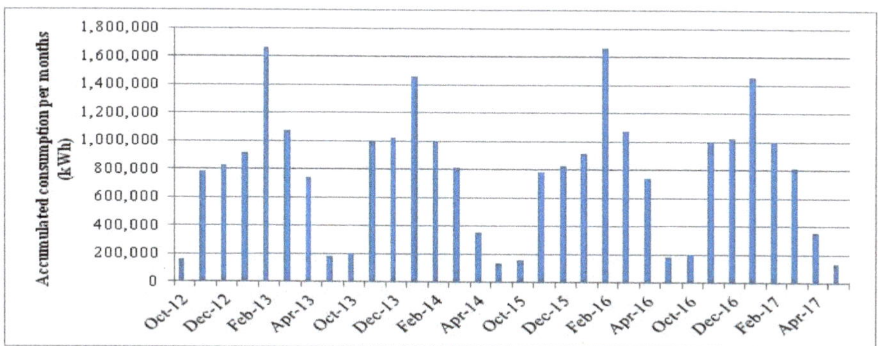

Figure 5. Total heat consumption of Miguel Delibes campus buildings during the reference and study period.

2.3. Statistical Analysis of Variables Correlated to Consumption

The statistical study was performed using SPSS software [44], and statistical inference techniques were used throughout the process, establishing a 95% trust level.

A first step was to determine the independent climatic variables for each building and with specific weight in the regression analysis, the dependent variable being the consumption of each building during the period from October 2012 to May 2014. Using the stepwise method, the independent variable is chosen which, in addition to meeting the highest input tolerance (its significance level is ≤ 0.05), correlates in absolute value with the dependent variable (has the highest absolute value of the partial correlation). The independent variable is then chosen which, in addition to meeting the input tolerance, has the next highest partial correlation coefficient (in absolute value). Each time a new variable is included in the model, the previously selected variables are re-evaluated to determine whether they still meet the output tolerance (with the lowest regression coefficient in absolute value, level of significance ≥ 0.1). If a chosen variable meets the output tolerance, it is eliminated from the model, since the regression or elimination is already explained by the rest of the variables and lacks a specific contribution of its own. The process stops when there are no variables that meet the input tolerance and the variables chosen do not meet the output tolerance [45].

The D3 building model has been built in a single step (Table 2) by entering variable GD15 with $t = 8.851$, a partial correlation of 0.921 and a level of a significance (Sig.) = 0.000 (≤ 0.05). As the remaining variables do not meet the tolerance input of Sig. ≤ 0.05, no more variables could be introduced into the model.

- The statistic t and its meaning (Sig.) are used to check that the regression coefficient equals zero in the model. Sig. > 0.05 implies that the slope of the independent variables in the regression model is equal to zero, and does not meet the input tolerance in the model.
- Partial correlation studies the relation between two quantitative variables by controlling for or eliminating the effect of third variables in the linear regression model. The higher the absolute value, the greater the relation between the dependent variable and the independent variable.
- Tolerance is a collinear statistic that looks for a relation between independent variables. If the tolerance is less than 0.1, there is a high degree of collinearity and the variable must be removed from the model.

Table 2. Inputs and deleted variables in the model D3 Building.

Variables Entered					
Model		**t**	**Sig.**	**Partial Correlation**	**Collinearity Statistics**
					Tolerance
1	DD15	8.851	0.000	0.921	1.000
Variables Removed					
Model		**t**	**Sig.**	**Partial Correlation**	**Collinearity Statistics**
					Tolerance
	DD20	−1.245	0.235	−0.326	0.004
	T_average	0.006	0.995	0.002	0.017
	T_average_day	−0.712	0.489	−0.194	0.033
	T_max	−0.951	0.359	−0.255	0.050
	T_min	0.872	0.399	0.235	0.075
	RH (Relative Humidity)	0.797	0.440	0.216	0.419
	RH_average_day	0.878	0.396	0.237	0.374
	RH_max	0.527	0.607	0.145	0.573
1	RH_min	0.992	0.339	0.265	0.346
	Radiation	−0.230	0.822	−0.064	0.392
	V	1.493	0.159	0.383	0.890
	V_day	1.325	0.208	0.345	0.958
	V_night	1.705	0.112	0.428	0.770
	V_max	1.097	0.293	0.291	0.985
	Wind_distance	1.493	0.159	0.383	0.890
	Accumulated_Precipitation	−0.032	0.975	−0.009	0.996
	Sun_hours	−0.512	0.618	−0.140	0.417

Dependent variable: kWh_D3, Predictors: DD15.

2.4. Obtaining Regression Models

The objective is to find some regression models that represent the consumption trends of each building, verifying the statistical hypotheses of the simple and multiple linear regression. There are a great number of studies that also use this kind of model [46–51].

In one-variable models, simple linear regression is (2):

$$kWh = c + \beta 1 \times \text{Variable} \tag{2}$$

For multivariable or multiple regression models that contain more than one or regression, the equation is (3):

$$kWh = \beta 0 + \beta 1 \times \text{Variable1} + \beta 2 \times \text{Variable2} \tag{3}$$

Once the regression model for predicting consumption has been obtained, the hypotheses of the model should be tested:

- Linearity of the variables.
- Normality of variables residues using the Shapiro-Wilk test for small samples.
- Independence of the residues using the Durbin-Watson statistic.
- Homogeneity of variance, checking the absence of correlation between residues, predictions and independent variables. The multiple linear regression models also prove this.
- Lack of multicollinearity in independent variables, analysing condition indeces, according to collinearity diagnoses.

An example is given below, showing compliance of the assumptions for the simple linear regression model for building D3, which is (4):

$$kWh\_D3 = -6854.944 + 192.51 \, GD15 \tag{4}$$

Table 3 shows the slope (B) obtained a value of Sig. = 0.000, which indicates that the null hypothesis that the slope is equal to zero is rejected and evidences the linearity between the dependent variable (kWhD3) and the independent variable (GD15). The positive value of the slope indicates a direct relation between consumption and GD15.

The statistics of the Shapiro-Wilk test for small sizes ($n < 30$) and the statistics of the residuals show a value of Sig. > 0.05 (Table 4), which allow us to accept the null hypothesis of the normality of variables.

Table 5 shows the Durbin-Watson statistic to determine the presence of autocorrelation between the residual corresponding to each observation and the previous one. According to Savin and White [52], for a sample size of 16 observations if the test statistic is greater 1.37092, there is no correlation.

Table 3. Compliance with linearity assumption and coefficients of the simple linear regression model. D3 building.

Model		B	t	Sig.
1	(Constant)	−6854.944	−1.324	0.207
	DD15	192.510	8.851	0.000

Table 4. Compliance with normality assumption (Shapiro-Wilk).

Variables and Residuals	Shapiro-Wilk		
	Statistics	df	Sig.
kWh_D3	0.931	16	0.251
DD15	0.953	16	0.541
Unstandardized Residual	0.945	16	0.414
Standardized Residual	0.945	16	0.414

Table 5. Compliance with the assumption of no autocorrelation.

Model Summary [b]				
Model	R	R Square	Adjusted R Square	Durbin-Watson
1	0.921 [a]	0.848	0.838	2.559

[a] Predictors: (Constant), DD15; [b] dependent variable: kWh_D3.

R: Pearson linear correlation coefficient measures the degree of linear relations between variables. Values of $R > 0$ indicate a direct linear relation between variables. Values of $R < 0$ indicate an inverse linear relation between variables. Values close to the unit indicate almost perfect correlations, whereas values close to zero indicate the variables are not correlated.

R^2: the linear determination coefficient measures the part of the variation of the dependent variable that can be explained by variations of the independent variables.

R^2 adjusted: linear determination coefficient over the number of independent variables included in the model and the sample size. It is used to compare regressions of the same sample size but with different number of regressors. Reduces the coefficient for very small samples with many independent variables (5).

$$R^2 \text{ adjusted} = 1 - [(N - 1)(1 - R^2)/(N - k - 1)] \tag{5}$$

where: N is the sample size and k the number of regressors

In order to check homoscedasticity, the linear determination coefficient (R^2) between residuals and predictions ($R^2 = 0$) and between residuals and the independent variable ($R^2 = 3.33 \times 10^{-16}$) is calculated. As shown in Figure 6, these are close to zero.

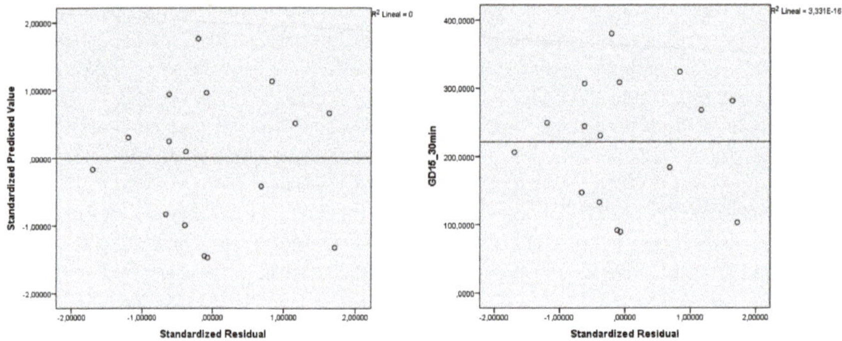

Figure 6. Compliance of homoscedasticity of residuals. D3 building.

The scatter plot is a totally random point cloud, which shows neither trends nor patterns in the graphical representation. Consequently, the hypothesis of linearity and homoscedasticity is accepted.

For building D1, the supposed lack of multiple collinearity in the independent variable was also tested in the multiple linear regression model, represented by the Equation (6):

$$kWh_{D1} = 271{,}370.906 - 20{,}045.184 \, T\_average + 50{,}568.513 \, V\_night \tag{6}$$

This model corresponds to model 2, shown in Table 6.

If two independent variables are closely correlated with each other and included in the model, certainly neither is likely to be statistically significant. However, if only one of them is included, it could prove to be statistically significant. To assess whether the model becomes unstable when a new variable is introduced, collinearity indices are evaluated. Following the studies of Belsley, Kuh, and Welsch, both with observed and simulated data, the problem of multicollinearity is severe when the condition index takes a value between 20 and 30 [53].

Table 7 shows the condition index for model 2, which is multiple linear regression, is 12.192, below 20. In addition, the tolerance shown in Table 6 takes a value of 0.738, close to the unit (the higher the tolerance, the lower the collinearity), so that it can be deduced that there is no multiple collinearity between the two independent variables.

Table 6. Linear regression models. D1 building.

		Coefficients [a]				
		Non-Standardized Coefficients				**Collinearity Statistics**
	Model		t	**Sig.**		
		B				**Tolerance**
1	(Constant)	391,037.854	13.525	0.000		
	$T\_average$	−23,838.939	−7.408	0.000		1.000
2	(Constant)	271,370.906	4.659	0.001		
	$T\_average$	−20,045.184	−6.154	0.000		0.738
	V_night	50,568.513	2.278	0.042		0.738

[a] Dependent Variable: kWh_D1.

Table 7. Verification of the assumption of lack of multicollinearity between variables. D1 building.

	Collinearity Diagnostics [a]	
Model	**Dimension**	**Condition Index**
1	1	1.000
	2	4.823
2	1	1.000
	2	3.915
	3	12.192

[a] Dependent Variable: kWh_D1.

2.5. Prediction of Expected Consumption

For the periods following construction of the network, using regression models and climate variables for subsequent periods, consumption without the district heating was predicted, and the accumulated consumption was calculated. This consumption was compared to the actual accumulated consumption. Table 8 shows the current consumption of building D3, corresponding to the period of the district heating from November 2015 to May 2017 and the values foreseen for the same period using the linear regression equation shown in Table 3.

Table 8. Current and predicted consumption for building D3, from November 2015 to May 2017.

Date	DD15	Real Value (kWh)	Predicted Values (kWh)
15 November	208.36	28,683	33,257
15 December	300.09	50,191	50,916
16 January	271.93	38,231	45,495
16 February	266.36	68,430	44,423
16 March	276.34	41,775	46,343
16 April	181.04	24,655	27,998
16 May	99.34	7268	12,268
16 October	97.34	3,711	11,883
16 November	245.69	35,452	40,443
16 December	337.25	45,615	58,069
17 January	386.98	47,454	67,643
17 February	226.76	33,104	36,799
17 March	203.17	26,179	32,257
17 April	129.72	16,672	18,118
17 May	54.86	12,122	3706
Total		479,543	529,618
Mean		31,969.4667	35,307.8667

It can be seen in Table 8 and Figure 7 how the 15-month average of the actual consumption is 31,969 kWh during the two seasons following the installation of the district heating, while the 15-month average of expected consumption for these seasons if installation had not been built, was 35,307 kWh, 3338 kWh higher than actual consumption, representing a saving of 9.5%. However, as will be seen below, the priority saving is in fact not statistically significant.

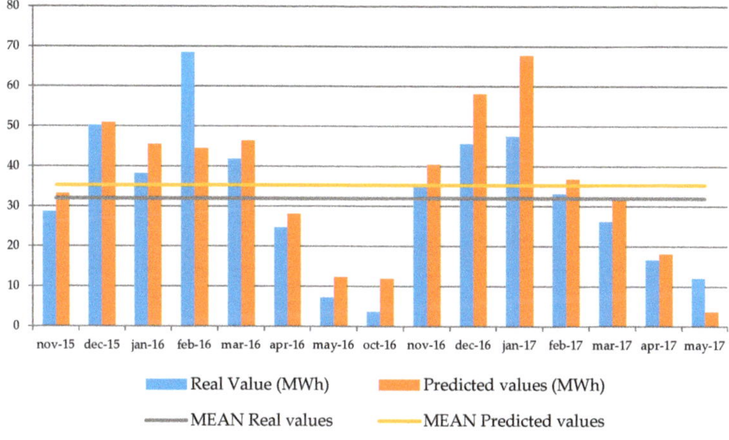

Figure 7. Mean real values and mean predicted values of consumption D3 building.

2.6. Statistical Verification of Significant Differences

An analysis was carried out to ascertain whether the difference between the predicted consumption, had the network not been built, and the actual consumption after it had been built, was statistically significant, at a confidence level of 95%.

The t-Student test was used in the study for related samples, which is considered particularly suited to compare the means of two groups when there is some relation between the individuals in the two groups. In this study, the relation was that consumption was associated to the same facility but during different periods of time.

Therefore, if the variables are distributed normally and the statistical significance is 0.05, it can be said that there are some significant differences. On the contrary, the null hypothesis that the two means are equal is not rejected, and the differences found are not considered statistically significant and do not go beyond what could be expected at random [54]. All of this assumes accepting a 5% error risk or, put differently, a confidence level of 95%.

The following shows how the differences found between the actual and predicted consumption of building D3 are not significant.

In line with the Shapiro–Wilk normality test, (Table 9) both the variables that represent actual consumption and those representing predicted consumption are distributed according to a normal distribution as a result of Sig. > 0.05, such that the null hypothesis of normality is accepted.

Table 10 shows that the forecast average consumption is 35,308 kWh, which compares with the actual average consumption of 31,969 kWh, which may lead one to believe that there is a difference of 3338 kWh between the averages.

Table 11 shows that the difference found is not significant (value of Sig. > 0.05), such that we accept the null hypothesis of equal means. However, we are not in a position to say whether or not the difference found is due to more than mere chance.

Table 9. Test of normality for real and predicted consumption D3 building.

kWh	Kolmogorov-Smirnov [a]			Shapiro-Wilk		
	Statistic	df	Sig.	Statistic	df	Sig.
kWh_real	0.085	15	0.200 *	0.980	15	0.969
kWh_predicted	0.100	15	0.200 *	0.976	15	0.933

* This is a lower bound of the true significance; [a] Lilliefors significance correction.

Table 10. Actual and predicted average consumption D3 building.

	Paired Samples Statistics				
	kWh	Mean	N	Std. Deviation	Std Error Mean
Par 1	kWh_predicted	35,307.8667	15	18,097.26058	4672.69259
	kWh_real	31,969.4667	15	17,668.79051	4562.06209

Table 11. Paired samples test D3 building.

	Paired Samples Test					
kWh	Paired Differences			t	df	Sig. (2-Tailedl)
	Mean	Std. Deviation	Std. Error Mean			
kWh_predicted—kWh_real	3338.40000	9703.90931	2505.53861	1.332	14	0.204

2.7. Exploring the Possible Causes to Explain the Results Obtained

As a final step in the study, an analysis was carried out of the possible causes that might justify the statistical results of the existence or other wise of significant differences before and after the district heating works.

3. Results

The regression models found, which allow the consumption of each building to be explained in terms of the explanatory independent variables, are shown in Table 12. For building D1 (Apartments), a multiple regression model was found that predicted expected consumption with greater correlation. Table 12 includes the independent terms and the slopes of the regression variables (β), the Pearson linear correlation coefficients (R), and the linear determination coefficients between the variables (R^2).

In seven of the 12 buildings, in other words 58.3% of the campus buildings, the explanatory variable found was degree-day on a 15 °C basis; the remaining 41.7% correlated better with mean temperature, and in building D1, in the regression multiple model, a new variable is introduced: nocturnal wind speed.

The absolute correlation value between the independent variables was between 0.896 and 0.999, and was positive for the models explained with the grade-day variable on a basis of 15 °C and negative for the mean temperature variables. The determination coefficient was between 0.802 and 0.998, so, accepting a risk error of 5%, the models found a predicted expected consumption with an accuracy probability greater than 80%., Figure 8 shows the linear regression model for building D3 (CTTA), displaying the consumption in kWh of the study period when the district heating was operating, and the straight line of consumption obtained with the explanatory variable.

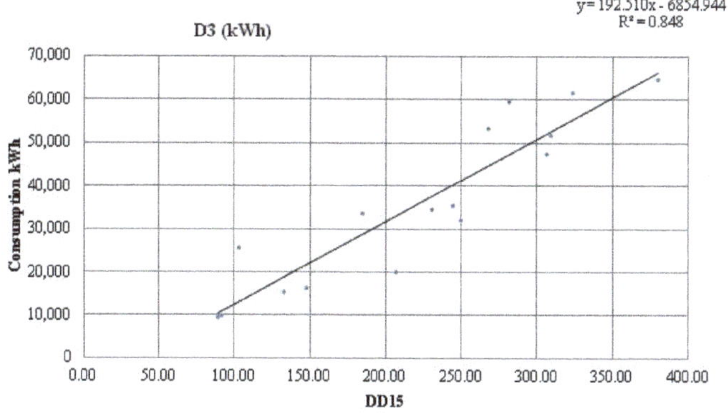

Figure 8. Linear regression model. D3 building (Centre for the Transfer of Applied Technologies (CTTA)).

From the baselines of the regression models found and the climatic variables for the 2015–2017 periods, the expected or predicted consumption was obtained and compared with the actual consumption obtained in those periods. Figure 9 shows the results of actual consumption and forecast consumption (dashed line) for the periods following the installation of the district heating in several of the buildings on the Miguel Delibes campus. Building D3 (CTTA), where hardly any variation is observed, building D5 (University Institute of Applied Ophthalmology (IOBA)), where an increase in real consumption is observed with respect to the forecast, and building D1 (Apartments), where energy savings are observed.

Table 12. Regression models.

Ref.	Simple Model					Multivariable Model						
	Variable	Constant	β	R	R^2	Variable_1	Variable_2	Constant	$\beta 1$	$\beta 2$	R	R^2
D3	DD15	−6854.944	192.510	0.921	0.848							
D5	DD15	−653.894	56.535	0.993	0.986							
D10	T_average	66,791.063	−4393.782	−0.896	0.802							
D1	T_average	391,037.854	−23,838.939	−0.899	0.808	T_average	V_night	271,370.906	−20,045.184	50,568.513	−0.931	0.866
D2	DD15	−663.338	38.971	0.985	0.970							
D4	T_average	271,046.123	−17,453.026	−0.940	0.884							
D6	DD15	−6540.157	656.818	0.995	0.989							
D8	T_average	91,484.52	−5414.100	−0.906	0.821							
D9	DD15	−654.602	139.864	0.999	0.997							
D11	T_average	187,840.052	−12,648.948	−0.898	0.807							
D12	DD15	−663.549	173.214	0.999	0.998							
D7	DD15	−6543.072	733.335	0.996	0.991							

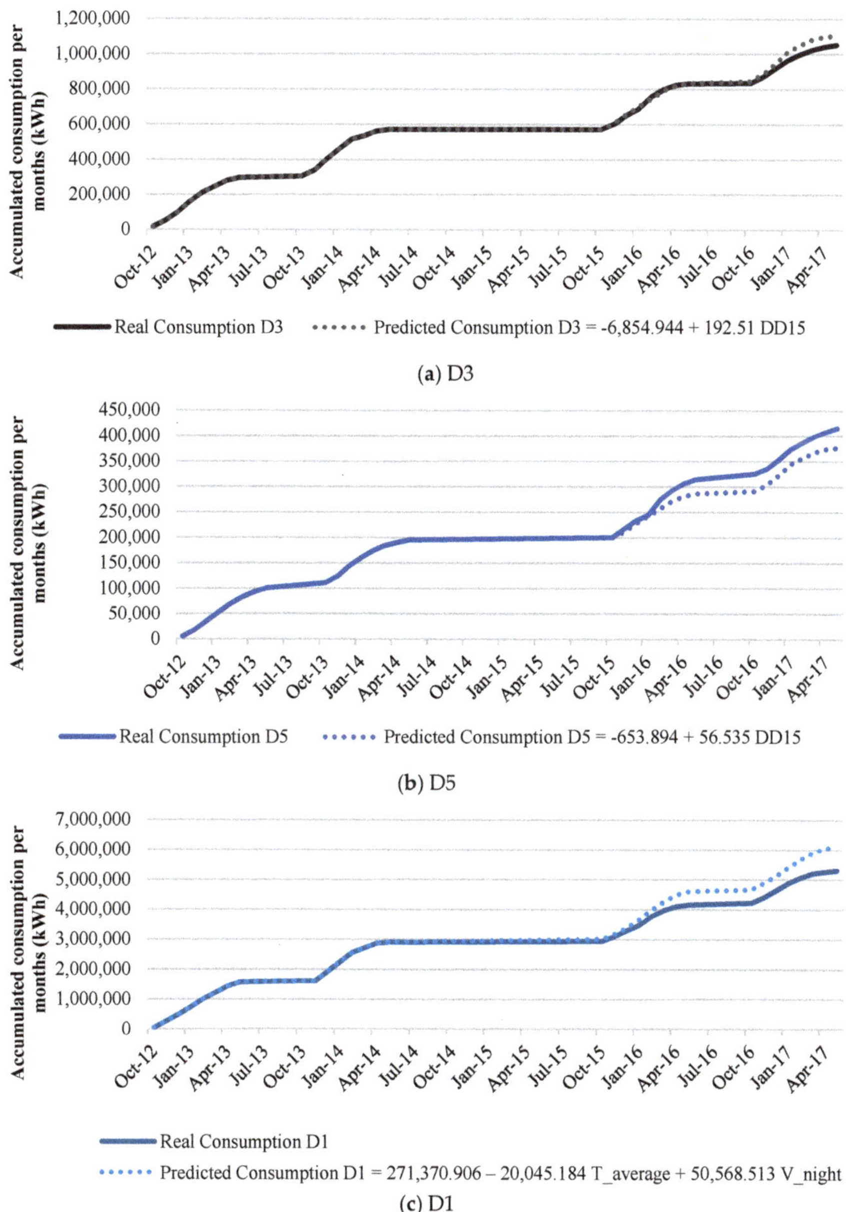

(a) D3

(b) D5

(c) D1

Figure 9. Graphs of real and predicted consumption. D3, D5 and D1 buildings.

The differences found for each building and each of the facilities are shown in Tables 13 and 14. The differences that turned out to be statistically significant are shaded. The savings that appear as negative are the increases in consumption.

Table 13. Simple linear regression model.

Ref.	Independent Variables	% Savings 2016–2017	Significant	% Savings 2016	Significant	% Savings 2017	Significant
D3	DD15	9.5%	NO	0.6%	NO	18.1%	NO
D5	DD15	−16.1%	YES	−20.1%	NO	−12.0%	NO
D10	$T\_average$	5.2%	NO	5.8%	NO	4.5%	NO
D1	$T\_average$	20.0%	YES	18.5%	YES	21.6%	YES
D2	DD15	30.6%	YES	27.6%	NO	34.4%	YES
D4	$T\_average$	15.7%	YES	19.4%	NO	11.7%	NO
D6	DD15	−11.1%	NO	−16.0%	NO	−6.4%	NO
D8	$T\_average$	−11.9%	NO	−1.8%	NO	−23.0%	YES
D9	DD15	0.9%	NO	3.7%	NO	−1.9%	NO
D11	$T\_average$	13.6%	NO	12.9%	NO	14.4%	NO
D12	DD15	18.9%	YES	12.2%	NO	25.7%	YES
D7	DD15	10.9%	NO	7.8%	NO	13.9%	NO

Table 14. Multiple regression model.

Ref.	Independent Variable 1	Independent Variable 2	% Savings 2016–2017	Significant	% Savings 2016	Significant	% Savings 2017	Significant
D1	$T\_average$	V_night	24.8%	YES	26.5%	YES	22.8%	YES

The significant savings differences in the simple linear regression analysis appear in buildings D1 (Apartments), D2 (Apartment Library), D4 (Teaching Block Library) and D12 (Research and Development (R&D) building), obtaining average savings of 21.3% (20.0% in D1, 30.6% in D2, 15.7% in D4, and 18.9% in D12). When applying the multiple regression model to the D1 building, significant savings in consumption continue to emerge, going from 20% in the simple model to 24.8% in the multiple model.

In two buildings: D5 (IOBA) and D8 (Building of Fine Chemistry and Advanced Materials (QUIFIMA)), statistically significant increases in consumption are observed; 16.1% for the 2016 season and 2017 in D5 and 23% for D8, although only in the 2017 season, reflecting an average increase in consumption of 20% for these buildings.

For the other six buildings, the differences found are not statistically significant.

Buildings D1, D2, D4 and D12 are buildings that are not used for teaching, so the time of use is extended to seven days a week and even to holiday periods in some cases.

4. Discussion

All the buildings studied used natural gas boilers as an initial system and in 50% of them no statistically significant differences were found between consumption before and after the installation of the heating network. Although this analysis did not reveal significant energy savings in all buildings, what is undeniable are the economic savings and emission savings achieved by substituting natural gas by biomass. This study analyses the viability of district heating in terms of predicting the expected heat demand, using as independent variables only the climatic data of the moment, and with a risk error equal to or less than 5%.

The results for the two heating seasons obtained to develop the baseline models (16 months) indicate that the regression mainly uses degree-days (58.3%) and mean temperature (41.7%) to establish its model, without taking into consideration the remaining variables such as: relative humidity, radiation, rainfall or wind speed. According to Granderson [55], a reference period of over 12 months does not guarantee a lower error.

The appropriateness of using district heating could be discussed depending on the initial system in operation to meet demand. As reflected in Ulseth's work [56], this information is important in order to plan a district and conducting the financial feasibility study without making too many mistakes. In Norway, the use of heating networks to heat houses with heat pumps that were also used to meet the demand for domestic hot water was questioned. Other studies have even assessed the feasibility of using electricity as a source for a municipal district heating system with low local emissions, although this would clearly not be economically feasible given the current price of electricity [57].

The next step being taken in the UVA network is the use of solar thermal energy for domestic hot water (DHW), as Winterscheid presented in his studies [58]. This would certainly improve the energy efficiency of the grid and would also be the logical step for the gradual conversion to a fourth generation grid, as Pavicevic explained in his work on a district heating system in Zagreb [59].

Given the layout of university and regional administration buildings that are close to the grid, but not yet connected, and that there are heating systems with different thermal jumps, new buildings can be cascaded, rather than parallel, as in the case of the proposal presented by Mertz [60]. Buildings that require lower temperatures (buildings with all-air systems) could be connected to the exit of buildings that require higher temperatures (hospitals or clinics).

5. Conclusions

The case study provides an example of how district heating by biomass can improve a city's environmental performance. The level of energy efficiency can still clearly be improved, since the savings obtained to date in the district heating system of the Miguel Delibes campus, where 12 buildings have been evaluated, is higher than 21% in 33% of the buildings. Overall, the district heating has achieved a significant reduction in CO_2 emissions (6800 tons of CO_2) according to the UVA, having changed natural gas for biomass (wood chips), also obtaining a significant economic saving of more than 30%.

The UVA district heating has achieved a reduction in installed power, going from 59 natural gas boilers with a total installed power of 27.4 MW to three boilers of 4.7 MW each, representing a total power of 14.1 MW.

A consumption model has been obtained for the Miguel Delibes university campus, which consists of 12 buildings and has an initial installed power of 9 MW, with a risk of error of 5%, and which has exceeded all statistical requirements to validate this type of model.

The response variable used to generate the models was natural gas consumption data for the 12 buildings from October 2012 to May 2014 and the climatic conditions in the area were used as an explanatory variable.

In 58% of the buildings, the variable that best correlates in the model found for the baseline was the grade-day, while for the remaining 42% it was the mean temperature. Variables such as relative humidity, rain, radiation or wind speed were not significant in the simple linear regression models found.

The absolute value of the correlation between the independent variables was between 0.896 and 0.99, and was positive for the models with the base explanatory variable of 15 degree-days and negative for the models with the mean temperature explanatory variable.

For 33% of the campus buildings, the savings obtained to date with this third generation is significant and above the 21% average. For 17% of the campus buildings, there is a significant increase in consumption, with an estimated average of around 20%. For the remaining 50% of the buildings, no significant differences were found in consumption before and after the installation of district heating.

Author Contributions: J.F.S.J.A. and F.J.R.M. conceived and designed the experiments; A.M.M.D. performed the experiment and analyzed the data; J.M.R.-H. wrote the paper and R.M.C. contributed simulation tools.

Funding: This research received no external funding.

Acknowledgments: The work presented in this article has been made possible thanks to the support of the University of Valladolid (UVA), the cooperation of the Public Infrastructures and Environment Company at the Regional Government of Castilla y León (SOMACYL), Institute of Advanced Production Technologies (ITAP) and RETO GIRTER Project (New Intelligent Manager for Thermal Networks), project funded by the European Regional Development Fund through the "2016 RETOS COLABORACIÓN" Program of the Ministry of Economy, Industry and Competitiveness of the Government of Spain.

Conflicts of Interest: The authors declare no conflict of interest.

References

1. International Energy Agency (IEA). *Market Report Series: Energy Efficiency 2017. Analysis and Forecasts to 2022*; IEA: Paris, France, 2017.
2. Communication from the Commission to the European Parliament, the Council, the European Economic and Social Committee and the Committee of the Regions. Energy Roadmap 2050. Available online: https://eur-lex.europa.eu/legal-content/EN/TXT/?uri=celex%3A52007DC0575 (accessed on 18 October 2018).
3. IPCC. Climate Change 2014: Synthesis Report. Contribution of Working Groups I, II and III to the Fifth Assessment Report of the Intergovernmental Panel on Climate Change. Available online: http://www.ipcc.ch/pdf/assessment-report/ar5/syr/SYR_AR5_FINAL_full_wcover.pdf (accessed on 18 October 2018).
4. *Effort Sharing Regulation, 2021–2030. Limiting Member States' Carbon Emissions*; European Parliament: Bruxelles, Belgium. Available online: http://www.europarl.europa.eu/thinktank/en/document.html?reference=EPRS_BRI(2016)589799 (accessed on 18 October 2018).
5. Ministerio de Fomento, Gobierno de España. *Actualización De la Estrategia a Largo Plazo para la Rehabilitación Energética en el Sector de la Edificación en España (ERESEE 2017)*; Ministerio de Fomento: Madrid, Spain, 2017.
6. IDAE. *PROYECTO SECH-SPAHOUSEC Análisis del Consumo Energético del Sector Residencial en España*; IDAE: Madrid, Spain, 2011.
7. Directive 2009/28/EC of the European Parliament and of the Council of 23 April 2009 on the Promotion of the Use of Energy from Renewable Sources and Amending and Subsequently Repealing Directives 2001/77/EC and 2003/30/EC. Available online: http://data.europa.eu/eli/dir/2009/28/oj (accessed on 18 October 2018).
8. United Nations Environment Programme (UNEP). District Energy in Cities: Unlocking the Potential of Energy Efficiency and Renewable Energy. Available online: http://hdl.handle.net/20.500.11822/9317 (accessed on 18 October 2018).
9. The European Parliament and the Council of the European Union DIRECTIVE (EU) 2018/844 of the European Parliament and of the Council of 30 May 2018 Amending Directive 2010/31/EU on the Energy Performance of Buildings and Directive 2012/27/EU on Energy Efficiency. *Off. J. Eur. Union* **2018**. Available online: https://eur-lex.europa.eu/legal-content/EN/TXT/PDF/?uri=CELEX:32018L0844&from=IT (accessed on 18 October 2018).
10. Van Deventer, J.; Derhamy, H.; Atta, K.; Delsing, J. Service Oriented Architecture enabling the 4th Generation of District Heating. *Energy Procedia* **2017**, *116*, 500–509. [CrossRef]
11. Lee, J.S.; Kim, H.C.; Im, S.Y. Comparative Analysis between District Heating and Geothermal Heat Pump System. *Energy Procedia* **2017**, *116*, 403–406. [CrossRef]
12. Ericsson, K.; Werner, S. The introduction and expansion of biomass use in Swedish district heating systems. *Biomass Bioenergy* **2016**, *94*, 57–65. [CrossRef]
13. Sandvall, A.F.; Ahlgren, E.O.; Ekvall, T. Cost-efficiency of urban heating strategies—Modelling scale effects of low-energy building heat supply. *Energy Strateg. Rev.* **2017**, *18*, 212–223. [CrossRef]
14. Lund, H.; Möller, B.; Mathiesen, B.V.; Dyrelund, A. The role of district heating in future renewable energy systems. *Energy* **2010**, *35*, 1381–1390. [CrossRef]
15. Delmastro, C.; Mutani, G.; Schranz, L. Advantages of Coupling a Woody Biomass Cogeneration Plant with a District Heating Network for a Sustainable Built Environment: A Case Study in Luserna San Giovanni (Torino, Italy). *Energy Procedia* **2015**, *78*, 794–799. [CrossRef]
16. Wojdyga, K.; Chorzelski, M. Chances for Polish district heating systems. *Energy Procedia* **2017**, *116*, 106–118. [CrossRef]
17. Kolokotsa, D. The role of smart grids in the building sector. *Energy Build.* **2016**, *116*, 703–708. [CrossRef]
18. Laakkonen, L.; Korpela, T.; Kaivosoja, J.; Vilkko, M.; Majanne, Y.; Nurmoranta, M. Predictive Supply Temperature Optimization of District Heating Networks Using Delay Distributions. *Energy Procedia* **2017**, *116*, 297–309. [CrossRef]
19. Bush, R.E.; Bale, C.S.E. The role of intermediaries in the transition to district heating. *Energy Procedia* **2017**, *116*, 490–499. [CrossRef]
20. Ziemele, J.; Gravelsins, A.; Blumberga, A.; Blumberga, D. The Effect of Energy Efficiency Improvements on the Development of 4th Generation District Heating. *Energy Procedia* **2016**, *95*, 522–527. [CrossRef]

21. Lake, A.; Rezaie, B.; Beyerlein, S. Review of district heating and cooling systems for a sustainable future. *Renew. Sustain. Energy Rev.* **2017**, *67*, 417–425. [CrossRef]
22. Werner, S. District heating and cooling in Sweden. *Energy* **2017**, *126*, 419–429. [CrossRef]
23. Wouters, C. Towards a regulatory framework for microgrids—The Singapore experience. *Sustain. Cities Soc.* **2015**, *15*, 22–32. [CrossRef]
24. Schmidt, D.; Kallert, A.; Blesl, M.; Svendsen, S.; Li, H.; Nord, N.; Sipilä, K. Low Temperature District Heating for Future Energy Systems. *Energy Procedia* **2017**, *116*, 26–38. [CrossRef]
25. Averfalk, H.; Werner, S. Essential improvements in future district heating systems. *Energy Procedia* **2017**, *116*, 217–225. [CrossRef]
26. Rämä, M.; Sipilä, K. Transition to low temperature distribution in existing systems. *Energy Procedia* **2017**, *116*, 58–68. [CrossRef]
27. Wahlroos, M.; Pärssinen, M.; Manner, J.; Syri, S. Utilizing data center waste heat in district heating—Impacts on energy efficiency and prospects for low-temperature district heating networks. *Energy* **2017**, *140*, 1228–1238. [CrossRef]
28. Persson, U.; Münster, M. Current and future prospects for heat recovery from waste in European district heating systems: A literature and data review. *Energy* **2016**, *110*, 116–128. [CrossRef]
29. Nord, N.; Schmidt, D.; Kallert, A.M.D. Necessary Measures to Include more Distributed Renewable Energy Sources into District Heating System. *Energy Procedia* **2017**, *116*, 48–57. [CrossRef]
30. Flores, J.F.C.; Lacarrière, B.; Chiu, J.N.W.; Martin, V. Assessing the techno-economic impact of low-temperature subnets in conventional district heating networks. *Energy Procedia* **2017**, *116*, 260–272. [CrossRef]
31. Lund, H.; Werner, S.; Wiltshire, R.; Svendsen, S.; Thorsen, J.E.; Hvelplund, F.; Mathiesen, B.V. 4th Generation District Heating (4GDH): Integrating smart thermal grids into future sustainable energy systems. *Energy* **2014**, *68*, 1–11. [CrossRef]
32. Council Directive 2010/31/EU of 19 May 2010 on the Energy Performance of Buildings, in 2010 OJ L153/13. Available online: http://data.europa.eu/eli/dir/2010/31/oj (accessed on 18 October 2018).
33. Directive 2012/27/EU of the European Parliament and of the Council of 25 October 2012 on Energy Efficiency, Amending Directives 2009/125/EC and 2010/30/EU and Repealing Directives 2004/8/EC and 2006/32/EC. Available online: http://data.europa.eu/eli/dir/2012/27/oj (accessed on 18 October 2018).
34. Communication From the Commission to the European Parliament, the Council, the European Economic and Social Committee and the Committee Of The Regions, an EU Strategy on Heating and Cooling. COM/2016/051 Final. Available online: https://eur-lex.europa.eu/legal-content/en/TXT/?uri=CELEX%3A52016DC0051 (accessed on 18 October 2018).
35. Werner, S. International review of district heating and cooling. *Energy* **2017**, *137*, 617–631. [CrossRef]
36. Sathaye, J.; Murtishaw, S.; Price, L.; Lefranc, M.; Roy, J.; Winkler, H.; Spalding-Fecher, R. Multiproject baselines for evaluation of electric power projects. *Energy Policy* **2004**, *32*, 1303–1317. [CrossRef]
37. UNE-EN ISO 15927-6:2009. Hygrothermal Performance of Buildings—Calculation and Presentation of Climatic Data—Part 6: Accumulated Temperature Differences (Degree-Days) (ISO 15927-6:2007). Available online: https://www.une.org/encuentra-tu-norma/busca-tu-norma/norma/?c=N0043434 (accessed on 18 October 2018).
38. Golden, A.; Woodbury, K.; Carpenter, J.; O'Neill, Z. Change point and degree day baseline regression models in industrial facilities. *Energy Build.* **2017**, *144*, 30–41. [CrossRef]
39. Meng, Q.; Mourshed, M. Degree-day based non-domestic building energy analytics and modelling should use building and type specific base temperatures. *Energy Build.* **2017**, *155*, 260–268. [CrossRef]
40. Verbai, Z.; Lakatos, Á.; Kalmár, F. Prediction of energy demand for heating of residential buildings using variable degree day. *Energy* **2014**, *76*, 780–787. [CrossRef]
41. Atalla, T.; Gualdi, S.; Lanza, A. A global degree days database for energy-related applications. *Energy* **2018**, *143*, 1048–1055. [CrossRef]
42. De Rosa, M.; Bianco, V.; Scarpa, F.; Tagliafico, L.A. Heating and cooling building energy demand evaluation; A simplified model and a modified degree days approach. *Appl. Energy* **2014**, *128*, 217–229. [CrossRef]
43. Efficiency Valuation Organization (EVO). International Performance Measurement and Verification Protocol: Concepts and Options for Determining. 2010. Available online: https://evo-world.org/en/products-services-mainmenu-en/protocols/ipmvp (accessed on 18 October 2018).

44. IBM SPSS Software. Available online: https://www.ibm.com/analytics/spss-statistics-software (accessed on 15 September 2018).
45. Walpole, R.E.; Myers, R.H.; Myers, S.L. *Probabilidad y Estadística Para Ingeniería y Ciencias*, 9th ed.; Prentice Hall: Englewood Cliffs, NJ, USA, 2012; ISBN 9786073214179.
46. Bilous, I.; Deshko, V.; Sukhodub, I. Parametric analysis of external and internal factors influence on building energy performance using non-linear multivariate regression models. *J. Build. Eng.* **2018**, *20*, 327–336. [CrossRef]
47. Permai, S.D.; Tanty, H. Linear regression model using bayesian approach for energy performance of residential building. *Procedia Comput. Sci.* **2018**, *135*, 671–677. [CrossRef]
48. Roth, J.; Rajagopal, R. Benchmarking building energy efficiency using quantile regression. *Energy* **2018**, *152*, 866–876. [CrossRef]
49. Fumo, N.; Rafe Biswas, M.A. Regression analysis for prediction of residential energy consumption. *Renew. Sustain. Energy Rev.* **2015**, *47*, 332–343. [CrossRef]
50. Arregi, B.; Garay, R. Regression analysis of the energy consumption of tertiary buildings. *Energy Procedia* **2017**, *122*, 9–14. [CrossRef]
51. Aranda, A.; Ferreira, G.; Mainar-Toledo, M.D.; Scarpellini, S.; Llera Sastresa, E. Multiple regression models to predict the annual energy consumption in the Spanish banking sector. *Energy Build.* **2012**, *49*, 380–387. [CrossRef]
52. Savin, N.E.; White, K.J. The Durbin-Watson test for serial correlation with extreme sample sizes or many regressors. *Econom. J. Econom. Soc.* **1977**, 1989–1996. [CrossRef]
53. Belsley, D.A.; Kuh, K.; Welsch, R.E. *Regression Diagnostics: Identifying Influential Data and Sources of Collinearity*; John Wiley & Sons: New York, NY, USA, 1980; ISBN 0-471-05856-4.
54. Navidi, W. *Estadística Para Ingenieros y Científicos*; McGraw-Hill: New York, NY, USA, 2006; ISBN 970-10-5629-9.
55. Granderson, J.; Price, P.N. Development and application of a statistical methodology to evaluate the predictive accuracy of building energy baseline models. *Energy* **2014**, *66*, 981–990. [CrossRef]
56. Lindberg, K.B.; Georges, L.; Alonso, M.J.; Utne, Å. Measured load profiles and heat use for "low energy buildings" with heat supply from district heating. *Energy Procedia* **2017**, *116*, 180–190. [CrossRef]
57. Khabdullin, A.; Khabdullina, Z.; Khabdullina, G.; Lauka, D.; Blumberga, D. Demand response analysis methodology in district heating system. *Energy Procedia* **2017**, *128*, 539–543. [CrossRef]
58. Winterscheid, C.; Dalenbäck, J.-O.; Holler, S. Integration of solar thermal systems in existing district heating systems. *Energy* **2017**, *137*, 579–585. [CrossRef]
59. Pavičević, M.; Novosel, T.; Pukšec, T.; Duić, N. Hourly optimization and sizing of district heating systems considering building refurbishment—Case study for the city of Zagreb. *Energy* **2017**, *137*, 1264–1276. [CrossRef]
60. Mertz, T.; Serra, S.; Henon, A.; Reneaume, J.M. A MINLP optimization of the configuration and the design of a district heating network: study case on an existing site. *Energy Procedia* **2017**, *116*, 236–248. [CrossRef]

Article

Techno-Economic Analysis of Rural 4th Generation Biomass District Heating

Víctor M. Soltero [1,*], Ricardo Chacartegui [2], Carlos Ortiz [3] and Gonzalo Quirosa [1]

[1] Department of Design Engineering, Universidad de Seville, 41011 Seville, Spain; gonzaquirosa@gmail.com
[2] Department of Energy Engineering, University of Seville, 41092 Seville, Spain; ricardoch@us.es
[3] Department of Electronic and Electromagnetism, University of Seville, 41012 Seville, Spain; cortiz7@us.es
* Correspondence: vmsoltero@us.es; Tel.: +34-630-877-180

Received: 25 October 2018; Accepted: 21 November 2018; Published: 25 November 2018

Abstract: Biomass heating networks provide renewable heat using low carbon energy sources. They can be powerful tools for economy decarbonization. Heating networks can increase heating efficiency in districts and small size municipalities, using more efficient thermal generation technologies, with higher efficiencies and with more efficient emissions abatement technologies. This paper analyzes the application of a biomass fourth generation district heating, 4GDH (4th Generation Biomass District Heating), in a rural municipality. The heating network is designed to supply 77 residential buildings and eight public buildings, to replace the current individual diesel boilers and electrical heating systems. The development of the new fourth district heating generation implies the challenge of combining using low or very low temperatures in the distribution network pipes and delivery temperatures in existing facilities buildings. In this work biomass district heating designs based on third and fourth generation district heating network criteria are evaluated in terms of design conditions, operating ranges, effect of variable temperature operation, energy efficiency and investment and operating costs. The Internal Rate of Return of the different options ranges from 6.55% for a design based on the third generation network to 7.46% for a design based on the fourth generation network, with a 25 years investment horizon. The results and analyses of this work show the interest and challenges for the next low temperature DH generation for the rural area under analysis.

Keywords: low temperature district heating system; biomass district heating for rural locations; 4th generation district heating; CO_2 emissions abatement

---

1. Introduction

Global warming is a major concern and urgent measures are required in a short time with the development of low carbon economy technologies [1]. At the same time, one of the key aspects to be developed by the scientific community in the coming years is an efficient energy distribution to cover the growth of the energy demand [2,3]. Building sector consumes 20% of the total energy in the world, with an expected increase ratio of 1.4% per year in the period between 2012 and 2040 [3]. By other side, of all new power generation plants installed worldwide in 2015, a 61% were based on renewable energies and an increase of this ratio is expected in the coming years [2].

District heating (DH) systems are considered as one of the most effective tools for an efficient and sustainable thermal energy distribution [4]. DH systems are centralized generation facilities, where thermal energy is distributed through a pipe network, which connects the thermal generation facility with the different consumption nodes at the buildings integrated in the system [5]. Thermal generation is centralized, and efficient and less contaminant heat generation equipment can be used, taking advantage of scaling up in the incorporation of combustion and emissions abatement technologies. It makes possible the efficient use of sustainable alternative energy sources. The energy efficiency of the global system is higher than the obtained with individual heating equipment [6] and, in addition, if renewable energy sources are used for thermal generation, a remarkable reduction in greenhouse gases emissions can be obtained [7].

European Directive 27/2012 [8] establishes a common framework to promote energy efficiency in Europe, in which DH networks have a relevant role. In previous research works methodologies to evaluate the potential of carrying out a massive implementation of different DH systems at a regional level have been developed [9,10]. Their results show how high impacts can be obtained if adequate methodologies for the selection, evaluation and implementation of district heating networks are used.

Distributed heating and cooling systems have a high potential for combining different renewable energy sources [10,11]. In heating systems, from the thermal point of view the most direct renewable energy sources to be integrated are biomass, solar energy [12–14] and geothermal energy [7,15]. Among them, the most direct integration is to use biomass as fuel, either renewable forestry products or sustainable agriculture biomass, with a higher potential in rural areas, where local biomass resource is available and the access to other primary energy sources is limited (i.e., natural gas).

The sustainability of biomass district heating rural networks will depend on the local biomass availability. The price of biomass depends on the proximity between the energy resource and the location of use [10,16,17]. The compensations that could arise in the local economy must be identified to ensure a stable and balanced development of heating networks in rural areas, the sector and the entire value chain [18]. Forestry activities can be a positive development in the economy of the area. Production of biomass for thermal generation can be a way to preserve rural areas, improve the quality of life of the municipality and to support the development of the territory that surrounds these activities. In this line, in reference [19], the impacts caused by the whole chain of biomass exploitation on the use of resources, human health and the environment are evaluated. The management of biomass, its transport, production of shavings, its treatments and its combustion are taken into account. It also analyzes the reduction of local pollution and GHG emissions that involves the displacement of a set of systems based on fossil fuels to a centralized production based on local biomass.

Among renewables, wind energy and photovoltaic can also be integrated for thermal integration of systems [7]. In reference [20] was shown how integration strategies between pipelines and buildings heat storage can be utilized to break the strong linkage of electric power and heat supply of CHP units, improving the operational flexibility of the system and reducing total operational costs.

DH networks are well established technologies, with an evolution along the year to incorporate advances in the operational strategies to increase their efficiency, to reduce distribution losses and to optimize their management [21]. The different generations of DH can be classified in terms of their operation temperatures, where operating temperatures have been lowered to reduce distribution heat losses [22,23].

There are different technologies that combined with DH systems improve efficiency and energy savings of the global system. Among the most interesting solutions is the combination of CHP biomass heating stations with DH systems. In reference [24] is shown how these installations are viable if the heating demand is high along the year, and it develops a methodology to evaluate the interest of these systems in order to improve the accuracy of the evaluation of the heating demand. It corrects the estimated demand for more accurate forecasting of heating demand that could lead to inadequate designs based on wrong performance parameters. The use of energy storage in these systems can reduce emission, operational cost and to enlarge the lifetime of existing CHP installations [25].

In reference [26] the different generations of DH are described. The third DH generation, and the most widely implemented nowadays, was developed in the period between 1980 and 2015. It is based on the distribution of liquid water at medium temperature, with temperatures below 100 °C. They use preinsulated and flexible pipes to facilitate their installation and to reduce installation and substitution costs. The fourth DH generation is currently under development. It is expected to have its maximum implementation in the period between 2020 and 2050. It is based on the distribution of water at a lower temperature, in the range between 30 °C and 60 °C. With this reduced operating temperatures, the thermal gradient between the heat transfer fluid and the environment is reduced and therefore the insulation requirements for controlling the thermal losses during distribution. Previous studies show the interest of low temperature district heating networks from the energy efficiency point of view [27]. The latest generation networks also make it easier to integrate these systems into Smart Energy Systems [23,28]. At the same time, the energy requirements of the buildings must be improved, improving their thermal behavior according to their location and energy demand characteristics. Internal equipment, as radiators, and end users habits, must be adapted for satisfying demand with these lower temperatures [29], increasing the available surface and adapting the heat demand profile.

Recent research activity in District Heating systems is oriented to integrate renewable energy sources with these systems and those that focus on the analysis or study of District Heating systems of low temperature, that is, investigate the applicability and viability of the heating urban low and very low temperature for buildings with high thermal performance. On this line different following European Commission funded projects: TEMPO [30], COOL DH [31], RELaTED [32], FLEXYNETS [33]. Among these European projects, the Project CELSIUS, develops Smart energy solutions. It deals with technical, social, economic and policy barriers for the development of new energy solutions. It shows the relevant role that DHC must have in the current and future sustainable Smart Cities [34,35].

In recent works the use of excess heat in industry as resource for district heating is analysed [36] with spatial and thermodynamic criteria. The analysis is expanded including the requirements linked to capital and operational expenditures.

This paper analyzes the application of a biomass fourth generation district heating, 4GDH, in a rural municipality. The heating network is designed to supply 77 residential buildings and eight public buildings, to replace the current individual diesel boilers and electrical heating systems. The development of the new fourth district heating generation, 4GDH, implies the challenge of combining using low or very low temperatures in the distribution network pipes and delivery temperatures in existing facilities buildings. A methodology is developed with the intention of comparing the operation of a biomass third generation or medium temperature network and a fourth generation or low temperature network. The aim is to compare operation modes estimated economic results. The evaluation methodology has common steps for both DH generations, so a common framework is established for a clear evaluation of results. The methodology is composed of two stages. The first stage carries out a technical analysis of the operation of the network with the objective of evaluating the total energy demand of the system, thermal load profile and pumping power consumption. The second stage performs an economic analysis of the operation of the network, based on the results obtained in the former stage, in order to determine the viability of the DH systems. The results show the viability of these DH and how it is affected by the operation with variable supply temperature.

2. Methodology

The objective of this methodology is to compare the viability and performance of two biomass District Heating networks (BioDHs), operating at two different levels one at medium temperature, with a design equivalent to the widely extended third generation DH design and other at low temperature equivalent to the fourth generation DH concept. The methodology has common stages for both cases but some steps differ according to the operation parameters of each one type of DH. The evaluation besides is extended to include the evaluation of the integration of another renewable source, wind energy, for providing electricity for pumping.

The proposed methodology comprises two different stages. In the first stage, the technical analysis of the operation of the network is performed. It drives to the sizing of the network and the definition of the energy profiles, thermal loads and pumping work. In the second stage, the economic analysis of the operation of the network is carried out, through a cost/benefit analysis. Stage one departs from the definition of the location of the system. Stage two departs from the results obtained in the previous stage and from the energy market characteristics, electricity and fuel costs definition.

2.1. Stage 1: Technical Analysis of DH Network Operation

Step 1: The first step defines the location of BioDH network, definition of the buildings to be supplied, and topographic conditions.

Step 2: After establishing the buildings to be supplied, the number of network users can be determined. On the other hand, it is also key to the determination of climatological data, mainly related to the minimum temperature, the degree days and the characterization of the wind resource throughout the year. The number of buildings included in the BioDH system will determine the total area to be heated, on which the heating demand depends proportionally, while the number of users will determine the SHW production requirements according to local regulations.

Step 3: Design of the network layout. Definition of its length, supply branches and integrated structure. It is defined as function of the supply zone characteristics: roads, altitude profiles, available area, other facilities, etc. The design minimizes network length, when possible, minimizing investment and civil works. The methodology of this stage is presented in Figure 1, while Figure 2 presents an example of the layout definition for the case study developed in Section 3.

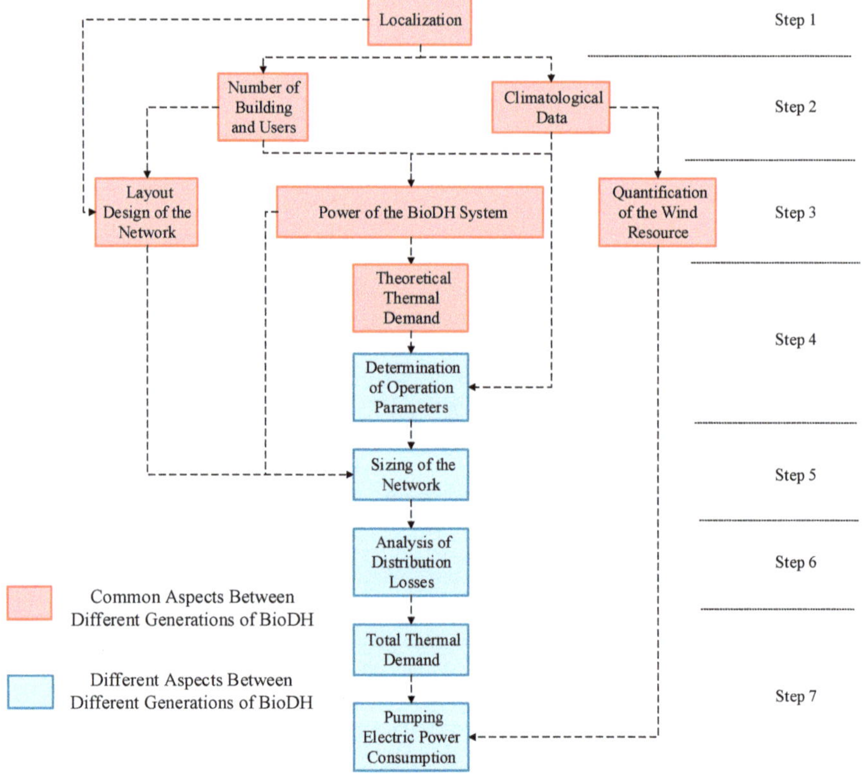

Figure 1. Stage I methodology.

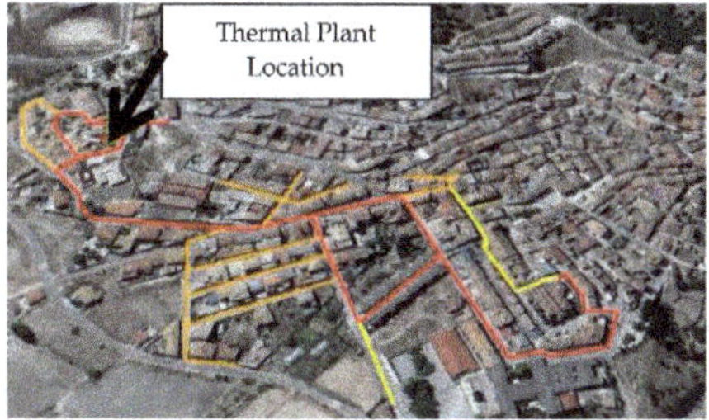

Figure 2. Network layout. Main distribution network (red lines) and secondary distribution network (yellow and orange lines).

With the heating demand, evaluated from the area to be heated, estimated from number of buildings and the demand of SHW, coming from the number of users, BioDH system heating power can be established, based on the minimum temperature and the thermal constant of the building, $K_{building}$ (1):

$$P = K_{building} \cdot (T_{c\_indoor} - T_{out}) = (\sum_{i=1}^{i=n} A_i U_i + V_{ainf}\rho \cdot C_p) \cdot (T_{c\_indoor} - T_{out}) \tag{1}$$

where P is the heat demanded (W), A_i is the surface (m^2) of the outer wall (i), $K_{building}$ is the building's thermal constant, $V_{a\,inf}$ is the infiltration air flow (m^3/s), U_i is the global heat transfer coefficient (air/air) of this wall (W/ m^{2}K), ρ is the density of the outside air (kg/m^3), C_p is the average specific heat of the outside air (Ws/kgK), T_{c_indoor} is the comfort indoor temperature (base temperature, 20 °C), and T_{out} is the outside temperature [9].

After determining the heating demand of the system, it is necessary to apply a simultaneity factor of the use of the network (2):

$$P_{design} = FS \cdot \sum P \tag{2}$$

Simultaneity factor is a key parameter for an adequate sizing of any system with multiple users. The value of the simultaneity factor used in this work is based on an expression derived from the analysis of the surveys done with end users, although there are different methodologies to obtain it, and it can vary in function of the integrated systems [37].

In this step, the wind resource is also characterized, to evaluate the subsequent incorporation into the electric power for pumping. To do this, the equivalent wind hours of the wind turbine implantation area will be established, correcting the wind speed with the height.

Step 4: The theoretical thermal demand is established. Based on the climatological data of the location and the power of the network established in the previous section, the annual theoretical thermal demand that the network will have can be determined.

Up to this point, all the previous steps are common for the definition of a third generation medium temperature BioDH network and a fourth generation low temperature BioDH network. Up to this point, the energy requirements to be met by the BioDH system have been determined without taking into account the operating characteristic parameters of the network.

At this point, the operating parameters of each system are determined. These parameters are: supply temperature, return temperature, circulation flow and thermal drop in substations. The combination of these parameters will lead to different results in the technical analysis depending on the characteristics of the DH network. In this study, are compared BioDH networks of low and medium temperature where the fourth generation low temperature DH operates at temperatures below 60 °C and the third generation BioDH operates with temperatures in the range between 60 and 90 °C.

In this BioDH, the emissions in the biomass boilers are a main constraint to the design of the network and thermal power station. By other side, in the rural areas under analysis, most of the heating systems are based on individual biomass heaters and gasoil boilers. With the development BioDH the current heating systems are displaced by a more efficient heating systems that uses thermal generation technologies of higher efficiency and with more advanced and efficient emissions abatement technologies, as particles filters. It also includes the required maintenance operations in the boiler not always followed by small individual heating systems. The commercial biomass boilers proposed for this study case produce the following emission levels al 98% de rated power under EN (European Standard) 303-5:2013: 66 mg/Nm^3 Dust, 62 mg/Nm^3 of CO, 157 mg/Nm^3 of NO$_x$ and 1 mg/Nm^3 of Org. C.

Therefore the effective implementation of these BioDH systems will improve significantly the air quality and efficiency of the heating systems. However they have an important challenge in the social acceptance of biomass thermal stations within an urban location [10].

Another relevant aspect is the typology of the terminal heating devices already installed in houses. In the location under study, with a high level of energy poverty, of those users with heating by individual boilers, a 45% have modern low temperature radiators, 35% of houses have mid temperature old radiators, and 20% have radiating floor. For the implementation of low temperature DH, the substitution of these indoor terminal radiators prepared for high temperature implies an additional cost in comparison with higher temperature DH.

Step 5: In this step the size of the BioDH network is set. Based on the data obtained from the previous steps, the design of the layout gives the length of the network and pressure drops meanwhile the network power and the operating parameters provide mass flow and fluid velocity to meet energy requirements and sizing pipes. With the use of Equation (3), the flow rate can be obtained:

$$Q = \frac{P_{design}}{c \cdot \rho \cdot (T_{flow} - T_{ret})} \tag{3}$$

where Q is the flow rate (m^3/s), T_{flow} is the flow temperature (K), T_{ret} is the return temperature (K), P_{design} is the design power (W), c is the heat capacity of the fluid (J/(kg·K)) and ρ is the density (kg/m^3).

As the operation parameters are specific for each type of BioDH system, the sizing of the network will also depend on the generation of DH that is being studied. Each mode of operation, therefore, will entail different pipe dimensions, an associated civil works, linked directly to required investments.

Step 6: After sizing the system, it is possible to evaluate the thermal losses along the distribution network. They depend directly on the diameter of the pipes and the average operating temperature of the system. The greater the diameter of the pipe, the lower the heat losses percentage produced per unit of flow; while at a lower average operating temperature, there is a lower thermal gradient between transport fluid and the external environment of the area, thus reducing thermal distribution losses significantly. To evaluate distribution thermal losses, pipes insulation is evaluated from the value of global heat transfer coefficient, U_i (W/(m^{2}K)) provided in catalogue by pipe manufacturers.

To obtain the thermal distribution losses, Equations (4) and (5) can be applied:

$$\Delta\vartheta = \frac{T_{flow} + T_{ret}}{2} - T_{ground} \tag{4}$$

$$Tlosses = u \cdot \Delta\vartheta \cdot L \tag{5}$$

where $\Delta\vartheta$ is the thermal gradient (K), T_{flow} is the flow temperature (K), T_{ret} is the return temperature (K), T_{ground} is the ground temperature (K), T_{losses} is the thermal losses (W), u is the insulating capacity characteristic of the pipe W/(m*K)) and L is the length of the network (m).

Step 7: Finally, the total thermal demand and pumping power consumption are evaluated. To determine the total thermal demand, the theoretical thermal demand must be taken into account from heating and SHW demand, and corrected to compensate thermal losses. In this way, the total energy requirement from the generation thermal plant is established.

Once the total thermal demand and flows are known, the pumping costs can be determined, applying the electricity prices in each hourly section. If the use of wind power is evaluated to be used for supplying this pumping power, as included in this study integrating a microwind turbine, then this value is corrected with the quantification of the wind resource, data already obtained in Step 3, so that the net electric power consumption of the pumping group from the grid can be determined.

Both the total thermal demand and the pumping power consumption depend on the operation parameters of the network, therefore, these final results obtained in stage I, depend on the generation of BioDH that is under study.

For the low temperature DH, additional considerations must be taken into account, as the required maintenance and control for avoiding *Legionella* growth. It grows in humid environments with temperatures in the range el 20–50 °C, that correspond with the existing in low temperature DH. To control I, thermostatic valves in the primary network are required, and the maximum volume of water in heat exchangers is limited to three litres in order than the system operates below of 50 °C without using a recirculating external treatment [38]. Other possibility is to stress the network and tanks to periods of thermal shock to generate temporary adverse environments for *Legionella* growth. Besides, the use of decentralized substations can contribute with a higher control of temperature distribution along the network [39]. In reference [40] different models and solutions for low temperature DH are studied form the energy and economic points of view.

2.2. Economic Analysis of DH Network Operation

At this stage the economic analysis of the network operation is carried out, the scheme of the proposed methodology is shown in Figure 3. As data from previous stage are used, the final results will depend on the network characteristics allowing direct comparisons between different temperature designs and operation strategies. Likewise, there are data required for this step which are common for both types of networks, such as price of energy, the cost of the thermal plant and the indirect operating costs. As the purpose of this paper is to carry out a comparative study between the implantation of two different generation of DHs, the work will focus on what differs between both typologies of DH.

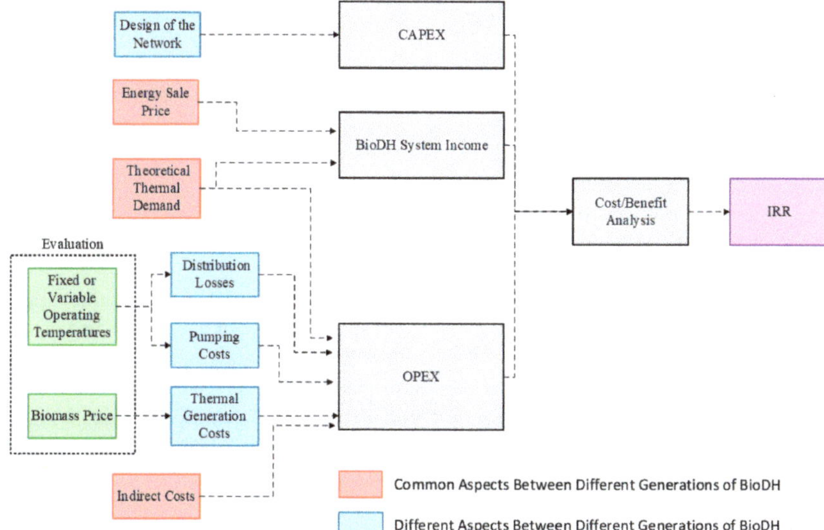

Figure 3. Stage II methodology.

- Capital Expenditure *(CAPEX)*. The initial investment to carry out the implementation of the network depends on many factors, but the most relevant, for a defined location and thermal plant, is the diameter of the pipes, which affects to the required investment.
- Operating Expenditure *(OPEX)*. Operating costs depend on several parameters that vary directly with some operation parameters. The total thermal demand consists of: the theoretical thermal demand, which does not depend on the mode of operation, and the distribution heat losses that do vary with them. Thus, the thermal energy to be supplied by the station is function of the temperature of the DH network and therefore of the DH generation. This variation in total thermal demand will affect to fuel costs, biomass in this case, and pumping costs. So, the higher the requirement for thermal generation, the higher these costs will be. These costs will vary markedly throughout the year, due to the large variation in the demand of these types of heating networks. The indirect costs will not depend, in principle, on the fact that a specific BioDH generation design is being used, therefore, although it is an important aspect in the quantification of operating costs, it is not an aspect required for the comparison study.
- Revenues from the BioDH System. The incomes generated by these systems come from the sale of thermal energy to users, therefore, the parameters to be taken into account in this section is the unit price of thermal energy sale and theoretical thermal demand. None of these aspects depends on the mode of operation of the BioDH system, therefore, the revenues structure generated by the two networks under the same coverage of demand will be the same. Therefore network profitability comparison involves the analysis of the operating costs and the investment.
- Cost/Benefit Analysis. Once the investment, operating costs and generated incomes have been determined, the economic study of the network operation can be carried out, in order to analyze the cash flows that occur throughout the life cycle of the installation including taxes and other economic aspects. Different economic indexes can be used. In this work, for the comparison of DH networks is used the Internal Rate of Return. Investment return rate (IRR) is chosen as a parameter for evaluation and comparison of the projects (6) and it is defined as:

$$-CAPEX + \sum_{i=1}^{LC} \frac{I_{ni} - OPEX}{(1 + IRR)^i} = 0 \tag{6}$$

where I_{ini} is the initial investment, and LC is the life cycle of the system.

3. Results

In this section are presented the main results obtained of applying the methodology previously exposed, in order to design, size and evaluate the performance of two district networks under different technological design criteria that belong to different DH generations, so that afterwards, a comparative analysis between the results obtained for each one can be performed. In this way, it will be possible to discuss, in a real application case, the advantages and disadvantages of each type of technology proposed by the different generations of district heating. Simulations where implemented in MatLab (MathWorks, Natick, MA, USA).

The third generation BioDH works at medium temperature, with approximate supply and return temperatures of 85 °C and 60 °C respectively [41], in the case of facilities in Spain, with the consideration of the current typologies of radiators. The fourth generation BioDH works at a lower temperature level, with a network supply temperature of 55 °C and a return temperature at 25 °C [42].

For the case analysis proposed for implementation of the methodology, 85 buildings, belonging to Santiago de la Espada (Spain), will be integrated in the network, and the DH network will be composed of 173 stretches of pipes arranged in a branched way. Heating in buildings is provided with hot water radiators. In both cases is assumed that the life cycle of both DH facilities is the same, 25 years.

Table 1 shows the overall parameters and results of the BioDH system obtained from stage one. The power of the system is 1.9 MW. With a demand of 3.68 GWh per year, combining SHW and heating. The total length of the network is 1504 m to supply 85 buildings.

Table 1. Common characteristics of both DH networks.

Parameter	Value
Thermal power generated (kW)	1909
Heating demand (MWh)	3434
SHW Demand (MWh)	248.4
Number of buildings	85
Surface of the buildings (m^2)	9078
Network length (m)	1504
Minimum temperature of the area (99) (°C)	−9.3
Setpoint temperature in buildings (°C)	20
Average temperature of the ground (°C)	8.9
Thermal constant of the building (kW/°C)	88.38
Equivalent wind hours	2462
Electricity cost, period 1, peak hours (€/kWh)	0.138
Electricity cost, period 2, regular hours(€/kWh)	0.104
Electricity cost, period 3, off-peak hours(€/kWh)	0.069
Biomass Cost (€/MWh)	30.00

The cost of the electricty required for pumping varies along the day. The electricity tariff changes depending on the time slot: peak demand period, from 18:00 to 22:00 h, regular demand period, from 8:00 to 18:00 and from 22:00 to 24:00 h, and off-peak demand period from 0:00 to 8:00 h.

Figure 4a shows the flow of the medium and low temperature DH systems according to the demand of the BioDH system. Figure 4b relates the flows of the networks for the different months of the year.

In the low temperature network, the mass flow is smaller due to a higher temperature drop in the substations than the one in higher temperature networks. With a higher temperature drop a reduced mass flow is required to provide the same heating power.

Figure 4b shows that the difference between flows supplied by each type of network is more evident in the months with higher thermal demand. The results show that for the low temperature network the minimum and maximum flows throughout the year are 5.4 and 31.7 m^3/h, respectively. However, in the case of the medium temperature network, the values of flow throughout the year are between the limits of 7.1 and 38.7 m^3/h. In Figure 4 can be verified that the hottest periods correspond to the lower flow rates, that is, when the demand for heating is minimal or nonexistent and only SHW is required. This value is obtained for impulsion and return temperatures of the network fixed throughout the year. The mean average operating flow in the network throughout the year; for the case of the low temperature network it is 15.2 m^3/h and for the medium temperature is 18.8 m^3/h.

In this case a smart strategy control hasn't been considered as operation temperatures cannot further reduced due to constrains associated with the installations of the buildings and to cover the comfort requirements for SHW and heating. These installation were developed for individual use and not considering their integration in DH networks.

Figure 5a shows the influence of the demand on the thermal losses produced during the distribution of the heat transfer fluid.

The results show a difference between thermal losses in each type of network. Thermal losses along the year in the medium temperature network are approximately the double than those in the low temperature network. In this way, for the period of maximum demand, and lower external temperature, the thermal losses for the low and medium temperature networks are 2.63% and 4.74%, respectively; these are the minimum values. The maximum values, in percentage, appear when the demand is minimal, thermal losses values are 7.22% and 19.52% for a low and medium temperature network, respectively. This difference between the losses is due to the difference in operating temperatures of each network.

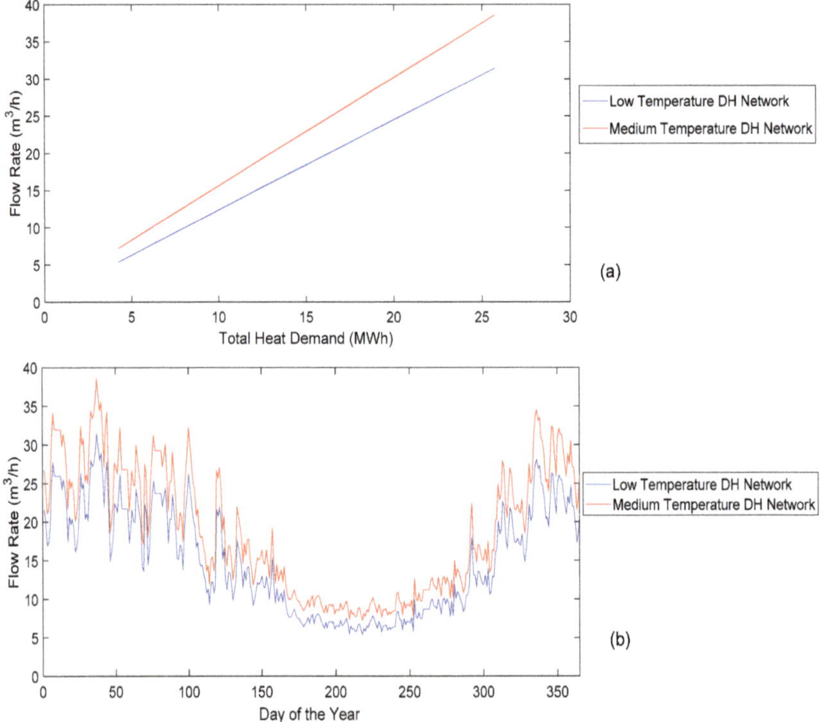

Figure 4. (**a**) Flow rate of medium and low temperature DH networks as function of thermal demand. (**b**) Flow rate for mean and low temperature DH networks along the year.

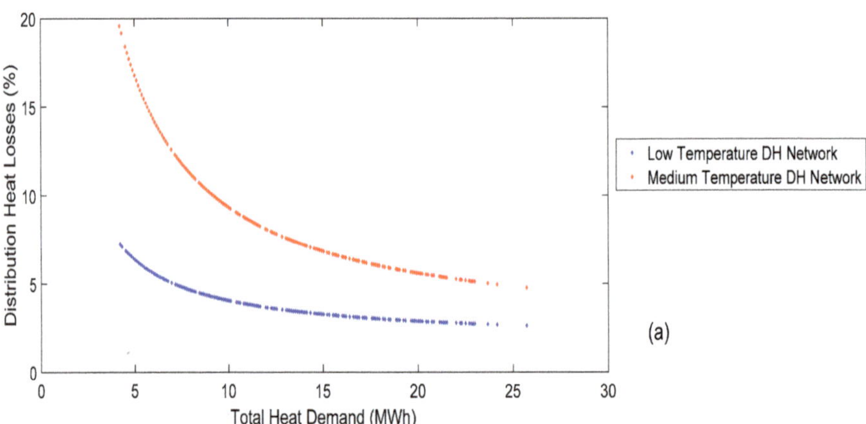

Figure 5. *Cont.*

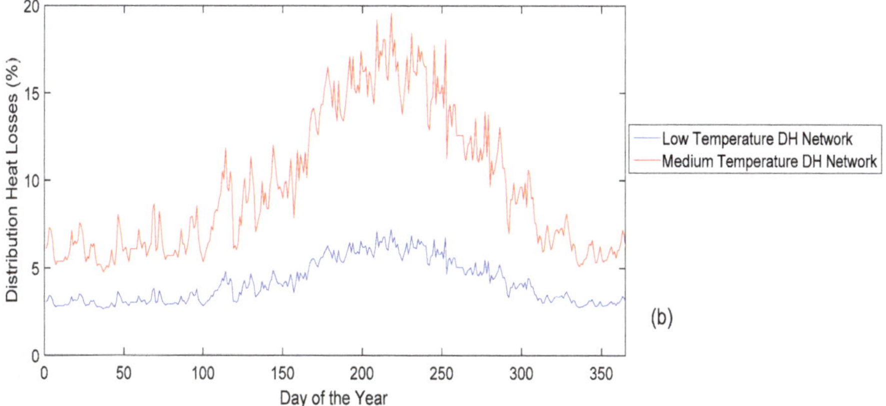

Figure 5. (**a**) Percentage of distribution heat losses for medium and low temperature DH networks as function of thermal demand, (**b**) Percentage of distribution thermal losses in the medium and low temperature DH networks along the year.

The losses that occur in the network depends on the thermal gradient that occurs between the operating temperatures of the network, the temperature of the ground around the pipe, insulation characteristics of the pipes and their size.

Ground temperature depends primarily on the ambient temperature and depth. From real experimental data of the temperature of the ground, measured at 20 cm of depth, an expression can be determined that relates the external ambient temperature to the temperature of the ground. These experimental data have been collected in an area close to the case study, within this project, it has not been validated for other geographical areas with different climatological and ground characteristics. In Figure 6 the relationship obtained from the experimental measurements is shown.

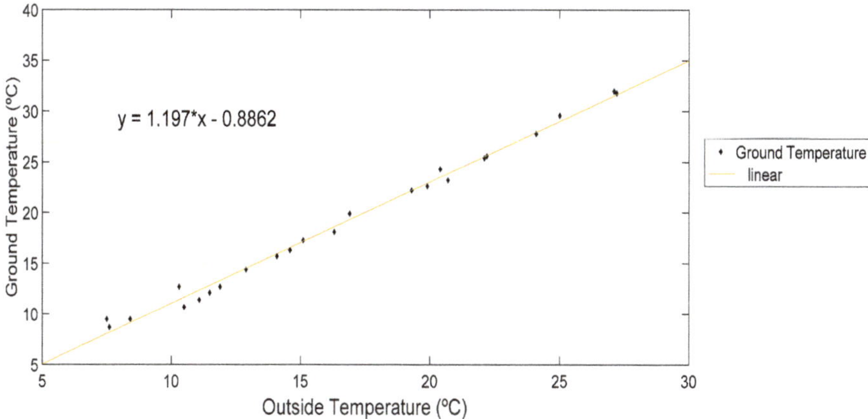

Figure 6. Relationship between ambient temperature and terrain temperature measured at 20 cm depth.

Another way to analyze thermal losses is by analyzing their total values with respect to the total demand. The variation of the thermal losses that occur in the low and medium temperature networks throughout the year are represented in Figure 7.

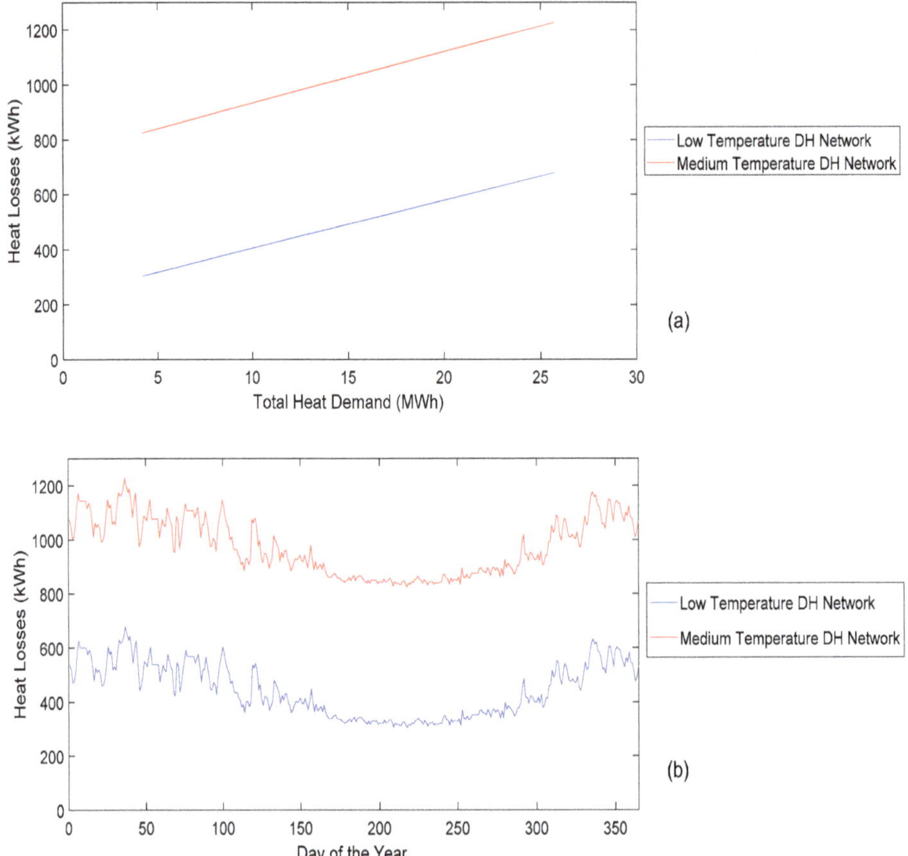

Figure 7. (a) Distribution heat losses for medium and low temperature DH networks as function of thermal demand, **(b)** Distribution thermal losses in the medium and low temperature DH networks along the year.

The evaluation of the investment associated to the different subsystems has been done according to [16,17], based on new construction systems. DH incomes will be the result of the energy sold by the sale price. For this case, end user sale price is set in 85 €/MWh. It implies to the end user 18% of savings compared with their current energy bills based on gasoil and wood in chimneys. This value has been set as function of the base line estimated as shown in [43]. Investment in DH is directly affected by pipes mean diameter. In the case of the low temperature network, due to a higher temperature drop in the thermal stations, mass flow circulating by pipes is lower and heating requirement can be maintained with smaller diameters. For the case under study it allows a reduction in pipes investment of 18.72%, and a reduced investment in the low temperature DH.

In Table 2 a comparison of the main results obtained in the design and calculation of the two district network designs is presented. The effect of installing a microwind turbine is also included to evaluate the effect on the total performance and system viability.

Table 2. Main operating parameters for both DH networks.

Main operating Parameters	Third Generation Network. Medium Temperature	Fourth Generation Network. Low Temperature
Supply Temperature	85 °C	55 °C
Return temperature	60 °C	25 °C
Flow rate at design conditions	44.00 m³/h	36.10 m³/h
Average distribution pipes diameter	101.27 mm	89.72 mm
Pressure drop at rated conditions	7.141 bar	8.699 bar
Total Heat Demand	4038 MWh	3845 MWh
Annual thermal losses	355.6 MWh	161.9 MWh
Rated operation thermal losses (%)	9.66%	4.40%
Annual pumping costs	14,964 €	14,246 €
Annual wind turbine savings	2198 €	2093 €
Wind turbine savings (%)	12.20%	12.81%
Annual biomass cost	121,145 €	115,335 €
Equivalent performance hours	2115	2014
Total inversion	1,080,898 €	1,036,678 €
Annual total operational costs	136,741 €	130,200 €
Payback$_{25\ years}$	15.34 years	13.16 years
IRR$_{25\ years}$	6.55%	7.46%

4. Discussion

To complete the study, parameters with a direct influence in the viability of the biomass DH system are analyzed. One parameter that can affect to the performance and viability of the district heating is to consider operation at constant network temperature or at variable temperature. The capacity of operating at variable temperature depends on the DH network technology. In the case of low temperature networks, which supply homes with radiators operating at low temperatures, the limit is to avoid using any auxiliary generation equipment. It constrains the supply and the minimum return temperature of the network, taking into account an indoor comfort temperature of 20 °C. On the other hand, for medium temperature networks, the limit is in the range of operation of the radiators, which in this case, operates with medium temperature. Then the limit can be obtained by analyzing the radiating surface and design temperatures of the radiators and comparing their capacity with the energy requirements that must be covered throughout the year. Therefore the variation range of operating temperatures is more limited for the case of low temperature DH networks.

If medium temperature the DH system operates with lower temperatures, *Legionella* growth risk increases, and control methodologies must be included in the medium temperature network in the same way that in the low temperature network.

By other side, the constraints associated with operating DH with variable temperature are not exclusively technological. Restrictions can be derived from legal aspects like the clauses in legal contracts related to energy supply, where it is usual to set a minimum supply temperature. For the development of variable operation DH this kind of clauses must be revised to guarantee the quality of heat supply and the optimal operation of the system. Besides end user awareness and formation about the most efficient use of indoor thermal facilities are required for a successful development. In [44] a thermos-hydraulic model for thermal networks is presented taking into account decentralized heating and cooling sources, variable temperature and mass flows and flow inversion.

The comparison of the losses that occur in a network with these designs, considering that the temperature of operation in the network is fixed or it can be variable is presented in Figures 8 and 9. Figure 8 shows the results obtained at fixed and variable network temperatures for a low temperature network and Figure 9 shows the same analysis for a medium temperature network.

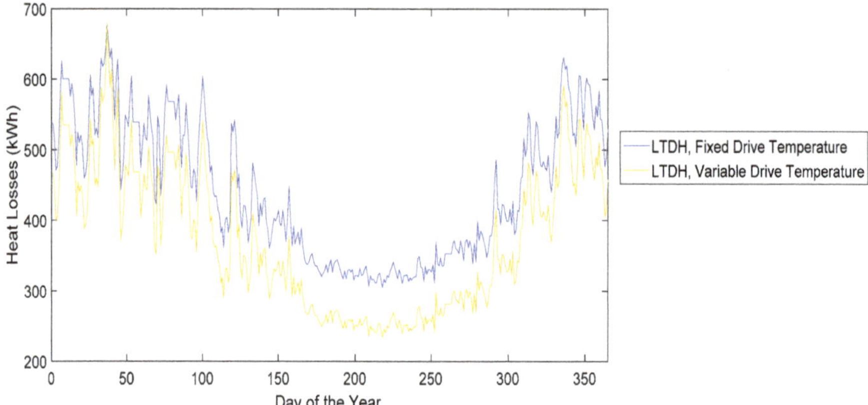

Figure 8. Comparison of losses in a low temperature network along the year with fixed and variable network temperatures.

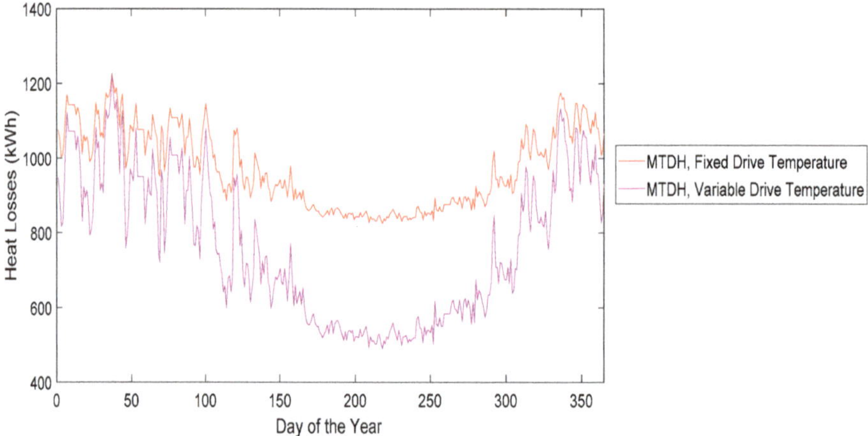

Figure 9. Comparison of losses in a medium temperature network along the year with fixed and variable network temperatures.

From these figures can be observed that in a medium temperature network, the supply and return temperatures can be lowered in a greater proportion than in a low temperature network, the losses that occur throughout the year in a medium temperature network can also be reduced in greater proportion. In the same way, the low temperature network is less sensitive to consider a variable network temperature.

Figure 10 analyzes losses reduction along the year. Energy savings, linked to losses reduction along the year, with a variable temperature operation in the low temperature network are 25.5 MWh meanwhile that in a medium temperature they are 38.6 MWh.

Another factor that affects to the network viability is the variation in the price of biomass. The price of biomass is a value that can vary quite significantly depending on the location where the DH network is located and constrains the viability of the project. Its effect is analyzed for the medium and low temperature networks for both operation strategies, with of fixed and variable supply temperatures.

Figure 11 shows the influence of the cost of biomass on the viability of the project. The best IRR values are provided by the low temperature network with variable working temperature;

the worst values are provided by the medium temperature network with fixed values for supply and return temperatures.

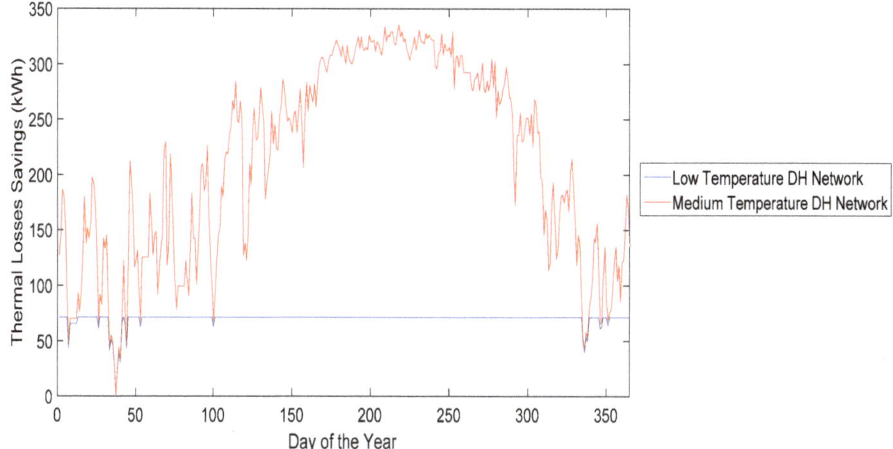

Figure 10. Energy savings in low and medium temperature networks operating at variable network temperatures.

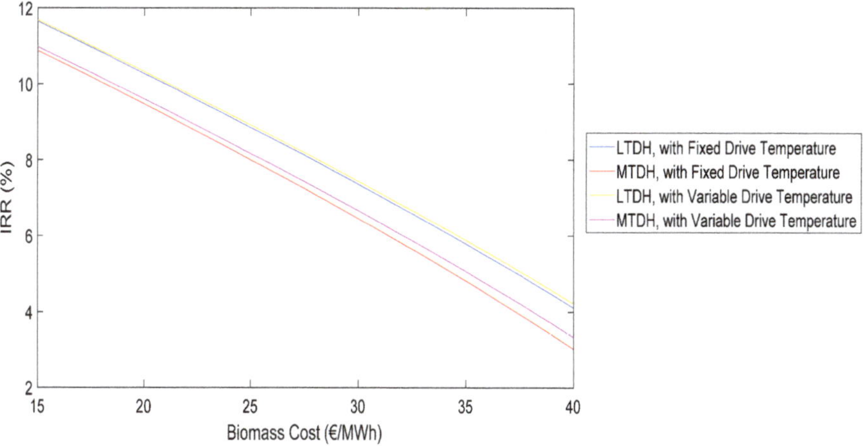

Figure 11. Influence of the cost of biomass on DH viability with different operation strategies.

In Figure 12 is presented the difference of IRR values between both networks as function of biomass price under the two different operation strategies, variable and fixed network temperature.

The energy savings that can be obtained in a medium temperature network, when considering variable working temperatures, is greater than in the case of a low temperature network. Therefore the IRR value will improve both for a medium temperature and low network, however, the IRR for medium temperature will be improved in a higher proportion reducing the differences between both designs.

In Figure 12 it is observed that on the one hand the IRR provided by a low temperature network is always greater than that which would be provided by a medium temperature network working under the same conditions, but that difference between IRR values is more noticeable when the fuel price is high and almost insignificant when the price is low. This is logical since in the low temperature network there are lower energy losses throughout the year, therefore, if the operating costs are high, this energy saving will be more relevant.

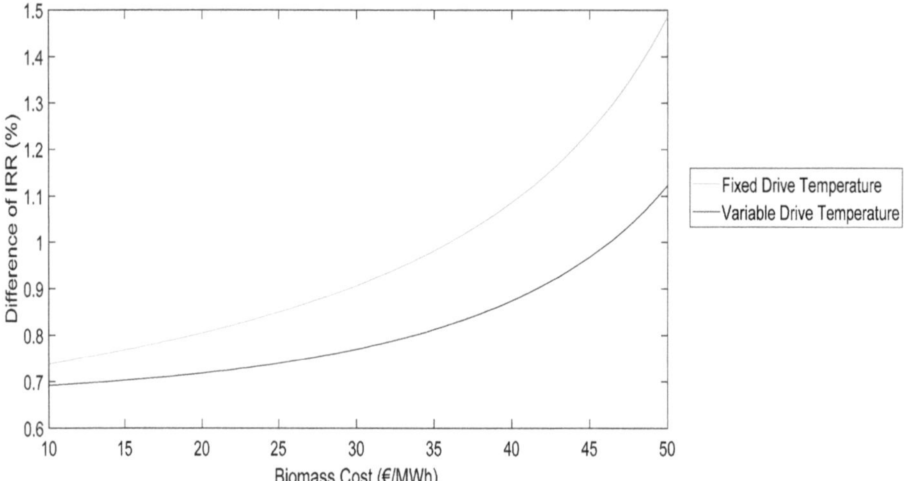

Figure 12. Influence of the cost of biomass on IRR difference between a network of low and medium temperature, with fixed and variable operating temperatures of the network.

It can also be observed that the IRR difference, in the case that the networks work at variable impulsion and return temperatures, is lower. This is due to what has been said above that energy savings because of losses reduction that can be obtained in a medium temperature network, when considering variable working temperatures, is greater than in the case of a low temperature network. Then the difference between IRR values is smaller. Since low temperature networks have higher IRR values, these can be installed in areas where the price of biomass is high and where medium temperature networks would not provide viable IRR values.

5. Conclusions

In this work the differences between two generations of district heating applied to a specific case are analysed. The third generation of district heating networks uses higher supply and return temperatures, but with lower temperature drops compared to that produced in the fourth generation of district heating network systems. One of the most relevant characteristic of operation are thermal losses from the network to the terrain surrounding the pipes. With networks operating at lower temperatures, there is a reduced gradient, with the consequent reduction of thermal losses in the hot water distribution.

Regarding the total circulation flow, because the temperature drop in the substations is greater in the case of the fourth generation network, it is required that a lower flow rate circulates through them to supply the same thermal power. Since the flow rate through the network is lower, the diameters of the pipes can be reduced, thus causing a small increase in the pressure drop during operation.

This results in very similar pumping costs. This is due to the fact that in the fourth generation DH, lower flow rates are used in the operation of the system, but nevertheless, if smaller diameters are selected for reducing investment costs, pressure drops are greater.

These relationships between diameters, flows and heat losses are linked to the specific case study. Variations in network distribution and external conditions can lead to different ratios and a general rule of thumb cannot be derived.

Depending on the location characteristics, other renewable energy sources can be integrated. For the DH analyzed in this work the effect of integrating a microwind turbine is considered. As pumping cost are quite similar the savings obtained in both cases are similar.

The difference in the investment cost to carry out the projects is mainly due to the reduction in piping costs (materials and insulation) and civil works in the fourth generation DH network. This reduction is due to the fact that smaller dimensions of pipes and fittings are used, since it smaller circulation flow allows reduced diameters with controlled pressure drops. For the application case the investment savings are estimated in 44,220 €.

The reduction in operating costs that occurs in the district network of fourth generation is mainly due to the fact that the fuel required for the operation of the boiler is lower, since there are less thermal losses in the hot water distribution. When a fourth generation district network is implemented instead a third generation, as distribution thermal losses are reduced, there is a fuel saving of 4.8%.

Since the investment to be made and the operation costs of the fourth generation district network are lower, the payback period is lower, obtaining a more interesting investment profile as long as the income generated by both types is maintained. In this sense, the advantages of low temperature DH networks are clear when deciding from the investor's point of view, in this case study. However, the development of heating networks in consolidated areas from the urban point of view involves the replacement of existing heating systems that currently operate at medium or high temperature. In these cases the application of the network at low temperature should consider the design of the interior installation of the houses and the adequateness of the network for maximum demand periods. The results of applying the methodology in a specific location and under operation conditions that allow to evaluate the differences in applying third and fourth DH network designs.

Analyzing the influence of operation with variable supply and return temperatures in low and medium temperature networks, it has been shown that energy losses are considerably reduced throughout the year. It has been shown, from the economic point of view, that it is always better when operating temperatures are variable, but it has also been seen that the variation suffered by the IRR of the project is not excessively high either. The price of biomass has been shown as key for viability of the project. The price of biomass can strongly vary different between projects with different locations, because this price depends strongly on the geographical area. When analyzing the profitability between the different types of networks, in the cases with a high cost of biomass, more interesting is to install a District Heating system with low temperature and with variable flow and return temperatures.

Author Contributions: Conceptualization, V.M.S.; Data curation, C.O.; Formal analysis, G.Q. Investigation, V.M.S. and G.Q.; Methodology, V.M.S.; Supervision, R.C.; Validation, C.O.; Writing—original draft, V.M.S. and R.C.

Funding: This research received no external funding.

Conflicts of Interest: The authors declare no conflict of interest.

Nomenclature

z	Density (kg/m^3)
$\Delta\vartheta$	Thermal gradient (K)
A_i	Surface (m^2) of the exterior wall (i)
c	Specific heat, J/(kg K)
C_p	Average specific heat of the exterior air (Ws/kgK)
$CAPEX$	Capital Expenditure
DH	District Heating
EN	European Standard
FS	Simultaneity factor
GDH	Generation of District Heating
I_{ini}	Initial investment (€)
$K_{building}$	Thermal constant of the building

L	Length of the network (m)
LC	Life cycle of the system (years)
$OPEX$	Operating Expenditure
P	Power demanded (kW)
P_{design}	Design power (kW)
Q	Flow rate (m^3/s)
SHW	Sanitary Hot Water
T_{c_indoor}	Comfort temperature selected for the indoor space (base temperature, 20 °C)
T_{flow}	Flow temperature (K)
T_{ground}	Ground temperature (K)
T_{losses}	Thermal losses (W)
T_{out}	Temperature of the exterior air.
T_{ret}	Return temperature (K)
u	Insulating capacity characteristic of the pipe W/(m*K))
U_i	Global heat transfer coefficient (air/air) of this wall (W/ m^{2}K)
V_{ainf}	Flow of infiltration air (m^3/s)

References

1. Lionello, P.; Scarascia, L. The relation between climate change in the Mediterranean region and global warming. *Reg. Environ. Chang.* **2018**, *18*, 1481–1493. [CrossRef]
2. REthinking Energy 2017: Accelerating the global energy transformation. Available online: http://www.irena.org/publications/2017/Jan/REthinking-Energy-2017-Accelerating-the-global-energy-transformation (accessed on 25 October 2018).
3. U.S. Energy Information Administration (EIA). 2018. Available online: https://www.eia.gov/ (accessed on 25 July 2018).
4. Lake, A.; Rezaie, B.; Beyerlein, S. Review of district heating and cooling systems for a sustainable future. *Renew. Sustain. Energy Rev.* **2017**, *67*, 417–425. [CrossRef]
5. Frederiksen, S.; Werner, S. *District Heating and Cooling. Studentlitteratur*; Studentlitteratur AB: Lund, Sweden, 2013.
6. Sayegh, M.A.; Danielewicz, J.; Nannou, T.; Miniewicz, M.; Jadwiszcak, P.; Piekarska, K.; Jouhara, H. Trends of European research and development in district heating technologies. *Renew. Sustain. Energy Rev.* **2017**, *68*, 1183–1192. [CrossRef]
7. Østergaard, P.A.; Mathiesen., B.V.; Möller, B.; Lund, H. A renewable energy scenario for Aalborg Municipality based on low-temperature geothermal heat, wind power and biomass. *Energy* **2010**, *35*, 4892–4901. [CrossRef]
8. EUR-Lex-32012L0027-EN-EUR-Lex. Available online: https://eur-lex.europa.eu/eli/dir/2012/27/oj (accessed on 25 July 2018).
9. Soltero, V.M.; Chacartegui, R.; Ortiz, C.; Velázquez, R. Evaluation of the potential of natural gas district heating cogeneration in Spain as a tool for decarbonisation of the economy. *Energy* **2016**, *115*, 1513–1532. [CrossRef]
10. Soltero, V.M.; Rodríguez-Artacho, S.; Velázquez, R.; Chacartegui, R. Biomass universal district heating systems. *E3S Web Conf.* **2017**, *22*, 00163. [CrossRef]
11. Østergaard, P.A.; Lund, H. A renewable energy system in Frederikshavn using low-temperature geothermal energy for district heating. *Appl. Energy* **2011**, *88*, 479–487. [CrossRef]
12. Carpaneto, E.; Lazzeroni, P.; Repetto, M. Optimal integration of solar energy in a district heating network. *Renew. Energy* **2015**, *75*, 714–721. [CrossRef]
13. Østergaard, P.A.; Andersen, A.N. Booster heat pumps and central heat pumps in district heating. *Appl. Energy* **2016**, *184*, 1374–1388. [CrossRef]
14. Lizana, J.; Ortiz, C.; Soltero, V.M.; Chacartegui, R. District heating systems based on low-carbon energy technologies in Mediterranean areas. *Energy* **2017**, *120*, 397–416. [CrossRef]
15. Aslan, A.; Yüksel, B.; Akyol, T. Effects of different operating conditions of Gonen geothermal district heating system on its annual performance. *Energy Convers. Manag.* **2014**, *79*, 554–567. [CrossRef]

16. Soltero, V.; Chacartegui, R.; Ortiz, C.; Jesus, L.; Gonzalo, Q. Biomass District Heating Systems Based on Agriculture Residues. *Appl. Sci.* **2018**, *8*, 476. [CrossRef]

17. Soltero, V.M.; Chacartegui, R.; Ortiz, C.; Velázquez, R. Potential of biomass district heating systems in rural areas. *Energy* **2018**, *156*, 132–143. [CrossRef]

18. Fagarazzi, C.; Tirinnanzi, A.; Cozzi, M.; Di Napoli, F.; Romano, S. The Forest Energy Chain in Tuscany: Economic Feasibility and Environmental Effects of Two Types of Biomass District Heating Plant. *Energies* **2014**, *7*, 5899–5921. [CrossRef]

19. Neri, E.; Cespi, D.; Setti, L.; Gombi, E.; Bernardi, E.; Vassura, I.; Passarini, F. Biomass Residues to Renewable Energy: A Life Cycle Perspective Applied at a Local Scale. *Energies* **2016**, *9*, 922. [CrossRef]

20. Li, P.; Wang, H.; Lv, Q.; Li, W. Combined Heat and Power Dispatch Considering Heat Storage of Both Buildings and Pipelines in District Heating System for Wind Power Integration. *Energies* **2017**, *10*, 89. [CrossRef]

21. Connolly, D.; Lund, H.; Mathiesen, B.V.; Werner, S.; Moller, B.; Persson, U.; Boermans, T.; Trier, D.; Qustergaard, P.A.; Nilesen, S. Heat Roadmap Europe: Combining district heating with heat savings to decarbonise the EU energy system. *Energy Policy* **2014**, *65*, 475–489. [CrossRef]

22. Ziemele, J.; Gravelsins, A.; Blumberga, A.; Blumberga, D. Combining energy efficiency at source and at consumer to reach 4th generation district heating: Economic and system dynamics analysis. *Energy* **2017**, *137*, 595–606. [CrossRef]

23. Ziemele, J.; Gravelsins, A.; Blumberga, A.; Blumberga, D. The Effect of Energy Efficiency Improvements on the Development of 4th Generation District Heating. *Energy Procedia* **2016**, *95*, 522–527. [CrossRef]

24. Sartor, K.; Quoilin, S.; Dewallef, P. Simulation and optimization of a CHP biomass plant and district heating network. *Appl. Energy* **2014**, *130*, 474–483. [CrossRef]

25. Sartor, K.; Dewallef, P. Optimized Integration of Heat Storage Into District Heating Networks Fed By a Biomass CHP Plant. *Energy Procedia* **2017**, *135*, 317–326. [CrossRef]

26. Lund, H.; Werner, S.; Wiltshire, R.; Svendsen, S.; Thorsen, J.E.; Hvelpuld, F.; Mathiesen, B.V. 4th Generation District Heating (4GDH). *Energy* **2014**, *68*, 1–11. [CrossRef]

27. Li, P.; Nord, N.; Ertesvåg, I.S.; Ge, Z.; Yang, Z.; Yang, Y. Integrated multiscale simulation of combined heat and power based district heating system. *Energy Convers. Manag.* **2015**, *106*, 337–354. [CrossRef]

28. Li, H.; Wang, S.J. Challenges in Smart Low-temperature District Heating Development. *Energy Procedia* **2014**, *61*, 1472–1475. [CrossRef]

29. Østergaard, D.S.; Svendsen, S. Replacing critical radiators to increase the potential to use low-temperature district heating-A case study of 4 Danish single-family houses from the 1930s. *Energy* **2016**, *110*, 75–84. [CrossRef]

30. TEMPerature Optimisation for Low Temperature District Heating across Europe | Projects | H2020 | CORDIS | European Commission. Available online: https://cordis.europa.eu/project/rcn/212364_en.html (accessed on 25 July 2018).

31. Cool Ways of Using Low Grade Heat Sources from Cooling and Surplus Heat for Heating of Energy Efficient Buildings with New Low Temperature District Heating (LTDH) Solutions. Available online: https://cordis.europa.eu/project/rcn/212356_en.html (accessed on 25 July 2018).

32. REnewable Low TEmperature District. Available online: https://cordis.europa.eu/project/rcn/212357_en.html (accessed on 25 July 2018).

33. Fifth Generation, Low Temperature, High EXergY District Heating and Cooling NETworkS. Available online: https://cordis.europa.eu/project/rcn/194622_en.html (accessed on 25 July 2018).

34. Borelli, D.; Devia, F.; Lo Cascio, E.; Shenone, C.; Spolandore, A. Combined Production and Conversion of Energy in an Urban Integrated System. *Energies* **2016**, *9*, 817. [CrossRef]

35. Perez, A.; Stadler, I.; Janocha, S.; Fermnando, C.; Bonvicini, G.; Tillmann, G. Heat recovery from sewage water using heat pumps in cologne: A case study. In Proceedings of the International Energy and Sustainability Conference, Koln, Germany, 30 Jun–1 July 2016; pp. 1–7.

36. Bühler, F.; Petrović, S.; Holm, F.M.; Karlson, K.; Emegaard, B. Spatiotemporal and economic analysis of industrial excess heat as a resource for district heating. *Energy* **2018**, *151*, 715–728. [CrossRef]

37. Winter, W.; Haslauer, T.; Obernberger, I. Simultaneity surveys in district heating networks: Results and project experience (Untersuchungen zur Gleich-zeitigkeit in Nahwärmenetzen: Ergebnisse und Projekterfahrungen). *Euroheat Power* **2001**, *30*, 42–47. (In German)

38. Ianakiev, A.I.; Cui, J.M.; Garbett, S.; Filer, A. Innovative system for delivery of low temperature district heating. *Int. J. Sustain. Energy Plan. Manag.* **2017**, *12*, 1–28. [CrossRef]

39. Yang, X.; Li, H.; Svendsen, S. Decentralized substations for low-temperature district heating with no Legionella risk, and low return temperatures. *Energy* **2016**, *110*, 65–74. [CrossRef]

40. Yang, X.; Li, H.; Svendsen, S. Energy, economy and exergy evaluations of the solutions for supplying domestic hot water from low-temperature district heating in Denmark. *Energy Convers. Manag.* **2016**, *122*, 142–152. [CrossRef]

41. Asociacion Técnica Española de Climatización y Refrigeracion (ATECYR). *Guía Técnica de Instalaciones de Calefacción Individual*; Institute for the Diversification and Saving of Energy: Madrid, Spain, 2010.

42. Lund, R.; Østergaard, D.S.; Yang, X.; Mathiesen, B.V. Comparison of Low-temperature District Heating Concepts in a Long-Term Energy System Perspective. *Int. J. Sustain. Energy. Plan. Manag.* **2017**, *12*, 5–18. [CrossRef]

43. Nielsen, S.; Grundahl, L.; Nielsen, S.; Grundahl, L. District Heating Expansion Potential with Low-Temperature and End-Use Heat Savings. *Energies* **2018**, *11*, 277. [CrossRef]

44. Van der Heijde, B.; Fuchs, M.; Ribas Tugores, C.; Schweiger, G.; Sartor, K.; Basciotti, D.; Muller, D.; Nutsch-Geusen, C.; Wetter, M.; Helsen, L. Dynamic equation-based thermo-hydraulic pipe model for district heating and cooling systems. *Energy Convers. Manag.* **2017**, *151*, 158–169. [CrossRef]

 *energies*

Article

Analysis of the Methodology to Obtain Several Key Indicators Performance (KIP), by Energy Retrofitting of the Actual Building to the District Heating Fuelled by Biomass, Focusing on nZEB Goal: Case of Study

Rosaura Castrillón Mendoza [1,2], Javier M. Rey Hernández [2,3,*], Eloy Velasco Gómez [2], Julio F. San José Alonso [2] and Francisco J. Rey Martínez [2]

[1] Department of Energy and Mechanics, University Autónoma de Occidente Cali (UAO), Cali 760030, Colombia; rcastrillon@uao.edu.co
[2] Department of Energy and Fluid Mechanics, School of Engineering (EII), University of Valladolid (UVa), 47002 Valladolid, Spain; eloy@eii.uva.es (E.V.G.); julsan@eii.uva.es (J.F.S.J.A.); rey@eii.uva.es (F.J.R.M.)
[3] Higher Polytechnic College, European University Miguel de Cervantes (UEMC), 47012 Valladolid, Spain
* Correspondence: javier.rey@uva.es; Tel.: +34-983-423-366

Received: 20 November 2018; Accepted: 21 December 2018; Published: 28 December 2018

Abstract: In order to achieve the objectives of the European 20/20/20 strategy, and to obtain a greater energy efficiency, integration of renewable energies and the reduction of carbon emissions, a District Heating (DH) system has been designed by the University of Valladolid (UVa), Spain, one of the most important DH fed by biomass fuel in Spain, supplying heating and domestic hot water (DHW) to 31 buildings in Valladolid, the majority of them, educational buildings on the University Campus. The aims of this paper were to study the change from an energy system fueled by natural gas to District Heating by biomass in a building on the campus of the University of Valladolid—the School of Engineering (EII)—studying its consumption from its connection to the District Heating system. An energy management methodology such as ISO 50001 is carried out, applied to efficiency systems in buildings, thus establishing new criteria of sustainability and economic value. In this paper, energy management will also be analyzed in accordance with the proposed tools of an Energy Management System (EMS) applied to the EII building, through the measurement of energy parameters, calculation of thermal consumption, thermal energy savings as a result of the change from system to District Heating by biomass, economic savings, reduction of environmental impact and indicators of thermal efficiency I_{100} and CUSUM indicator. Finally, the primary renewable and non-renewable energy efficiency indicators for the new District Heating system will be determined. The concept of the near Zero Energy Buildings is defined in the European Union (EU) in order to analyze an approach to an nZEB which results from replacing the natural gas heating system by a biomass District Heating system.

Keywords: district heating; biomass; energy management in renovated building; nZEB

1. Introduction

The spread of university campuses in recent years has led to a significant increase in energy consumption. The total amount of energy consumption includes lighting, DHW, heating or air conditioning systems in all campus buildings.

Due to the large amount of energy consumption in these buildings, a policy of promoting energy savings, energy management, supply of energy useful to the final energy consumption, perfectly planned at a local scale, can maintain a standard of energy consumed on campuses in a controlled environment. The University as an institution is one that can best stimulate energy saving among other institutions by setting an example of its actions, pertaining to the consumption that it requires (Figure 1).

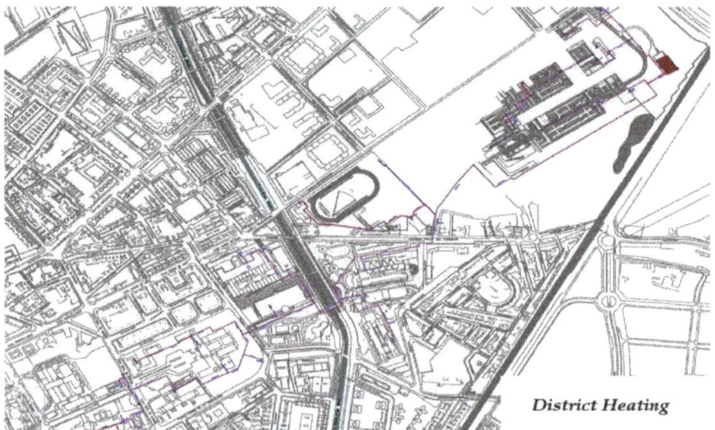

Figure 1. Plan view of connected buildings.

Among the different actions carried out by the University of Valladolid to reduce its carbon footprint, is the implementation of a DH system, making it to also the most important biomass DH in Spain, supplying heating and DHW to 31 buildings.

The thermal biomass generation plant has a power of 14 MW with a total distribution network of 12 km in length, to supply an annual energy consumption of 22 million kWh. The expected economic saving is 30% and the reduction in CO_2 emissions of 6800 $\frac{tonCO_2}{year}$. This represents a reduction in CO_2 emissions of around 30% of all our energy consumption (electricity and thermal) and an increase of 40% of energy production through renewables in our facilities.

The main purpose of this project is to achieve energy efficiency objectives and reduction of carbon emissions, through the implementation of the DH. The DH uses forest biomass as fuel, wood residues from near forests, a renewable energy source with low greenhouse gas emissions, which is cheaper than conventional fossil fuels, and allows us to achieve the objectives of the European strategy 20/20/20.

The use of renewable energy is an important commitment by the University of Valladolid to help control its carbon footprint through a significant reduction in CO_2 emissions. It also means reducing energy dependence on fossil fuels. The current natural gas boilers will be maintained to ensure the supply of heating in the event of a disconnection of District Heating, thus ensuring thermal comfort in buildings.

The biofuel to be used are wood chips, with a grain size range from G50 to G100 and humidity between 20% and 40%. Consumption is 7886 tons/year, of which the University will consume 6140 $\frac{tonCO_2}{year}$ (77.87%), 183.74 tons per year (2.33%) by buildings of the City Council of Valladolid, and 1561.43 $\frac{tonCO_2}{year}$ (19.80%) by buildings of the Government of Castilla y León.

The annual energy generation in 2016–2017 was 22,069 MWh per year to the whole of the network, of which 17,188 MWh correspond to the consumption of the UVa (77.87%), 515,180 kWh correspond to buildings of the City Council of Valladolid (2.33%) and 4,366,685 kWh belong to buildings of the Junta de Castilla y León (19.80%).

Some current research focuses on the study and application of the Thermal District concept. Mazhar et al. [1] conducted a review of district heating systems, which highlights the importance of modernizing heating systems to reduce greenhouse gas emissions and contribute to sustainability, for which district heating (DH) systems are considered the future trend. In the view of Gao et al. [2], District Heating offers greater advantages than traditional heating systems, including energy savings, regulation and control, which represents a great potential for development and a broad market outlook for the future. Regarding DH development, Lund et al. [3] presented a vision of the future of district heating systems and technologies and their role in a sustainable energy future. Wener [4] presented the background and current position of district heating and cooling in Sweden. The review structure considers market, technical, supply, environmental, institutional and future contexts. Lygnerud and Werner. [5] concluded in their study of 107 heat recovery systems in Swedish industry that the recovery of excess industrial heat using DH systems can be characterized by high political interest, high potential, low utilization and often high profitability. A paper submitted by Schmidt [6] exposes the cooperative work carried out together with the International Energy Agency (IEA) on thermal districts, determined that low-temperature district heating is a key technology that allows to increase the integration of renewable and residual energy for heating and cooling. Turski-Michal and Sekret-Robert [7] analyzed 63 heating stations in Poland in order to determine the energy effect of using buildings and the district heating network as thermal energy storage to compensate for the reduced heat output of the district heating system. In the investigation some parameters are often underestimated, such as the effect of the incidence of external temperature and the duration of outdoor temperature on the production of heat from DH systems. In Vienna, Fallahnejad et al. [8] studied the impacts of key DH parameters (ceilings, network costs, DH participation and extension), in which they determined that increasing DH market share in HD areas under a given network cost limit significantly reduces average network costs. Furthermore, that expanding the DH system without increasing market share in the DH areas does not effectively increase the DH share of the total heat demand and leads to higher average network costs. Restoration of existing buildings is considered a way to reduce energy use and carbon emissions in building stock. In [9] the impact of renovating a single-family home with different district heating systems under restoration packages chosen to represent typical but innovative ways to improve the energy efficiency and indoor climate of a single-family home was studied. In Bavaria, the performance of six small-scale district heating systems (DH) was monitored for 12 months in order to identify typical optimization potentials and develop standards for performance monitoring, analyze extensive operational data and evaluate Key Performance Indicators (KPIs). The KPI showed strong fluctuation and variation between different DH indicating that the main potential was found in the control of DH [10].

One of the challenges for near Zero Energy Consumption Buildings (nZEB) is thermal conditioning systems, Wu et al. [11] compared the energy performance of HVAC systems for a zero-grid energy residential building (nZEB) in different climate zones in the United States, where air conditioning represented 23.8% to 72.9% of the total energy of the building, depending on the air conditioning option and climate zone. In [12] whether nZEB standards applied to Italian school buildings guarantee good thermal conditions inside and which building configuration is the most favored was studied. Nielsen and Möller [13] examined the excess heat production of nZEBs in DH, determined that the excess heat of nZEBs can benefit DH systems by decreasing the production of fuel-using production units. In DH areas where the heat demand in the summer months is already covered by renewable energy, adding seasonal heat storage is essential to achieve nZEB status in terms of efficiency indicators as an alternative to track changes in efficiency and savings obtained according to the improvements implemented. Sekki et al. [14] showed that energy efficiency can be measured using alternative indicators and confirmed that different indicators have a different impact on the results showing efficiency. In the cases studied, energy savings can be achieved by investing in technical measures or operating the building automation system based on actual occupancy. On the other hand, Abu et al. [15] reviewed the Energy Efficiency Index (EEI) as an indicator to measure the performance of energy consumption in a building, with the forecast of energy consumption in a building being an important strategy to achieve the goal of reducing energy demand as well as improving energy efficiency. A review of how to quantify the Environmental Building Performance (EBP) is conducted by Maslesa et al. [16] through a systematic review of the literature where the information obtained is valuable to decision makers and facility managers in the process of implementing an environmental strategy and focusing on improving the EBP. It also concludes that buildings as products are complex structures with a long service life compared to most other products and induce considerable environmental impacts throughout their life cycle. As for the importance of integrating SGEn in the industrial and service sector, ISO 50001 [17] provides benefits for the industrial and service sectors. It is estimated that the standard, applied to different economic sectors, could influence savings of up to 60% in the world's energy consumption according to studies by Van der Hoeven [18]. In terms of methodologies to implement SGEn, Castrillón et al. [19] reported their results of the implementation of SGIE in a wet cement production facility, which showed an increase in energy efficiency associated with a reduction in electricity consumption of 4.6%, achieved without investment. This means that only through the innovation of processes through applied management technologies, in addition to the adoption of a culture of efficient energy management and continuous improvement. Benedetti et al. [20] also presented a new methodology for managing energy performance through the development, analysis and maintenance of energy performance indicators in manufacturing plants, taking into account the requirements of ISO 50001:2011 and ISO 50006:2014. The proposed methodology allows for the immediate identification of energy performance deviations from the manufacturing plant through the monitoring of Energy Performance Indicators over time and the identification of possible causes and responsibilities for such deviations. Saadi et al. [21] described the results of a study that has been carried out to reduce the energy consumption of a library building in the warm Omani climate. The project follows a systematic approach of collecting data from the maintenance department and projects, performing an energy audit and generating a building simulation model. To learn the real scenario of building energy management facilities, Afroz et al. [22] conducted a study in an institutional building at Murdoch University (Australia), which has incorporated leading-edge technologies over the past two decades. Through this case study analysis, relevant information is revealed that will bring benefits to energy management staff as well as researchers in this area, showing that an efficient energy management system in commercial or institutional buildings can reduce energy consumption and operating costs and provide a comfortable and healthy indoor environment. In terms of management tools and techniques as well as statistics, Castrillón and González [23] set out the procedures for validating the energy indicators and baselines needed in the energy planning of any institution and its subsequent implementation of an energy planning management system. This represents a reduction of 75% of emissions associated with the thermal supply of heating, allowing us to reduce our carbon footprint. This will also allow us to improve the energy certification of campus buildings. A building that is connected to a network improves its energy rating, an improvement that can be even greater when the network of biomass heaters is used, as in this case.

2. Case Study

The Engineering School (EII), is an educational building, mainly dedicated to teaching and research. It has a base area of 20,397 m^2 (16,691.4 m^2 useful), distributed over four floors and the orientation of the main façade is southwest. (Figure 2).

Figure 2. School of Industrial Engineering (EII) building.

Table 1 lists the features and facilities of the EII building.

Table 1. Features of the building.

System/Fuel	Lighting:	Heating:	HVAC	Other Equipment	Working Time and Set Point
Gas Boilers (NG)	LED and fluorescence with electromagnetic ballast T5	Natural Gas boilers of 540 kW "REMEHA" model "Gas-3B/13-Duo" $\eta = 0.78$	Individual Split-type equipment. 19 Air Handling Units (AHU)	Elevators, computers and laptops, printers, laboratory equipment, etc.	Weekdays (8 a.m. to 10 p.m.). Saturdays (9 a.m. to 2 p.m.). Set point (21 °C)
DH (Biomass)	LED and fluorescence with electromagnetic ballast T5	119,000 L of backup tank DH substation with flat plate heat exchanger Biomass Boilers 19 MW $\eta = 0.85$	Individual Split-type equipment 19 Air Handling Units (AHU)	Elevators, computers and laptops, printers, laboratory equipment, etc.	Weekdays (8 a.m. to 10 p.m.) Saturdays (9 a.m. to 2 p.m.) Set point (21 °C)

The heat exchange system between the DH and the EII is done through a secondary installation in the facilities of the building, formed by a flat plate exchanger, elements regulating and controlling the heating operation, and a discharge pump, which distributes the fluid to the different areas of the building (Figure 3).

Figure 3. Heat exchanger substation in the facilities at the EII Building.

The system consists of two circuits, the primary circuit, from the Central Heat Power (CHP) to the building facilities where the heat exchanger is located, and the secondary circuit, from the heat exchanger to the heat supply to the building.

The connection between the DH and the substation of the building is from underground pipes that lead to the installations of the building. Once the supply and return pipes reach the facilities at EII building, the pressure and temperature from the DH is regulated with the heat exchanger to the necessary conditions for the heating demand of the building. Both the supply and return temperatures of the system ought to be regulated in such a way that the temperature that reaches the heat exchanger from the central heating plant is 90 °C and the temperature at the outlet of the heat exchanger in the return pipe towards the central heating plant is close to 70 °C, with a thermal gap of approximately 20 °C.

In order to visualize, control and manage at any time the operation of the installation, a dynamic monitoring is developed through sensors and automatons that register data, which are exported and saved in a Supervisory Control And Data Acquisition (SCADA). In this way, any possible failure or warning can be read quickly and the exact location of the problem can be visualized (Figure 4).

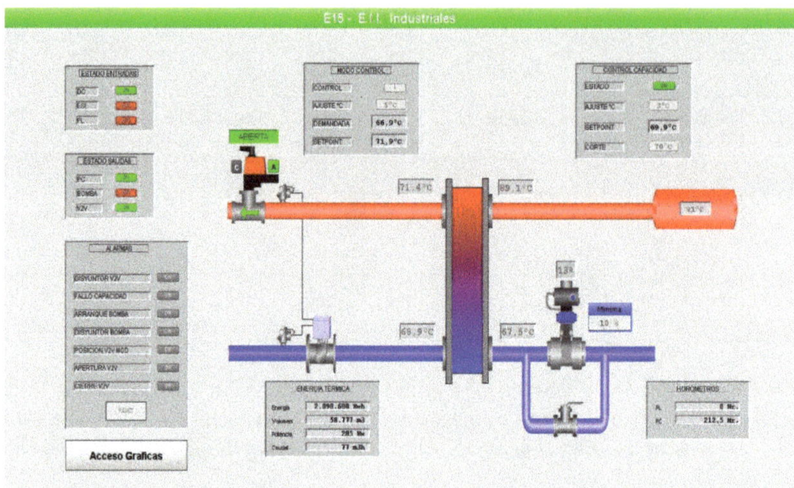

Figure 4. Screenshot of the EII Heat Exchanger substation SCADA.

The data provided by the implemented monitoring system is accessed via SCADA, which manages the DH generation and the heat exchange substations connected to DH. With the SCADA, all the energy parameters being measured by sensors can be visualized, represented and studied in depth. The consumption by the generation and energy demand are displayed in real time, in addition to being able to observe an instant control of the heat exchangers of each building.

The SCADA implemented in the system facilitates the analysis and studies of historical consumption and can generate reports on the development of each system together or separately, comparing values between the different buildings connected to the DH. In addition all this data can be exported to Excel files and simulation files, enabling their study through other software programs.

In addition, there are different sensors spread throughout the building that record the different energy and comfort parameters in real-time, delivered through a Modbus network to a dynamic monitoring tool, where are stored for subsequent study, using them as an energy management tool for the building.

The dynamic monitoring of energy parameters through SCADA, provides information on the use of energy consumed as an energy management tool. In this way, it has been possible to obtain the thermal and electrical energy consumption, where this consumption occurs and the conditions of the spaces where this energy is being supplied. Figure 5 shows the dynamic monitoring of the thermal parameters in the heating system, such as the power generated at the EII building during a period of nine standard winter days.

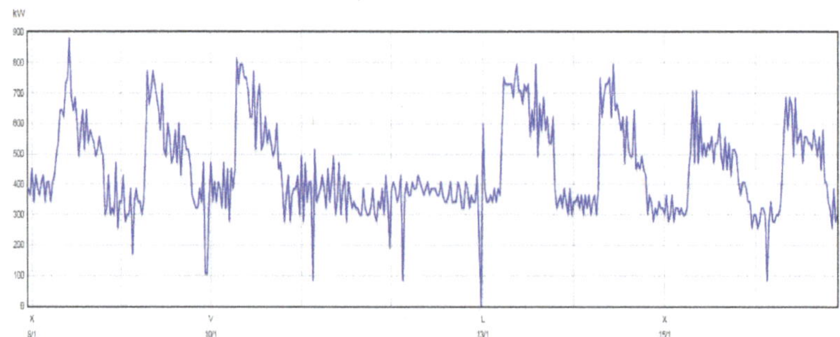

Figure 5. Screenshot of EII SCADA. Thermal power consumed in a winter period.

2.1. Thermal Consumption

The thermal primary energy consumption per year in kWh and the thermal efficiency indicator $\left(\frac{kWh}{m^2}\right)$ of Natural Gas in the EII Building at the UVa, between 2006 and 2017, are shown in Table 2, where it can be observed that in 2010 there was a maximum due to the extreme temperatures that winter, and the necessity of higher heating consumption. There is also a reduction in the thermal consumption of Natural Gas from 2015 to 2017, when the DH fueled by biomass begins to work, thus the NG boilers just work as backup.

Table 2. Natural Gas Consumption from 2006 to 2017.

Year	Natural Gas	
	KWh	**KWh/m^2**
2006	1,264,550	75.76
2007	1,415,661	84.81
2008	1,632,630	97.81
2009	1,585,339	94.98
2010	2,569,220	153.92
2011	1,639,719	98.24
2012	1,805,694	108.18
2013	1,510,476	90.49
2014	1,229,459	73.66
2015	1,131,687	67.8
2016	87,716	5.26
2017	14,912	0.88

In Table 3, the natural gas thermal consumption of in kWh and the thermal efficiency indicator $\left(\frac{kWh}{m^2}\right)$ are shown monthly during the last years, from 2014 to 2017.

Table 3. Natural Gas consumption and indicators from 2014 to 2017.

Month	2014		2015		2016		2017	
	kWh	kWh/m^2	kWh	kWh/m^2	kWh	kWh/m^2	kWh	kWh/m^2
January	258,821	15.51	377,448	22.61	33,617	2.01	2834	0.17
February	276,638	16.57	360,341	21.59	2294	0.14	0	0
March	208,852	12.51	243,979	14.62	12,292	0.74	851	0.05
April	116,325	6.97	95,293	5.71	12,395	0.74	5184	0.31
May	18,881	1.13	31,520	1.89	2912	0.17	4587	0.27
June	0	0	0	0	0	0	0	0
July	0	0.02	0	0	0	0	0	0
August	0	0	0	0	0	0	0	0
September	0	0	0	0	574	0.03	0	0
October	993	0.06	5905	0.35	11,345	0.68	672	0.04
November	173,818	10.41	13,498	0.81	8908	0.53	464	0.03
December	174,875	10.48	3703	0.22	3379	0.2	320	0.02

In the variation of the monthly NG thermal efficiency indicator $\left(\frac{kWh}{m^2}\right)$, of the EII in the years 2014 to 2017, it can be noted that the months of June, July, August and in some years the month of September, due to the fact that the external temperature is high and no heating is needed, there is no thermal consumption. However, there is a cost due to the payment of the fee for contracting this NG service. From September 2015, the DH was incorporated, this year still experimentally, which reduced the thermal consumption of NG as well as the cost of heating.

As for the cost, the annual consumption of NG for heating shows that the NG tariff was stable, with service contracting rates distributed as follows: $135.13\frac{\text{€}}{\text{kWh}}$ in 2014, $137.42\frac{\text{€}}{\text{kWh}}$ in 2015, $136.54\frac{\text{€}}{\text{kWh}}$

in 2016 and 108.66 $\frac{€}{kWh}$ in 2017. Another important consideration is that in September 2015, both the energy and cost decreased thanks to the supply of the DH network.

The biomass District Heating (DH) system began to provide heating service to the entire campus of the UVa and therefore the EII building in 2015/2016, with the year 2016/2017 being the second year of operation of the DH network. Table 4 shows the variation in thermal consumption of DH fed with biomass in kWh and the thermal efficiency indicator of biomass in the EII during the different months of the year.

Table 4. Primary energy consumption and energy efficiency indicator for DH fueled by biomass from 2015 to 2017.

Month	2015		2016		2017	
	KWh	KWh/m^2	KWh	KWh/m^2	KWh	KWh/m^2
January	0	0	167,500	10.04	283,300	16.97
February	0	0	224,400	13.44	205,600	12.32
March	0	0	164,700	9.87	169,800	10.17
April	0	0	126,700	7.59	53,800	3.22
May	0	0	18,900	1.13	13,800	0.83
June	0	0	0	0	0	0
July	0	0	0	0	0	0
August	0	0	0	0	0	0
September	0	0	0	0	0	0
October	53,000	3.18	28,700	1.72	8100	0.49
November	173,200	10.38	186,100	11.15	137,100	8.21
December	126,100	7.55	177,400	10.63	219,700	13.16

The evolution of thermal consumption of biomass, as well as the thermal efficiency indicator, shows the impact that the incorporation of viomass, through the DH system, has had for the EII as a substitute fuel for NG in recent years. Figure 6 shows the evolution of the thermal efficiency indicator for biomass as a function of the months in the annual period 2015–2017.

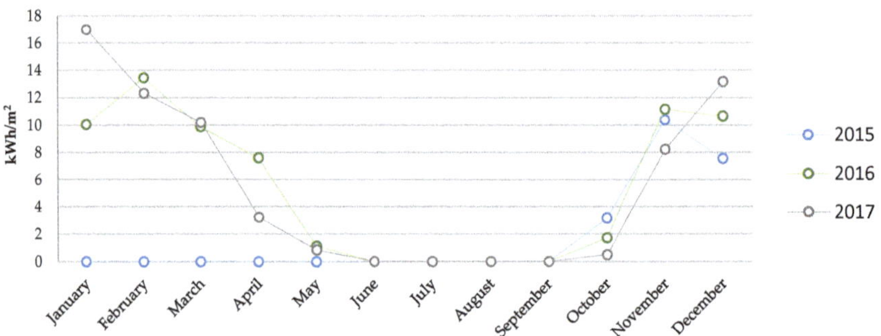

Figure 6. Evolution of the thermal efficiency indicator for biomass use 2015–2017.

With regards to the composition of the EII's energy matrix, in 2014 the NG heating system was installed, having only high thermal consumption in the months when the heating started. At that time the distribution of annual energy consumption corresponded to 56% thermal consumption of NG and the remaining 44% is a consumption of electricity.

In 2015, the implementation of the DH began, which began to heat the EII in September, however due to climate issues, heating consumption in this month was not necessary. Once the DH began to operate, there was a decrease in NG consumption, resulting in an annual distribution of thermal consumption for that year of 14% biomass, and 44% NG. Due to the fact that the DH was operating in experimental mode with frequent stoppages for verification and balancing, the EII was still using the old NG system to allow for proper heating. The remaining 42% corresponds to the electricity consumption of the EII.

In 2016, the DH fueled by biomass started to operate year round, with the support of the NG boiler system of the EII. The distribution of the energy matrix throughout this year corresponded to a thermal consumption of 51% of biomass consumption, and 4% of NG, and the remainder is electricity consumption. In this year and subsequent ones the percentage of NG consumption is very low compared to the biomass consumption. In 2017, the second year of operation of the DH network, the installation was already at an optimum point of operation due to its balance. In this way the distribution of energy consumption percentage of the EII, corresponds to a thermal consumption of 72% of biomass consumption and 1% of NG. As a result, NG consumption is negligible compared to biomass consumption. The annual analysis is shown in Table 5, where it is possible to decrease both the kWh/m^2 indicator and the economic expenditure €/m^2.

Table 5. Indicator of thermal efficiency $\left(\frac{\text{kWh}}{\text{m}^2}\right)$ and energy cost Indicator of biomass from 2015 to 2017.

Biomass	€	€/m^2	KWh	KWh/m^2
2015	28,359.93	1.7	352,300	21.11
2016	88,060.89	5.28	1,094,400	65.57
2017	87,803.40	5.26	1,091,200	65.43

Figure 7 shows the thermal consumption of the EII building from 2014 to 2017.

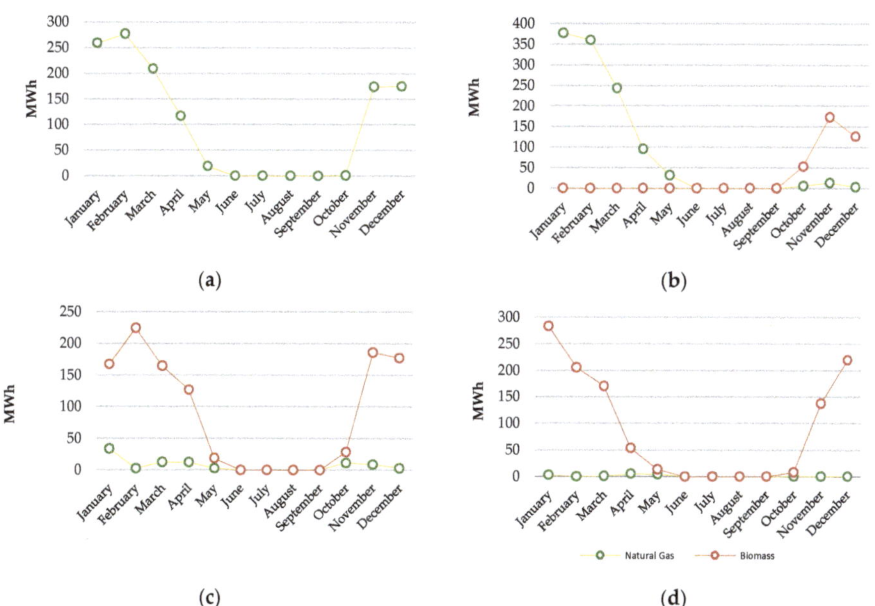

Figure 7. Thermal consumption of the EII building: (**a**) 2014; (**b**) 2015; (**c**) 2016; (**d**) 2017.

Figure 8 shows how from 2015 onwards, the energy cost of natural gas decreases and is replaced by the energy cost of biomass itself; each year the cost is lower as the operation of DH stabilizes.

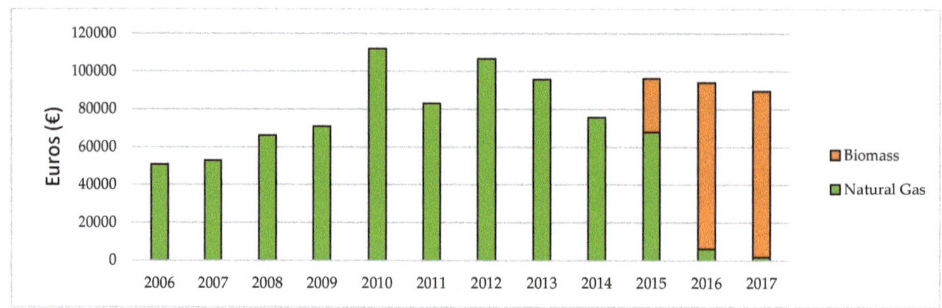

Figure 8. Evolution of the annual thermal cost of heating with NG and biomass DH.

Once the energy consumption, the indicators, their evolution over time have been analyzed, and establishing the energy saving by changing the heating system of the NG system for the new DH system with biomass, the methodology of the statistical model used in an energy management system ISO 50006 is proposed as an analysis. An energy consumption baseline is selected through a linear regression model, understanding that an Energy Baseline is a quantitative reference that provides the basis for comparing energy performance, allowing the evaluation of energy consumption. In this sense, with the aim of evaluating the impacts of both thermal consumption, economic cost and environmental impact referred to CO_2, in the replacement of NG by biomass for EII heating and starting from the Energy Baseline is possible to establish a procedure to predict the calculation of energy consumption of NG in a time period, where DH is used with biomass. The analysis, allows evaluating the energy saving between the District Heating (DH) with biomass and the previous system of NG in the temporary demonstrative period 2016–2017 for EII heating.

For the construction of the Energy Baseline, the NG energy consumption of the 2014/2015 academic year is obtained, the degrees day base 19, for that period of time reference calculated through the climatic data measured in Valladolid and selecting the days of each month where the NG heating system of the EII works. 19°C is experimentally justified because through dynamic monitoring with the SCADA system we are aware of the outdoor and indoor temperatures of the building. It also presents the monthly economic cost and the indicators $\left(\frac{\text{€}}{\text{m}^2}\right)$ and $\left(\frac{\text{€}}{\text{kWh}}\right)$. All of this data is presented in Table 6.

Table 6. Data for the calculation of the Energy Baseline.

Data	October 2014	November 2014	December 2014	January 2015	February 2015	March 2015	April 2015	May 2015
Degree days	29	138.1	137.4	227.8	238.6	166.6	59.3	43.3
Thermal Consumption of NG (kWh)	993	173,818	174,875	377,448	360,341	243,979	95,293	31,520
€	198.88	10,637.21	10,701.12	22,492.18	21,289.07	14,459.23	5421.08	1882.58
€/m²	0.01	0.64	0.64	1.35	1.28	0.87	0.32	0.11
€/kWh	0.2003	0.0612	0.0612	0.0596	0.0591	0.0593	0.0569	0.0597

Figure 9 represents the Energy Baseline for NG in this reference period, where we relate the thermal consumption of NG kWh versus Degree Days base 19.

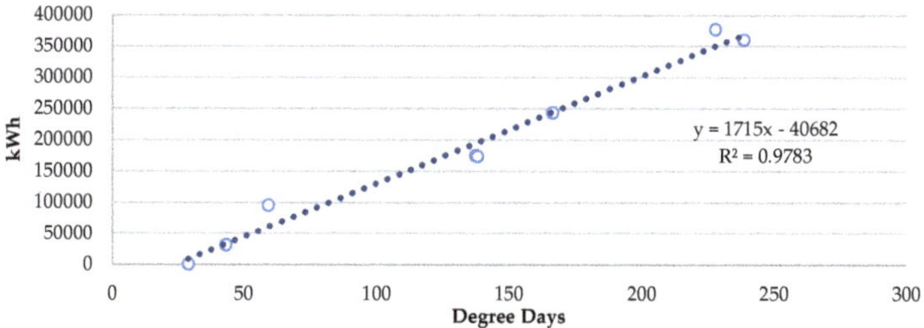

Figure 9. kWh vs. Degree days.

The interpretation of this Energy Baseline is based on the fact that thermal energy consumption is linearly related to the Degree Days, through the equation:

$$E = m \times (\text{Degree Days}) + E_0 \tag{1}$$

where there is a fixed energy consumption. E_0 does not depend on the calculated degrees days and another variable part required to carry out the heating process of the building, $m \times$ (Degree Days). From the regression model the linear equation of the energy baseline is obtained:

$$C_{(\text{Thermal Consumption})} = 1.715 \times (\text{Degree Days}) - 40.682 \tag{2}$$

In this way, it will be possible to obtain the thermal consumption of the monthly NG in a new temporary heating demonstration period. 2016–2017 can be simply substituted in this linear equation by the new base 19 Degree Days, obtained from a meteorological station in Valladolid for the year 2016/2017, taking the days of operation of the EII heating system.

It is important to interpret these results with respect to the negative E_0, i.e., the thermal consumption for Degree Days, where it can be observed that as the Degree Days is reduced, base 19 reaches a point at which, due to the inertia and insulation of the building, the thermal loads of heat transfer through it with the exterior are balanced with the internal thermal loads and radiation, so that the temperature of thermal comfort is kept constant and it is not necessary to provide heating for that building. As the base 19 Degree Days continues to be reduced to zero the energy consumption becomes negative, physically this means that the building generates more internal load and radiation than thermal loads by transfer through the building whose energy balance to maintain constant comfort temperature it is necessary to cool the building. Another important consideration is the correlation analysis that allows to measure the degree of relationship between two variables (E and Degree Days). The measurement of this intensity is evaluated through the correlation coefficient "R2", indicating that the closer to 1 as in our case, 0.9783 stronger will be the linear association between the two variables.

Once validated, the previous linear correlation equation through its Energy Baseline, (Figure 9), we are going to use it to calculate the thermal consumptions of heating in the EII of NG that we would have if the installation of NG continued in the hypothesis that would not have been changed by the new DH system with biomass. For this we use the Degree Days base 19 calculated from the meteorological data in Valladolid 2016/2017. Table 7 shows the results obtained, and also includes the monthly economic costs.

Table 7. Monthly data obtained from thermal NG consumption and economic cost with linear regression for the 2016/2017 demonstration period.

Data	October 2016	November 2016	December 2016	January 2017	February 2017	March 2017	April 2017	May 2017
Degree days	63.2	190.6	145.1	203.3	179.7	152.1	67.8	29.5
Thermal Consumption of NG (kWh)	67,706	286,197	208,164.5	307,977.5	267,503.5	220,169.5	75,595	9910.5
€	9583	41,477	35,389	53,808	22,139	56,641	10,996	1470

Table 8 shows the monthly data for the year 2016–2017, (demonstration period) of the real thermal consumption of NG and DH biomass for heating, with their respective cost. It can be seen that every month has a consumption of biomass for DH but also some thermal consumption of NG, because the DH system in days and hours in 2016–2017 has not been able to maintain all the thermal demand for heating of the EII Building. As can be observed, there are some thermal power peaks, except in February when the DH operated completely without the backup of NG boilers. However, in relation to the economic cost, despite the fact that the thermal consumption of NG in the month of February is non-existent, a fixed rate has to be paid to the NG distributor, the amount of this fee was 108.66 $\frac{€}{kWh}$.

Table 8. Monthly consumption data for DH fueled by biomass in the EII building from 2016 to 2017.

Data	October 2016	November 2016	December 2016	January 2017	February 2017	March 2017	April 2017	May 2017
NG Consumption (kWh)	11,345	8908	3379	2834	0	851	5184	4587
€ (NG)	692.99	574.22	302.57	260.61	108.66	148.5	325.05	300.78
€/kWh (NG)	0.06	0.06	0.09	0.09	0.00	0.17	0.06	0.07
Biomass (kWh)	28,700	186,100	177,400	283,300	205,600	169,800	53,800	13,800
€ (Biomass)	2309.34	14,974.54	14,274.49	23,445.41	17,016.28	14,053.33	4452.7	1142.14
€/kWh (Biomass)	0.08	0.08	0.08	0.08	0.08	0.08	0.08	0.08
€/kWh (TOTAL)	0.07	0.08	0.08	0.08	0.08	0.08	0.08	0.08

With all these data, will proceed to make economic and environmental savings calculations to compare for the EII building, the NG heating system against the DH system fueled by biomass.

In the regression, a total thermal consumption of NG of 1,443,223 kWh has been obtained for the annual demonstration period 2016-17 (Table 7) and in the DH system fueled by biomass plus NG backup system, it is 1,155,588 kWh (Table 8). Therefore, the annual thermal consumption savings and the percentage savings by comparing these two heating systems for the EII building are:

$$kWh_{NG\ (linear\ Regression)} = 1,443,223\ kWh \tag{3}$$

$$kWh_{16/17\ (Real)} = kWh_{DH} + kWh_{NG} = 1,118,500 + 37,088 = 1,155,588\ kWh \tag{4}$$

$$Thermal\ Savings = kWh_{NG\ (linear\ Regression)} - kWh_{16/17\ (Real)} = 1,443,223.5 - 1,155,588 = 287,635.5\ kWh \tag{5}$$

A 19.9 % savings in thermal consumption is obtained by supplying heating in the EII building by replacing the NG boilers with a DH system fueled by biomass. In relation to the economic costs per year, the same method of calculation is carried out as for energy. The data are obtained from Tables 7 and 8:

$$€_{NG\ (linear\ Regression)} = 231.507,87\ € \tag{6}$$

$$€_{16/17\ (Real)} = €_{DH} + €_{NG} = 91,668.23 + 2,713.38 = 94,381.61\ € \tag{7}$$

Therefore, the annual Economic Saving in heating comparing both energy systems and its percentage are:

$$Economic\ Savings = €_{NG\ (linear\ Regression)} - €_{16/17\ (Real)} = 231,507.87 - 94,381.61 = 137,126.26\ € \tag{8}$$

A 59.2% economic savings are obtained for the EII building with the DH system fueled by biomass, compared to the original energy system fueled by NG. In order to analyze the environmental impact, it is necessary to take into account the "F" factor of the type of fuel in current Spanish standards:

$$F_{NG\ (Thermal)} = 0.252\ \frac{kg\ CO_2}{kWh}$$

$$F_{Biomass\ (Thermal)} = 0.018\ \frac{kg\ CO_2}{kWh}$$

For the environmental impact assessment of the change of the DH system fueled by biomass versus the old energy system fueled by NG, Carbon emissions per year are calculated on the heating consumption at the EII building, between 2016 to 2017, for each energy system. If the energy system fueled by NG were working, its carbon emissions would be:

$$\text{kg}_{CO_2\,\binom{NG}{\text{(linear Regression)}}} = C_{\binom{NG}{\text{(linear Regression)}}} \times F_{NG} \tag{9}$$

$$\text{kg}_{CO_2\,\binom{NG}{\text{(linear Regression)}}} = 1,443,223.5 \text{ kWh} \times 0.252\,\frac{\text{kg } CO_2}{\text{kWh}} \tag{10}$$

$$\text{kg}_{CO_2\,\binom{NG}{\text{(linear Regression)}}} = 363,692.322 \text{ kg}_{CO_2} \tag{11}$$

To calculate the carbon emissions of the current energy system (DH fueled by biomass + energy system fueled by NG) it is carried out in the same way:

$$\text{kg}_{CO_2\,\binom{NG}{\text{(Real)}}} = \left[\text{kWh}_{DH} \times F_{Biomasa} + \text{kWh}_{NG} \times F_{NG}\right] \tag{12}$$

$$\text{kg}_{CO_2\,\binom{NG}{\text{(Real)}}} = \left[1,118,500 \text{ kWh} \times 0.018\,\frac{\text{kg } CO_2}{\text{kWh}} + 37,088 \text{ kWh} \times 0.252\,\frac{\text{kg } CO_2}{\text{kWh}}\right] \tag{13}$$

$$\text{kg}_{CO_2\,\binom{NG}{\text{(Real)}}} = 29,479.176 \text{ kg}_{CO_2} \tag{14}$$

The environmental impact resulting from the heating system replacement in the EII building, achieves for the 2016–2017 period an annual reduction of carbon emissions of:

$$\text{kg}_{CO_{2(Savings)}} = \text{kg}_{CO_2\,\binom{NG}{\text{(linear Regression)}}} - \text{kg}_{CO_2\,\binom{DH}{\text{(Real)}}} - \text{kg}_{CO_2\,\binom{NG}{\text{(Real)}}} \tag{15}$$

$$= 363,692.322 \text{ kg}_{CO_2} - 29,479.176 \text{ kg}_{CO_2}$$

$$\text{kg}_{CO_{2(Savings)}} = 334,213 \text{ kg}_{CO_2} \tag{16}$$

A 93.2 % savings in environmental impact with the current energy system compared to the previous one:

$$\% \text{ Increased Renewable Energy} = \frac{\binom{DH}{\text{(Biomass)}}}{\binom{NG}{\text{(linear Regression)}}} = \frac{1,118,500}{1,443,223.5} \times 100 = 77.5\% \tag{17}$$

Figure 10 shows the performance between DH fueled by biomass versus the NG energy system compared to environmental economic energy savings and increased renewable energy versus the natural gas system from 2016 to 2017.

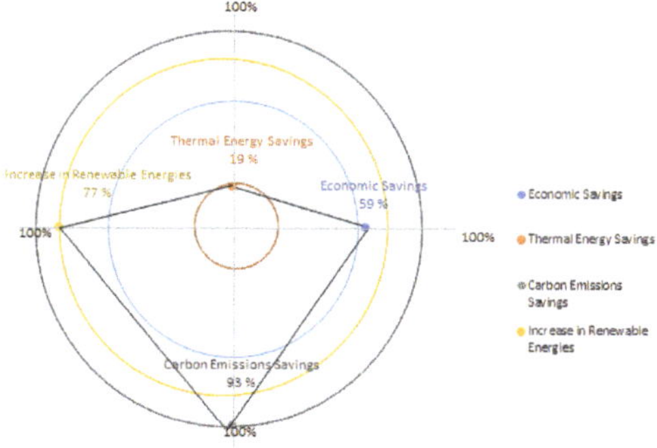

Figure 10. Performance between DH fueled by biomass versus NG energy system from 2016 to 2017.

2.1.1. Basic Efficiency Indicator I_{100}

In order to evaluate the energy efficiency of the new heating system for DH EII with biomass, the use of the base I_{100} is presented as a methodology since it is an energy management tool that allows comparing the performance of the results of the monthly thermal consumption of heating with DH of biomass, measured in a process during an operative period 2016–2017, with respect to the base or real energy consumption of the same, taking as a reference the dimensionless value of 100:

The Basic Efficiency Indicator 100 is:

$$I_{100} = \frac{E_{\text{Regression Baseline}}}{E_r} \tag{18}$$

where $E_{\text{Regression Baseline}}$: Energy that should have been consumed according to baseline:

$$E = m \times (\text{Degree Days}) + E_0$$

The result of the Basic Efficiency Indicator can be analyzed according to three states, which are within the numerical ranges <100, >100, =100. The interpretation of the efficiency indicator is shown in Table 9.

Table 9. Efficiency Indicator Interpretation.

Case	IDE	
1st Case	IDE = 100	Measured Consumption equal to projected with the base equation
2nd Case	IDE < 100	There is a decrease in energy performance since the energy consumption was higher than the baseline. the energy consumed or measured greater than the baseline energy calculated process with overconsumption or inefficient
3rd Case	IDE > 100	There is an improvement in energy performance since energy consumption was lower than the baseline.

In the DH system fueled by biomass, the monthly Basic Efficiency Indicator for the period 2016–2017 is obtained from the actual and base monthly thermal consumption by means of the Energy Baseline. The data are presented in Table 10.

Table 10. Monthly Basic Efficiency Indicator for the 2016 to 2017.

Efficiency Indicator	October 2016	November 2016	December 2016	January 2017	February 2017	March 2017	April 2017	May 2017
Actual thermal consumption (2016/2017)	40,045	195,008	180,779	286,134	205,600	170,651	58,984	18,387
Basic consumption (regression model)	67,706	286,197	208,164.5	307,977.5	267,503.5	220,169.5	75,595	9,910.5
Basic Efficiency Indicator	169.07	146.76	115.15	107.63	130.11	129.02	128.16	53.90

Figure 11 shows that most of the months analyzed correspond to values above 100, which indicate a compliance zone for energy performance (increased system efficiency). In May, the value of the indicator is less than 100, which indicates a decrease in efficiency that can be explained, insofar as the demand for heating is much lower than the power generated by the biomass boilers of DH, in addition to being a point of low energy consumption of heat.

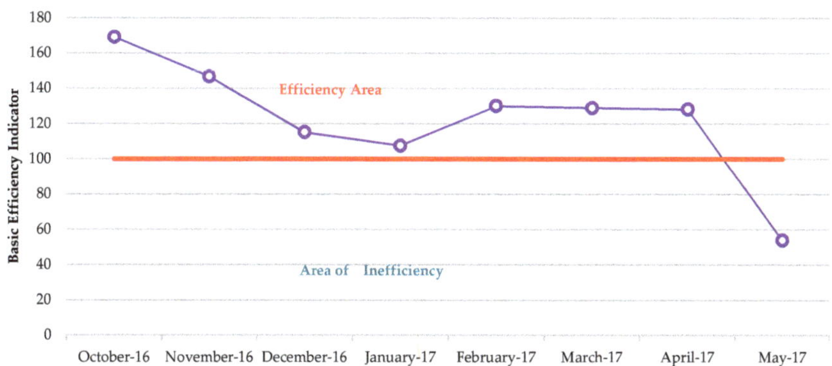

Figure 11. Basic Efficiency Indicator.

2.1.2. Cusum Indicator

With the aim of determining quantitatively the magnitude of the energy that has not been consumed, in other words, savings. Besides, the amount that has been overconsumed. The variation of this indicator for the DH system is presented, using for its construction the equation:

$$\text{Cusum Indicator} = (E_{\text{measured}} - E_{\text{trend}})_i + (E_{\text{measured}} - E_{\text{trend}})_{i-1} \tag{19}$$

Table 11 shows the cusum indicators for the period 2016 to 2017.

Table 11. Base data for indicator CUSUM for the period 2016 to 2017.

Month	Energy Linear Regression 2016/2017 [kWh]	Actual Energy [kWh]	Actual Energy-Energy Linear Regression [kWh]	Cusum [kWh]
October 2016	67,706	40,045	−27,661	−27,661
November 2016	286,197	195,008	−91,189	−118,850
December 2016	208,164.50	180,779	−27,386	−146,236
January 2017	307,977.50	286,134	−21,844	−168,079
February 2017	267,503.50	205,600	−61,904	−229,983
March 2017	220,169.50	170,651	−49,519	−279,501
April 2017	75,595	58,984	−16,611	−296,112
May 2017	9910.50	18,387	8477	−287,636

Therefore, this indicator shows any energy savings or over-consumption, or in economic terms related to energy, that the EII Building has had. The figure is used to track the trend of the process, in terms of the variation of its energy consumption, with respect to an energy period or baseline used as a reference. Figure 12 shows the CUSUM Indicator since 2016 to 2017.

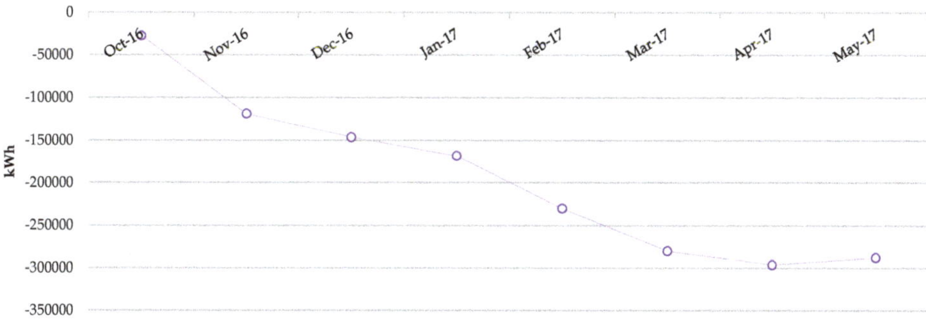

Figure 12. CUSUM indicator for the period 2016 to 2017.

If there is no change in consumption patterns, the CUSUM curve will simply fluctuate around a horizontal line. However, any change that produces savings or overconsumption will cause the curve to change direction: a downward trend in the curve means savings, while upward trend means losses.

2.1.3. The Primary Energy Indicator

All supplied and exported energies are shown in a unique indicator. The Primary Energy Indicator [24]. It is estimated from the energy supplied and exported using the national primary energy factor provided by Equation (20):

$$E_{p,nrem} = \sum_i (E_{del,i} f_{del,nren,i}) - \sum_i \left(E_{exp,i} f_{del,exp,i} \right) \tag{20}$$

$$EP_p = \frac{E_{p,nrem}}{A_{net}} \tag{21}$$

The near zero performance is determined by a Member State in terms of the achievable use of primary energy, the amount of primary energy supplied by renewable energies, the funding sources available for renewable energies or measures of energy efficiency, and the funding requirements as well as the successful rate of the definition.

The nZEBs such as a national use of energy which is both technologically and economically feasible of $> 0 \frac{kWh}{m^2 \cdot y}$, but no higher than the existing Member State threshold for primary non-renewable energy, generated by a mix of best-practice eco-efficiency measures plus renewable energy systems that might or might not be cost-optimal.

Spain has not yet laid down this limited value; however a draft document already exists for the Spanish Building Standard, in which bases for the nZEB definition are argued. December 2016, Spain launched a report to improve the evaluation methodology in high-efficiency buildings. This report suggests several predefined thresholds based on those defined in July 2016, in Commission Recommendation (EU) 2016/1318.

In Valladolid, which is marked by a continental climate, such upper limits are 85 to 100 $\frac{kWh}{m^2 \cdot y}$, of primary energy consumption in total on these buildings per year. At this point, the upper limits for non-renewable sources are between 45 and 55 $\frac{kWh}{m^2 \cdot y}$, and the contribution of renewable energies to primary energy must be greater than 45 $\frac{kWh}{m^2 \cdot y}$.

In order to assess compliance with upper limits for primary energy, primary energy intensity at the buildings is established. The methodology for the determination applies the approach defined in the general EPBD regulation. Primary energy is obtained as the biomass heat being multiplied by its respective Primary Energy Factor (PEF). Moreover, in Spain, these factors are established in its own normative, the Spanish Thermal Systems Standard. [25]. The Table 12 shows the primary energy factors by each energy sources.

Table 12. Spanish Building Standards. PEF established.

Energy	Source	Use	$f_{p,ren}$	$f_{p,nren}$
Electricity	Grid	Input	0.414	1.954
Biomass	On-site	Input	1.003	0.034
Natural Gas	On-site	Input	0.005	1.190

Non-renewable conversion factor f_i applied to the electricity, in Spain, is currently 1.954. Biomass conversion factor is 1. At the EII Building there is still no export energy, so the primary energy indicator is:

$$EP_p = 118.15 \frac{kWh}{m^2 \cdot y} \tag{22}$$

This is caused by the main factors established by Spain, which has a higher percentage of renewable energy (17%) from the main electricity grid. by calculating the on-site RES share; it is possible to obtain a high value of 91.21 $\frac{kWh}{m^2 \cdot y}$ of contribution of renewable primary energies in situ at EII Building. A primary energy comparison between the EII Building versus the assumptions of the European Union Directive is shown in Figure 13.

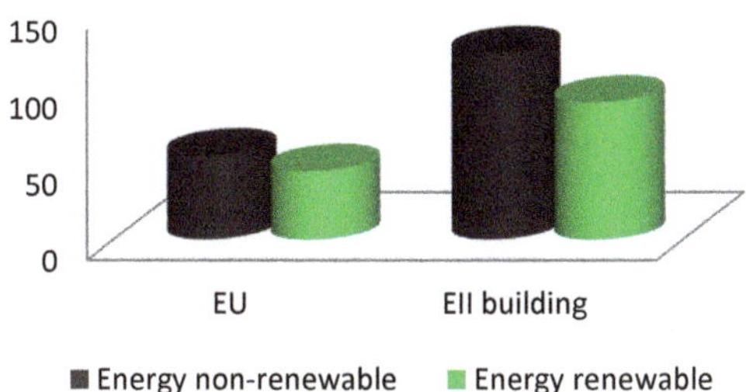

Figure 13. Primary energy comparison between EII Building and EU Directive.

The detailed boundary values of the system for determining the energy supplied and exported are added based on the reference limits of EN15603. As started in the EPBD recast, the positive influence of renewable energy produced on site is taken into account, so that it reduces the amount of energy supplied needed and can be exported when the demand of buildings is exceeded [26].

2.1.4. Renewable Energy Ratio

To calculate the amount of renewable energies used through the Renewable Energy Ratio (RER), all sources of renewable energy must be taken into account [24]. Solar thermal energy, photovoltaic energy, wind and hydro power, renewable energy from environmental heat sources such as heat pumps and free cooling, renewable biofuels and off-site renewable energy are included. The environmental heat sources used in heat pumps and free-cooling should be considered within the limit of the use of renewable energy system, because in the calculation of the RER, free cooling and heat pumps are not only accounted for in the calculation of the energy supplied on the basis of the COP, it is also taken into account in the energy obtained from the environmental heat sources.

The proportion of renewable energies are obtained in relation to the energy use at the building, as total primary energy. It keeps in mind that the exported energy compensates for the energy delivered. By default, the export energies are deemed to compensate for the grid mix or for thermal energy, the grid mix of district cooling or heating. For near and on-site renewable energies, the overall primary energy factor is 1.0, while the primary non-renewable energy factor is 0.

The total primary energy-based RER is given by the following Equation (23):

$$RER_p = \frac{\sum_i E_{ren,i} + \sum_i ((f_{del,tot,i} - f_{del,nrem,i}) E_{del,i})}{\sum_i E_{ren,i} + \sum_i (E_{del,i} f_{del,tot,i}) - \sum_i (E_{exp,i} f_{exp,tot,i})} \tag{23}$$

In order to get the RER_p for the EII building, $f_i = 1.954$ is taken for electricity and $f_i = 1.003$ for the biomass and $f_i = 0.005$ for natural gas:

$$RER_p = 0.424 \tag{24}$$

Thus, the RER, the third indicator, in the EII Building is 0.424, a figure that represents a pretty strong result for a nZEB.

3. Conclusions

A new analysis methodology is developed, based on the ISO 50001 standard, for the implementation of new thermal systems in building renovations using DH. Following the objectives proposed by the EU 20-20-20, in this research paper several indicators have been studied to analyze how the change from a NG heating system to a new heating system by a DH fueled by biomass on an educational building, Engineering School (EII) at the University of Valladolid in Spain.

The proposed indicators as well as the methodology carried out fulfill the established methodology in an SGEn ISO 50001, in order to analyze the reduction of thermal energy and the heating efficiency of the building with the new DH as well as the economic costs and environmental impact. Finally, three new indicators of non-renewable primary energy, renewable energy and RER are determined in accordance with the methodology proposed by EU EN15603 and EPBD 2018 on nZEB buildings.

With them it is possible to determine the level of impact in the change towards the DH, without altering the energy demand of the building, looking for this existing building to look like a building nZEB.

The results obtained show that the new DH system fueled by biomass in the period 2016–2017 achieved a percentage of thermal savings of 19.9 %, an economic saving of 59.2%, a carbon emissions saving of 93.2 %, and an increase in renewable energy of 77.5 %.

The I_{100} indicator has also been calculated and it can be seen that the DH system works in an energy efficiency zone, as well as the Cusum indicator where its graph shows that the DH over the entire time interval has a tendency to improve energy efficiency. The I_{100} indicator has also been calculated and it can be seen that the DH system works in an energy efficiency zone, as well as the Cusum Indicator where its graph shows that the DH over the entire time interval has a tendency to improve energy efficiency.

The non-renewable and renewable primary energy indicators of the new DH system fueled by biomass have been calculated at 118.15 $\frac{kWh}{m^2 \cdot y}$, 91.21 $\frac{kWh}{m^2 \cdot y}$, respectively. They have been compared by the values that the EU proposes for a city like Valladolid and it is observed that the non-renewable primary energy is of the order of the double, this is due that although the DH reduces in a 40% with respect to the old system of NG, to the lack of any improvement in the energetic demand of the building EII does that this value is still high.

On the other hand, however, the comparison of primary renewable energy from DH reaches a value of double, so that the integration of renewable energy by the change from NG to biomass gives a high RER of 0.4. Use of these indicators proposed by the EU enables us to conclude that the change in the EII building of the new DH system fueled by biomass achieves an approximation towards the nZEB building status.

Author Contributions: Conceptualization F.J.R.M., J.S.J.A. and E.V.G.; Methodology, F.J.R.M., R.C.M. and J.M.R.H.; Validation, F.J.R.M., J.S.J.A. and E.V.G.; Formal Analysis, R.C.M. and J.M.R.H.; Investigation, R.C.M. and J.M.R.H.; Resources, F.J.R.M., J.S.J.A. and E.V.G.; Data Curation, R.C.M.; Writing—Original Draft Preparation, R.C.M.; Writing—Review & Editing, J.M.R.H.; Visualization, F.J.R.M., J.S.J.A. and E.V.G.; Supervision, F.J.R.M.

Funding: This research received no external funding.

Acknowledgments: The support of La Dirección de Investigaciones y Desarrollo Tecnológico (DIDT) de la Universidad Autónoma de Occidente (UAO) in Cali, Colombia, and the RETO GIRTER Project (New Intelligent Manager for Thermal Networks), project funded by the European Regional Development Fund through the Ministry of Economy, Industry and Competitiveness of the Government of Spain, have made this research paper feasible.

Conflicts of Interest: The authors declare no conflict of interest.

Abbreviations

DH:	District Heating.
NG:	Natural Gas.
DHW:	Domestic Hot Water.
EII:	Engineering School of Valladolid.
UVa:	University of Valladolid.
EMS:	Energy Management System.
nZEB:	near Zero Energy Building.
IEA:	International Energy Agency.
KPI:	Key Performance Indicators.
HVAC:	Heat Ventilation Air Conditioning.
EEI:	Energy Efficiency Index.
EBP:	Environmental Building Performance.
LED:	Light Emitting Diode.
ISO:	International Organization for Standardization.
SCADA:	Supervisory Control And Data Acquisition.
EP_p:	Primary energy indicator ($\frac{\text{kWh}}{\text{m}^2 \cdot \text{y}}$).
$E_\text{p,nrem}$:	Primary Non-renewable energy ($\frac{\text{kWh}}{\text{y}}$).
$E_\text{del,i}$:	Energy supplied on-site or around the site ($\frac{\text{kWh}}{\text{y}}$).
$f_\text{del,exp,i}$:	Primary Non-renewable energy of energy supplied compensated by export-energy, which is by default the same as the factor of the energy supplied, if not defined at national level otherwise.
$f_\text{del,nren,i}$:	Primary Non-renewable energy for the energy supplied.
$E_\text{exp,i}$:	Export-Energy on site or around the site ($\frac{\text{kWh}}{\text{y}}$).
A_net:	Usable space (m^2).
RER_p:	Renewable Energy Ratio from primary energy.
$E_\text{ren,i}$:	Renewable energy generated on-site or around the site, ($\frac{\text{kWh}}{\text{y}}$).
$f_\text{del,tot,i}$:	Total primary energy factor for the delivered energy carrier i.
$f_\text{exp.tot,i}$:	Total primary energy factor of the delivered energy compensated by the exported energy for energy carrier i.

References

1. Mazhar, A.R.; Liu, S.; Shukla, A. A state of art review on the district heating systems. *Renew. Sustain. Energy Rev.* **2018**, *96*, 420–439. [CrossRef]
2. Gao, L.; Cui, X.; Ni, J.; Lei, W.; Huang, T.; Bai, C.; Yang, J. Technologies in Smart District Heating System. *Energy Procedia* **2017**, *142*, 1829–1834. [CrossRef]
3. Lund, H.; Duic, N.; Østergaard, P.A.; Mathiesen, B.V. Future district heating systems and technologies: On the role of smart energy systems and 4th generation district heating. *Energy* **2018**, *165*, 614–619. [CrossRef]
4. Werner, S. District heating and cooling in Sweden. *Energy* **2017**, *126*, 419–429. [CrossRef]
5. Lygnerud, K.; Werner, S. Risk assessment of industrial excess heat recovery in district heating systems. *Energy* **2018**, *151*, 430–441. [CrossRef]
6. Schmidt, D. Low Temperature District Heating for Future Energy Systems. *Energy Procedia* **2018**, *149*, 595–604. [CrossRef]
7. Turski, M.; Sekret, R. Buildings and a district heating network as thermal energy storages in the district heating system. *Energy Build.* **2018**, *179*, 49–56. [CrossRef]
8. Fallahnejad, M.; Hartner, M.; Kranzl, L.; Fritz, S. Impact of distribution and transmission investment costs of district heating systems on district heating potential. *Energy Procedia* **2018**, *149*, 141–150. [CrossRef]
9. Lidberg, T.; Gustafsson, M.; Myhren, J.A.; Olofsson, T.; Ödlund (former Trygg), L. Environmental impact of energy refurbishment of buildings within different district heating systems. *Appl. Energy* **2018**, *227*, 231–238. [CrossRef]
10. Bücker, D.; Jell, P.; Botsch, R. Performance monitoring of rural district heating systems. *Energy Procedia* **2018**, *149*, 5–14. [CrossRef]

11. Wu, W.; Skye, H.M. Net-zero nation: HVAC and PV systems for residential net-zero energy buildings across the United States. *Energy Convers. Manag.* **2018**, *177*, 605–628. [CrossRef]

12. Zinzi, M.; Pagliaro, F.; Agnoli, S.; Bisegna, F.; Iatauro, D. Assessing the overheating risks in Italian existing school buildings renovated with nZEB targets. *Energy Procedia* **2017**, *142*, 2517–2524. [CrossRef]

13. Nielsen, S.; Möller, B. Excess heat production of future net zero energy buildings within district heating areas in Denmark. *Energy* **2012**, *48*, 23–31. [CrossRef]

14. Sekki, T.; Airaksinen, M.; Saari, A. Effect of energy measures on the values of energy efficiency indicators in Finnish daycare and school buildings. *Energy Build.* **2017**, *139*, 124–132. [CrossRef]

15. Abu Bakar, N.N.; Hassan, M.Y.; Abdullah, H.; Rahman, H.A.; Abdullah, M.P.; Hussin, F.; Bandi, M. Energy efficiency index as an indicator for measuring building energy performance: A review. *Renew. Sustain. Energy Rev.* **2015**, *44*, 1–11. [CrossRef]

16. Maslesa, E.; Jensen, P.A.; Birkved, M. Indicators for quantifying environmental building performance: A systematic literature review. *J. Build. Eng.* **2018**, *19*, 552–560. [CrossRef]

17. International Organization for Standardization (ISO). Available online: https://www.iso.org/ (accessed on 14 September 2018).

18. Van der Hoeven, M. Energy Efficiency Indicators: Essentials for Policy Making. Available online: https://webstore.iea.org/energy-efficiency-indicators-essentials-for-policy-making (accessed on 5 November 2018).

19. González, A.J.; Castrillón, R.; Quispe, E.C. Energy Efficiency Improvement in the Cement Industry through Energy Management. In Proceedings of the 2012 IEEE-IAS/PCA 54th Cement Industry Technical Conference, San Antonio, TX, USA, 14–17 May 2012. [CrossRef]

20. Benedetti, M.; Cesarotti, V.; Introna, V. From energy targets setting to energy-aware operations control and back: An advanced methodology for energy efficient manufacturing. *J. Clean. Prod.* **2017**, *167*, 1518–1533. [CrossRef]

21. Al-Saadi, S.N.J.; Ramaswamy, M.; Al-Rashdi, H.; Al-Mamari, M.; Al-Abri, M. Energy Management Strategies for a Governmental Building in Oman. *Energy Procedia* **2017**, *141*, 206–210. [CrossRef]

22. Afroz, Z.; Higgins, G.; Urmee, T.; Shafiullah, G. Technological Advancement of Energy Management Facility of Institutional Buildings: A Case Study. *Energy Procedia* **2017**, *142*, 3088–3095. [CrossRef]

23. Castrillon, R.; González, A. *Metodología Para la Planificación Energética a Partir de la Norma ISO 50001*; Editorial Universidad Autónoma de Occidente: Cali, Colombia, 2018; ISBN 978-958-8994-59-8.

24. Rey-Hernández, J.M.; Velasco-Gómez, E.; San José-Alonso, J.F.; Tejero-González, A.; Rey-Martínez, F.J. Energy analysis at a near zero energy building. A case-study in Spain. *Energies* **2018**, *11*, 857. [CrossRef]

25. CTE (Spanish Technical Building Code). Available online: http://www.codigotecnico.org (accessed on 15 September 2018).

26. European Directives. Available online: http://eur-lex.europa.eu/ (accessed on 14 September 2018).

 *energies*

Review

Sustainable District Cooling Systems: Status, Challenges, and Future Opportunities, with Emphasis on Cooling-Dominated Regions

Valerie Eveloy * and Dereje S. Ayou

Department of Mechanical Engineering, Khalifa University of Science and Technology, P.O. Box 2533, Abu Dhabi, UAE; dere8020@gmail.com
* Correspondence: valerie.eveloy@ku.ac.ae; Tel.: +971-02-607-5533

Received: 7 December 2018; Accepted: 8 January 2019; Published: 13 January 2019

Abstract: A review of current and future district cooling (DC) technologies, operational, economic, and environmental aspects, and analysis and optimization methodologies is presented, focusing on the demands of cooling-dominated regions. Sustainable energy sources (i.e., renewable, waste/excess electricity and heat, natural/artificial cold) and cooling/storage technology options with emphasis on heat-driven refrigeration, and their integrations in published DC design and analysis studies are reviewed. Published DC system analysis, modeling, and optimization methodologies are analyzed in terms of their objectives, scope, sustainability-related criteria, and key findings. The current and future development of DC in the Gulf Cooperation Council (GCC) region, a major developing cooling-dominated market, is examined more specifically in terms of current and future energy sources and their use, and economic, environmental, and regulatory aspects, with potential technical and non-technical solutions identified to address regional DC sustainability challenges. From the review of published DC design and analysis studies presented, collective research trends in key thematic areas are analyzed, with suggested future research themes proposed towards the sustainability enhancement of DC systems in predominantly hot climates.

Keywords: district cooling; space cooling; air-conditioning; hot climate; thermally activated cooling; sustainable energy; Gulf Cooperation Council

---

1. Introduction

Energy sustainability, security and climate change are major threats of today's and future generations. Substantial reductions in anthropogenic greenhouse gas (GHG) emissions must be an integral part of sustainable energy production and use to limit the global rise in ambient temperatures over the coming decades [1]. Emissions reduction over the next two decades will be challenged by a projected 33% increase in global primary energy demand over the coming 25 years, 70% of which are anticipated to arise from the electricity sector [2]. This demand growth will be mainly contributed by the industrialization and electrification of non-OECD (Organization for Economic Cooperation and Development) economies, particularly China and India [2].

Buildings are responsible for approximately a third and 50% of global energy and electricity uses, respectively, and 20% of energy-related GHG emissions [1]. Approximately 99% of air-conditioning and refrigeration loads worldwide are met by electricity. The associated annual electricity demand (and carbon dioxide (CO_2) emissions) has tripled between 1990 and 2016, and currently represents 10% of global electricity use [3]. This demand is expected to triple again to 6200 TWh by 2050 in the baseline International Energy Agency (IEA) scenario, with 70% of this rise attributable to residential users, unless effective energy conservation and efficiency measures are adopted [3]. This growth will take place essentially due to rises in population in developing economies that are in hot-climate regions, and seek improved comfort (i.e., India, China, Indonesia, the Middle East) [3]. The cooling energy consumption of typical buildings in hot and hot/humid climates is up to three times higher than in moderate climates [1]. Furthermore, space cooling loads in hot climates are typically characterized by large seasonal and daily variations, that induce strain on electricity grids. Whereas air-conditioning represents on average 14% of peak electricity demand worldwide [3], in hot climates such as in for example the Gulf Cooperating Council (GCC) region, where the deployment of district cooling (DC) is discussed as part of this article, this demand represents approximately 50% [4,5] and up to 70% [3,5] of total and peak electricity consumptions, respectively. (The GCC comprises six Arabian states, namely Bahrain, Kuwait, Qatar, Oman, Saudi Arabia, and the United Arab Emirates (UAE). The GCC region extends over 2,500,000 km^2 area, with a population of 52.7 million in 2015 [6].) Similarly, in South Asian regions, where the energy demand of residential/service buildings accounts for approximately 60% of total energy consumption, approximately 44% and 50–57% of residential and commercial/office building electricity consumption is associated with space cooling [7]. Building cooling requirements will be exacerbated by climate change, as reflected by the measured augmentation in the number of cooling degree days in several regions [8]. Additional factors that contribute to a growing building cooling demand include building architectures, rising internal heat loads, and urban heat island effects.

At present, most air-conditioning loads worldwide are met by conventional on-site cooling systems (Figure 1a), consisting of either window units (i.e., split systems) applied in single rooms, apartment units or small buildings, or central air-cooled or water-cooled chillers, which tend to be located on the rooftop or basement of large buildings. In such systems cold energy is produced and distributed at the end-user's site [9]. The efficiency of on-site air-conditioning equipment varies widely depending on design and operating conditions but is generally half of that achievable with best commercially available technologies [3]. By contrast in DC systems (Figure 1b), cold production is centralized in a central chiller plant and delivered to end-users via a distribution network infrastructure and energy transfer stations (ETSs). The central chiller plant includes cooling equipment, pumps, heat-rejection equipment, a chemical treatment unit, controller, and other devices [10]. Indirect cold production by the central chiller can be supplemented by direct cooling provided by an available cold energy source. Other associated units include cold storage, pump stations and control systems. ETSs consist of heat exchangers and distribution/regulation valves [9].

DC systems can offer significant advantages over conventional standalone chiller plants installed in individual buildings, or industrial or commercial facilities. These advantages include (i) low energy requirements—for a given cooling demand, DC systems generally consume less energy than on-site cooling systems, mainly due to large-scale central water-cooled chiller plants being more efficient than on-site small-capacity air-cooled systems; (ii) efficient and flexible capacity use to fulfill load diversity and variability, as a result of DC system design and installation; (iii) peak-period saving potential; (iv) lower unit cost of cooling, due to lower energy-, maintenance- and construction costs; (v) reduced environmental impact—emissions are not only reduced but also more easily handled at a remote, centralized chiller plant than at individual building's air-conditioning systems; (vi) more reliable service (i.e., reliability in excess of 99.7%), because of high standard industrial equipment, backup chillers, and the availability of professional, ongoing operation and maintenance support, and longer life span (i.e., 25–30 years) than for conventional on-site air-conditioners (i.e., 10–15 years); (vii) space savings at the end-user site, since DC systems are remotely located [9–11]. The advantages of DC systems are most pronounced in dense districts exposed to hot climatic conditions throughout the year and characterized by rapid urbanization and building developments [9,12], as encountered in for example the GCC region.

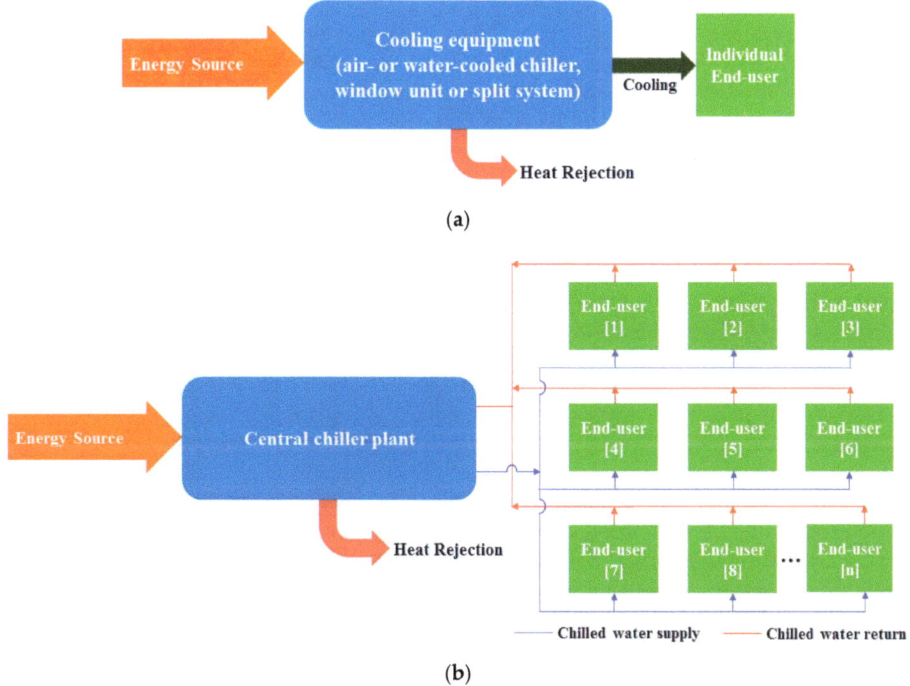

Figure 1. Building air-conditioning systems: (**a**) conventional on-site systems, (**b**) DC systems.

However, considering the substantial investments associated with double digit megawatt-cooling capacities and multi-kilometer distribution networks, DC systems should be carefully designed, evaluated, and optimized to permit a well-organized and cost-effective system to be ultimately deployed. Greater technology- and non-technology-related challenges exist for DC than district heating, with the former significantly less and later implemented than heating systems [8].

Historically, the United States of America (USA) were the main player in the early-stage implementation of contemporary commercial DC systems at district-scale (i.e., universities, airports, healthcare campuses, business districts) and then city-scale in the 1930's, the deployment of which grew significantly in the mid to late 1990s [9,13]. DC implementation in Canada followed a similar timeline [9,14], but at limited deployment scale due to affordable hydroelectricity and fossil energy, and lack of large dense districts [12]. The implementation of DC in Europe began in France and Germany in the 1960's and gradually spread to other European countries, mainly for summer air conditioning [15]. DC in Asia was introduced in Japan in 1970 where it expanded rapidly under government intervention towards higher efficiency and reduced environmental emissions [16]. After introducing DC in Beijing in 2004, China also actively deployed this technology [16]. Introduced in the GCC in 1999, DC implementation has since made significant progress, favored by rapid urbanization and building developments, particularly in the UAE [17]. Today, the Americas (led by the USA) hold 43% of the global installed DC capacity (i.e., 12.6 MRT or 44.3 GW_{th}), followed by the GCC (32%), Asia-Pacific and Africa (19%, led by Japan), and Europe (5%, led by Sweden and France) [14,17]. By 2019, the Middle East and North African (MENA) DC market is anticipated to overtake the American one and become the largest DC market [14].

To date, a limited number of reviews related to DC have been published [8,12,15,16,18–20]. Only Gang et al. [16] and Palm and Gustafsson [19] reviews were dedicated to DC, while [8,12,15,18,20] collectively considered district heating and cooling, with emphasis on heating. Gang et al. [16]

focused on the integration of renewable energy and combined cooling, heating, and power (CCHP) technologies, optimization, and projects in China. Palm and Gustafsson [19] analyzed technical, economic, environmental, and policy-related obstacles and enabling factors for the expansion of DC systems in Sweden. Rezaie and Rosen [12], Lake et al. [18] and Werner [15] reviewed district heating/cooling systems, in terms of technical, economic, environmental, and institutional/policy aspects. Werner [15] included some content specific to the DC market and cold sources, and highlighted the significantly smaller number of research publications and information on actual DC systems in comparison with district heating ones. Vandermeulen et al. [20] discussed control strategies for exploiting flexibility in district heating/cooling systems, to support increasing shares of fluctuating renewables in future energy systems. Pellegrini and Bianchini [8] defined the concept of cold district heating and cooling networks, that combine centralized energy distribution with minimized heat losses in supply, and their suitability for cold delivery.

The present article is intended to address the above gap through a review of DC design and analysis efforts that have aimed at improving the sustainability of cooling and dehumidification of buildings in cooling-dominated regions. These efforts are discussed in terms of DC cooling technologies and associated energy sources, cold energy distribution infrastructure, DC operation, analysis and optimization, economics, environmental impact, and challenges and opportunities.

This article is structured as follows. In Section 2, DC energy sources and associated cooling technologies, system configurations, cool thermal storage and cooling energy distribution network infrastructures are reviewed, with emphasis on heat-driven cooling technologies, and technologies suitable for high ambient temperature/humidity conditions. In Section 3, DC analysis, modeling and optimization methodologies are discussed. Considering the extreme and rising cooling demand of GCC countries, yet their limited exploitation of DC to date, Section 4 examines the current and future development of this technology specifically in the region, in terms of challenges, benefits, market, and potential solutions for improved sustainability. Based on the information compiled in Sections 2–4, collective research trends are identified in Section 5, leading to suggested future research themes in the design and analysis of sustainable DC systems for predominantly hot/humid climates. This article closes with concluding remarks in Section 6.

2. Sustainable District Cooling Systems

In the following sub-sections, DC cooling and thermal energy storage technologies, energy sources, distribution network infrastructure, operational aspects, and extension of DC end-uses to non-space cooling applications, are reviewed based on implemented DC technology information and published DC design and analysis research, focusing on the needs of cooling-dominated regions. In addition to space cooling technologies with present applicability to DC, alternative technologies in development are also discussed (Section 2.1.1). Renewable and waste/excess heat sources and their possible exploitations in DC systems are reviewed in Section 2.1.2. Natural and artificial cold energy sources and their potential applications are then contrasted in Section 2.1.3. Key aspects of cold energy storage and distribution, DC operation, and DC non-space cooling applications, are analyzed in Sections 2.2–2.5.

Published design and analysis DC studies are compiled in Table 1, in terms of geographical location, analysis timeline, energy sources (i.e., electrical, heat and cold), DC integration with other sectors, cooling and thermal energy storage technologies, cold energy distribution network, estimated energy savings and environmental benefits, and economics. This information is used in Sections 2.1–2.5 to analyze the potential of various energy sources and DC design/operational features to contribute to DC sustainability improvements, and to identify collective DC research trends and gaps in Section 5. The present emphasis is on analyses of DC systems with sustainable attributes and for cooling-dominated climates, complemented by DC studies considering other regions but with design/analysis features of value to hot climates. The studies in Table 1 are listed by chronological order of publication, to highlight developments in published DC research.

Table 1. Overview of published design and analysis studies of DC systems with sustainable energy attributes.

Source; Geographical Location; Analysis Timeline	Cross-Sectorial Integrations; Energy Sources	Cooling/Energy Storage Technologies and Capacities	Cold Energy Distribution Network Characteristics	Energy Savings and Environmental Benefits	Economic Benefits
Chan et al. (2006) [21]; Hong Kong (timeline N/R)	Integration: None (isolated DCS) Energy source: Grid electricity	Cooling: Seawater-cooled electrical vapor compression chillers (1×1500 RT and 7×5000 RT) with variable-speed seawater pumps Storage: Ice tanks, in either series storage-led, or series chiller-led arrangements with chillers	Primary (production) and secondary (distribution) loops with constant-speed/flow and variable-speed pumps, respectively	Energy savings: • DCS with ice storage and chiller-priority control consumes less electricity annually than for storage-priority control, regardless of storage fraction • Electricity consumption is minimum for ice storage with chiller-priority control at 40% partial storage Environmental benefits: N/R	Annual DCS electricity cost minimum for ice storage system with chiller-priority control at 60% partial storage under Guangdong Province Electricity Supply (GPES) tariff structure; however, prohibitive payback period
Chan et al. (2007) [22]; Hong Kong; (timeline N/R)	Integration: None (DCS distribution network only) Energy sources: N/R	N/R	Radial-shaped, tree-shaped, and mix of radial/tree-shaped networks	N/R	Piping network configuration optimization methodology developed to minimize piping investment and water distribution pumping energy cost
Trygg and Amiri (2007) [23]; Norrköping, Sweden (2004)	Integration: DCS, DHS and CHP Energy sources: • Waste-, biofuel-, rubber- and oil-fired CHP plant electricity and heat • Oil-fired boiler heat replaced by NG-fired CC heat for DHS • Free nearby river cooling water	Cooling: • Low-temperature absorption chillers (COP, 0.5–0.7) • Electrical vapor compression chillers (COP, 3–5) Storage: N/R	N/R	Energy savings: N/R Environmental benefits: Global CO_2 emissions reduced by up to 80% using absorption cooling only relative to compression cooling only	• Production cost of cooling (including investment and operating cost) reduced by 170% using absorption cooling only relative to compression cooling only • 302% reduction in system cost (i.e., capital, O & M, fuel, taxes, fees) due to electricity savings using absorption cooling only relative to compression cooling only
Söderman (2007) [24]; Finland (2006, 2020)	Integration: None (isolated DCS) Energy source: Unspecified electricity	Cooling: Electrical vapor compression chillers (COP, 3) • 2006: 1 centralized site, 8.3 MW$_{th}$ • 2020: 2 centralized sites, 42.4 MW$_{th}$ and 43.1 MW$_{th}$, and 6 distributed sites totaling 14 MW$_{th}$ Storage: Cold water in above-ground steel tanks or underground basins (2006: 3525 kW, 3650 m^3); 2020: 37 MW, 37,750 m^3)	Chilled water transported to consumer from either distributed (i.e., local) chiller plant or cold storage tank via direct pipeline, or from central DC chiller plant or cold storage tank via DC mainlines	N/R	Annual DC investment and operating costs minimized to identify optimum cooling plants locations/capacities, storage media/capacities and distribution pipeline routing

Table 1. *Cont.*

Source; Geographical Location; Analysis Timeline	Cross-Sectorial Integrations; Energy Sources	Cooling/Energy Storage Technologies and Capacities	Cold Energy Distribution Network Characteristics	Energy Savings and Environmental Benefits	Economic Benefits
Feng and Long (2010) [25]; China (timeline N/R)	Integration: None (isolated DCS) Energy sources: • Unspecified electricity • Free river water cooling	Cooling: Electrical vapor compression chillers (COP, 5.1); design cooling load, 18,989 kW Storage: N/R	Pipe network with 42 nodes, 65 branches and 24 heat exchanger substations, with pipe diameters to be optimized	N/R	DCS annual cost (i.e., annualized investment, O & M, amortization) reduced by 4.2% through optimization of pipe diameter, relative to conventional recommended-velocity pipe sizing method
Svensson and Moshfegh (2011) [26]; Södertälje, Sweden (2007 onwards)	Integration: DCS and DHS Energy sources: • Heat from waste-fired, peat/biomass-fired, oil-fired, and electric boilers • Coal-, oil- and waste-fired CHP plant electricity and heat • Free lake cooling water	Cooling: • Absorption chillers (COP, 0.7) • Electrical vapor compression chillers (COP, 2–5) Storage: N/R	N/R	Energy savings: • Electricity consumption reduced by ~40% through use of absorption chillers to replace electrical chillers and increased lake water cooling • Electricity/heat production-induced CO_2 emissions reduced through investments in condensing power	Investments in new absorption chillers to replace electrical chillers and increased lake water cooling optimized through minimization of system costs (including new absorption chillers, pipelines, pumps, electricity/heat production-induced CO_2 emissions); Investments profitable
Udomsri et al. (2011) [27]; Bangkok, Thailand (2009)	Integration: DCS and MSW-fired co-generation plants Energy sources: Waste heat and electricity from MSW-fired power plants	Cooling: • Single-effect (COP, 0.7–0.73), double-effect (COP, 1.2–1.3) and low-temperature (COP, 0.7) H_2O/LiBr absorption chillers (centralized: 14 × 10 MW; decentralized: 257 × 300 kW) • Electrical vapor compression chillers (centralized: COP, 2–4, 14 × 10 MW; decentralized: COP, 2–4, 257 × 300 kW) Storage: N/R	N/R	Energy savings: Cooling energy consumption reduced by 1 MW_{fuel}/$MW_{cooling}$ compared with electrical chillers-based cooling Environmental benefits: Emissions reduced by 0.13 kg_{CO2}/kWh of cooling (i.e., by 60%) using MSW co-generation coupled with absorption cooling compared with electrical chillers-based cooling	• Total investment and annual O&M costs of MSW co-generation with 14 × 10 MW-absorption chillers: M$149.49 and M$28.6, respectively; 4.7 years payback period • Total investment and annual O&M costs of MSW co-generation and 14 × 10 MW-electrical chillers: M$139.44 and M$32.04, respectively; 4.9 years payback period
Al-Qattan et al. (2014) [6]; Kuwait (timeline N/R)	Integration: DCS and SOFC-GT co-generation plant Energy sources: • Waste heat from NG-fired SOFC-GT power plant • Electricity from NG-fired SOFC-GT power plant	Cooling: • Double-effect H_2O/LiBr absorption chillers (2620 RT; COP, 1.3) • Electrical vapor compression chillers (20,680 RT; rated efficiency, 0.61 kW/RT) Storage: Cold tanks (20,868 RT h)	N/R	Energy savings: Fuel energy consumption reduced by 54%, peak electrical power consumption reduced by 57%, and fuel-to-cooling efficiency improved by 346%, all relative to PACUs Environmental benefits: Annual CO_2 emissions reduced by 50% relative to PACUs	Cost of per ton-hour of cooling reduced by 53% relative to PACUs

Table 1. *Cont.*

Source; Geographical Location; Analysis Timeline	Cross-Sectorial Integrations; Energy Sources	Cooling/Energy Storage Technologies and Capacities	Cold Energy Distribution Network Characteristics	Energy Savings and Environmental Benefits	Economic Benefits
Ondeck et al. (2015) [28]; Austin, USA (2011)	Integration: DCS, NG-fired district-level island-mode CHP system, and DHS Energy sources: • Waste heat from NG-fired GT; heat from auxiliary NG-fired boiler • Electricity from NG-fired GT and rooftop PVs	Cooling (48,000 tons/hrs total chilled water capacity): • Steam-driven absorption chillers • Electrical vapor compression chillers Storage: Chilled water system (39,000 ton/hr capacity)	N/R	N/R	Profit from electricity sales to grid and neighborhood, cooling sales and some heating sales to neighborhood, maximized at $421,434 for selected analysis week (i.e., July 1–7) via scheduling optimization; Profit halved without PV integration
Erdem et al. (2015) [29]; Turkey (timeline N/R)	Integration: DCS, DHS and coal-fired power plants Energy sources: Steam extracted from inlet stage of LP turbine in eight existing coal-fired power plants	Cooling: • Single-stage absorption chillers (COP, capacity N/R) • Two-stage absorption chillers (COP, 1.0) Storage: N/R	N/R	Energy savings: Cooling energy consumption reduced (not quantified); Use of 30% LP steam extraction from coal power plant for absorption cooling improves the modified first-law efficiency for co-generation of electrical power and cooling by ~3–5.5% Environmental benefits: emissions reduced (not quantified)	N/R
Marugán-Cruz et al. (2015) [30]; Spain (2012)	Integration: DCS and solar CSP tower Energy sources: • Excess heat from solar heliostats • Auxiliary heat from NG-fired burner	Cooling: Steam-driven double-effect H_2O/LiBr absorption chillers (8 × 11.63 MW; COP, 1.4) Storage: Chilled water tank	Rectangular grid-type district network including 10 km primary and 46.4 km secondary networks	Energy savings: N/R Environmental benefits: Annual CO_2 emissions reduced by 19,870 Tm/year due to use of excess solar heat for cooling instead of fossil-fuel-derived electricity	Total annual net earnings and net present value (TNPV) of €106 million and €38 million, respectively, for assumed economic parameters, due to substitution of fossil-fuel-derived electricity by excess solar heat; up to 75% savings on consumer electricity bills
Perdichizzi et al. (2015) [31]; Abu Dhabi, UAE (timeline N/R)	Integration: DCS, solar thermal PTC field and CC Energy sources: • Heat from CC (i.e., steam extracted from ST LP turbine) fed to absorption chillers • Heat from solar PTC fed to CC HRSG • Electricity from CC	Cooling: • Steam-driven double-effect H_2O/LiBr absorption chillers (4 × 20.7 MW; COP, 1.31) • Electrical vapor compression chillers (COP, 2.2–4.0) Storage: N/R	None	Energy savings: • Fossil-fuel savings of 26% and 33% in winter and summer days, relative to fossil-fueled electricity compression cooling only • Peak electricity demand reduced by ~8 MW (20%) and ~25 MW (31%) in winter and summer days, respectively, relative to fossil-fueled electricity compression cooling only Environmental benefits: N/R	Non-quantified economic savings associated with reduced electricity consumption, reduced peak demand charges, and reduced CC investment cost

Table 1. *Cont.*

Source; Geographical Location; Analysis Timeline	Cross-Sectorial Integrations; Energy Sources	Cooling/Energy Storage Technologies and Capacities	Cold Energy Distribution Network Characteristics	Energy Savings and Environmental Benefits	Economic Benefits
Khir and Haouari (2015) [32]; Qatar (timeline N/R)	N/R	Cooling: Unspecified, generic chiller technology and capacity Storage: Chilled water tank	Tree-type piping network with size and layout to be optimized	N/R	Optimization methodology developed to minimize DCS investment and operating cost through optimization of chiller/storage tank capacities, piping network size/layout, and hourly cooling energy production and storage fraction
Karlsson and Nilsson (2015) [33]; Sweden (timeline N/R)	Integration: DCS, biomass-based CHP and DHS Energy sources: Excess heat from CHP heat-driven flash pyrolysis process oil condenser with sequential vapor condensation	Cooling: Single-stage H_2O/LiBr absorption chillers (COP, 0.78) Storage: N/R	N/R	Energy savings Annual CHP and pyrolysis efficiency improved by 1.3% and 6%, respectively; annual CHP electrical power production increased by 8.6–18.7% Environmental benefits: N/R	N/R
Gang et al. (2016) [34]; Hong Kong (timeline N/R)	Integration: None (isolated DCS) Energy source: Grid electricity	Cooling: Electrical vapor compression chillers (Configuration-1: 5 × 17,500 kW and 2 × 8750 kW; Configuration-2: 7 × 15,000 kW) Storage: Ice system	Chiller (primary) and cooling water system (secondary) pumps with 40 m and 20 m heads, respectively	Energy savings: DCS energy consumption has 80% probability of being overestimated using conventional design method that does not account for uncertainties in outdoor/indoor conditions and building design/construction Environmental benefits: N/R	Annual operating cost of DCS with ice storage, optimized using uncertainty-based method (i.e., uncertainties in outdoor weather, indoor conditions and building design/construction), has 80% probability of being lower than that of DCS optimized using conventional method
Gang et al. (2016) [35]; Hong Kong (timeline N/R)	Integration: Isolated DCS Energy source: Grid electricity	Cooling: Electrical vapor compression chillers (either 7 chillers of equal capacities with 105 MW total capacity; or 6 chillers of equal capacities and 1 chiller with 50% smaller capacity) Storage: N/R	Chiller (primary) constant-speed pumps	Energy savings: Using an uncertainty-and reliability-based design method, the use of chillers of different capacities has little and less impact on DCS than ICS energy savings Environmental benefits: N/R	The impacts of uncertainties in outdoor weather, indoor conditions and building design/construction and reliability on the design optimization, hence total annual capital, operating and availability risk cost, is smaller for DCS than ICSs
Ameri and Besharati (2016) [36]; Tehran, Iran (timeline N/R)	Integration: DCS/DHS/CCHP and DCS/DHS/CCHP/PV (i.e., energy scenarios 3 and 4, respectively) Energy sources: • NG-fired GT electricity and waste heat • Heat from NG-fired auxiliary boiler • Grid electricity • Solar PV electricity (scenario 4)	Cooling: • Absorption chillers (3 × 1725 kW, 1 × 5332 kW; COP 0.8) • Electrical vapor compression chillers (1 × 637 kW, 1 × 2426 kW, 1 × 3025 kW, 3 × 3159 kW, 1 × 3635 kW; COP 4.0) Storage: N/R	Four distribution pipelines with optimized heat transfer capacities	Energy savings: PES of 17.1–38.7% for scenarios 3–4, respectively, compared to conventional system (scenario 1) Environmental benefits: CO_2 emissions reduced by 35.8–52.6% for scenarios 3–4, respectively, compared to conventional system (scenario 1)	• Energy supply system initial and operating costs minimized to identify optimum CCHP equipment capacities and operational strategies • Energy costs reduced by 29.4–40.8% for scenarios 3–4, respectively, compared to conventional supply system (i.e., scenario 1 with grid electricity, NG-fired boilers heating, individual electrical chillers), with payback periods of 4.8–9.7 years

Table 1. *Cont.*

Source; Geographical Location; Analysis Timeline	Cross-Sectorial Integrations; Energy Sources	Cooling/Energy Storage Technologies and Capacities	Cold Energy Distribution Network Characteristics	Energy Savings and Environmental Benefits	Economic Benefits
Kang et al. (2017) [37]; Hong Kong (2015)	Integration: DCS and DGs Energy sources: • Electricity from distributed electric generators (i.e., gas engines) • Waste heat from distributed electric generators • Grid electricity	Cooling: • Absorption chillers (6 × 2054 kW; COP, 1.2) • Electrical vapor compression chillers (6 × 3980 kW; annual average COP, 5.36) Storage: N/R	Star-shaped chilled water distribution network	Energy savings: DGs-DCS primary energy savings of 9.6% relative to CES Environmental benefits: N/R	• DGs-DCS capacity optimized through minimization of payback period. DGs-DCS capital and annual operation cost of M$20.7 and M$9.2, respectively • CES capital and annual operation cost of M$7.2 and M$16.6, respectively • DGs-DCS additional capital payback period (relative to CES) of 1.9 years
Yan et al. (2017) [38]; Hong Kong (year N/R)	Integration: None (isolated DCS) Energy sources: N/R	N/R	• Buildings groups formed based on a newly defined grouping coefficient, to minimize secondary network variable-speed pumps' energy consumption • Lorenz curve and Gini coefficient applied to quantify (i) spatial/temporal inequality in normalized DCS buildings' load distribution (normalized to buildings' rated loads), and (ii) deviation in chillers' load from full load	Energy savings: • Low grouping coefficients (i.e., even buildings' hydraulic pressure requirements) lead to higher annual secondary pumps energy savings (up to 50%) and higher DCS energy savings (up to 4%) relative to ICSs • Low Gini coefficients (i.e., even distribution in buildings cooling loads) correlate with higher annual DCS energy efficiency, and higher annual chiller plant COP, and tend to lead to higher annual DCS energy savings relative to ICSs • Annual DCS energy savings of up to 14% when both the grouping coefficient and Gini coefficients are low, essentially due to grouping coefficient (i.e., chilled water distribution) rather than Gini coefficient (i.e., building cooling loads) Environmental benefits: N/R	N/R

Table 1. *Cont.*

Source; Geographical Location; Analysis Timeline	Cross-Sectorial Integrations; Energy Sources	Cooling/Energy Storage Technologies and Capacities	Cold Energy Distribution Network Characteristics	Energy Savings and Environmental Benefits	Economic Benefits
Coz et al. (2017) [39]; Slovenia (year N/R)	Integration: None (DCS network only) Energy sources: N/R	N/R	Network with either (i) polyurethane pre-insulated steel pipes or (ii) non-insulated polyethylene pipes	N/R	• Exergoeconomic optimal pipe diameters and insulation thicknesses determined to minimize the cost of cold (i.e., exergoeconomic product) for cooling capacities of 50–1500 kW • For the assumed economic parameters, pre-insulated steel pipes have higher exergy efficiency and lower exergoeconomic cost of cold than non-insulated polyethylene pipes
Dominković et al. (2018) [40]; South-west and Northern parts (Woodlands) of Singapore (2014, 2050)	Integration: DCS, GT/PV power plant and LNG regasification terminal Energy sources: • Waste heat from waste incineration plants and NG/LNG-fired GT power plants • Cold energy from LNG regasification plants (0.23 kWh of cold/kg of LNG) • PV electricity	Cooling: • Single-effect H_2O/LiBr absorption chillers (COP, 0.7) • Electrical vapor compression chillers Storage: Ice (7.7 GWh_{cold}; 90,536 m^3)	54 and 61 transmission and distribution pipe sections, respectively, for Singapore south-west part 4 and 5 transmission and distribution sections for Singapore Northern part (Woodlands)	Energy savings: Annual PES of 7.3–19.5% in 2050 DC and DC-PV energy scenarios, respectively, compared to business-as-usual (BAU) scenario without DC Environmental benefits: Annual 2050 energy system CO_2 emissions reduced by 19.7% and 41.5% in 2050 DC and DC-PV scenarios, compared to BAU scenario without DC	Singapore 2050 energy system DC grid investment cost (M\$339) offset by annual socio-economic savings of 32.7% and 38.4% in DC and DC-PV scenarios, respectively, compared to BAU scenario
Gao et al. (2018) [41]; Hong Kong (2015)	Integration: DCS and DES Energy sources: • Electricity from distributed electric generators • Waste heat from distributed electric generators • Grid electricity	Cooling: • Double-effect absorption chillers (4 × 4000 kW; COP, 1.1) • Electrical chillers (4 × 4000 kW; COP, 5.5) Storage: N/R	Constant- and variable-speed pumps for primary and secondary chilled water networks, respectively	Energy savings: Monthly PES of 10–19% with DCS-DES depending on control strategy, compared with grid-powered DCS plus grid-powered individual cooling systems; Environmental benefits: N/R	• DCS-DES payback period of 6.4–10.4 years, depending on control strategy

Table 1. *Cont.*

Source; Geographical Location; Analysis Timeline	Cross-Sectorial Integrations; Energy Sources	Cooling/Energy Storage Technologies and Capacities	Cold Energy Distribution Network Characteristics	Energy Savings and Environmental Benefits	Economic Benefits
Franchini et al. (2018) [42]; Riyadh, Saudi Arabia (timeline N/R)	Integration: DCS and solar thermal PTC/ETC field integration; Energy sources: Thermal energy from solar PTCs and ETCs	Cooling: • PTC heat-driven double-effect H_2O/LiBr absorption chillers (2315 kW; COP, 1.39) • ETC heat-driven single-effect H_2O/LiBr absorption chillers (3250 kW; COP, 0.723) Storage: • Pressurized and non-pressurized hot water storage tanks for double- and single-effect absorption chillers, respectively • Cold tank (200 m^3)	District piping network with insulated pipelines and 16.8 km overall length; design	Annual primary energy and electricity consumptions, and CO_2 emissions reduced by ~500 toe (i.e., 70%) and 1400 tons per year relative to DCS with centralized electrical compression chiller cooling plant	• System components' sizes optimized by minimization of primary cost of solar field, storage tank and chillers • Overall cost of DCS with PTC-driven double-effect absorption chillers 30% lower than that of DCS with ETC-driven single-effect absorption chillers
Ravelli et al. (2018) [43]; Riyadh, Saudi Arabia (timeline N/R)	Integration: DCS and solar PTC/CRS co-generation plant; Energy sources: • Thermal energy from either North-south or East-west oriented PTCs, or CRS heliostats • Electricity and heat from PTC/CRS solar field-driven SRC, with steam extracted from SRC LP turbine to drive thermal chillers	Cooling: • Double-effect H_2O/LiBr absorption chillers (4 × 20.7 MW; COP, 1.31) • Electrical vapor compression chillers (50 MW; COP, 2.5–5.0) Storage: Two-tank molten salt	N/R	Energy savings of the three solar field configurations, the CRS system requires the lowest amount of solar energy and produces the least excess heat Environmental benefits: N/R	• Investment cost of solar and storage systems minimized through optimization of solar field aperture area and storage tank volume • Of the three solar field configurations, the CRS system leads to the lowest investment cost

Note: CES—centralized power plant and individual cooling systems. CC—combined cycle. CHP—combined heat and power. CCHP—combined cooling, heating, and power. CRS—central receiver system. CSP—concentrated solar power. DES—distributed energy system. DCS—district cooling system. DG—distributed generators. DHS—district heating system. ETC—evacuated-tube collector. GT—gas turbine. HP—high pressure. HRSG—heat recovery steam generator. ICS—individual cooling system. LNG—liquefied natural gas. NG—natural gas. LP—low pressure. MSW—municipal solid waste. N/R—not reported. O&M—operation and maintenance. PES—primary energy savings. PACU—packaged air-conditioning unit. PTC—parabolic trough collector. PV—photovoltaic or present value. RT—ton of refrigeration. SRC—steam Rankine cycle. SOFC—solid oxide fuel cell.

2.1. Cooling Technologies and Associated Energy Sources

Grid electricity, generally predominantly produced using fossil fuels, is currently the main energy source to produce cold using vapor compression chillers in DC systems. Figure 2 provides an overview of alternative sustainable energy sources and heat-activated cooling technology options, some of which have either been practically implemented in DC systems, previously evaluated in the DC literature, or are envisaged here. Sustainable energy sources include low-carbon (i.e., renewable, low-emission, waste/excess) electricity and heat to drive compression chillers and heat-activated chillers, respectively, as well as natural and artificial (e.g., cryogenic) cold for direct cooling. Key characteristics of sustainable heat-activated air-conditioning technologies presently applicable to DC central chiller plants, namely vapor compression and absorption refrigeration, are discussed in Section 2.1.1. Alternative refrigeration cycle options having features suitable for renewable/waste heat use and/or operation in high ambient temperature/humidity conditions are also identified, and their potential advantages and shortcomings discussed (Section 2.1.1). Sustainable heat sources for thermally activated cooling and cold sources, and their exploitations in published DC studies, are then reviewed in Sections 2.1.2 and 2.1.3, respectively.

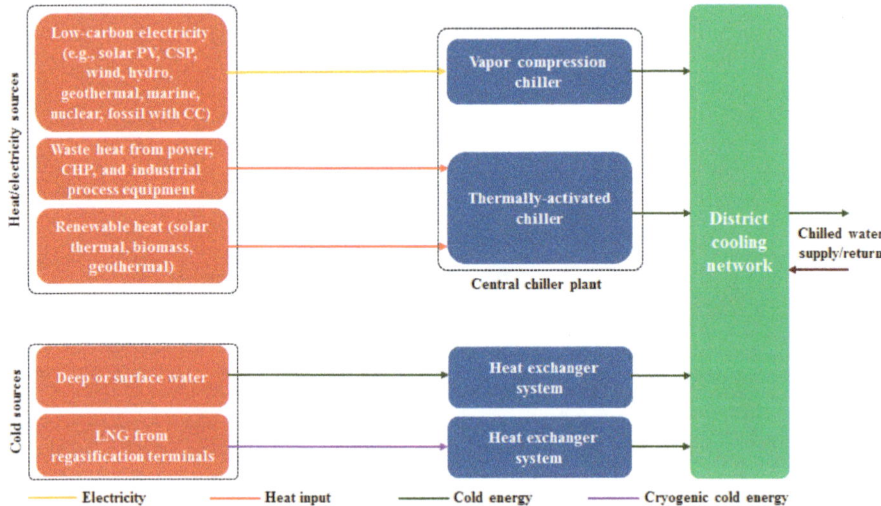

Figure 2. Sustainable cold production options for DC systems.

2.1.1. Cooling Technologies

Either a single or multiple cooling technologies may be incorporated in a DC central chiller plant, depending on available electrical/thermal energy sources and their production and demand profiles, and environmental and cost considerations. Technologies either implemented or evaluated in DC design/analysis studies mainly consist of vapor compression and single-/double-effect absorption refrigeration. This essentially reflects technology maturity and reliability, equipment availability at MW-scale capacity, and affordability. Among these technologies, electrical centrifugal water-cooled chillers are currently the most widely employed chillers in installed DC systems, because of their higher performance (COP > 7.0 10]) relative to their air-cooled counterparts [10], and reliable operation. Given the technology maturity of vapor compression refrigeration, the following sub-sections focus on thermally activated cooling technologies, which can be driven by renewable/waste heat and reduce the energy consumption of vapor compression chillers in a DC system and consequently peak electrical loads. In addition, some of the thermally activated cooling concepts discussed here have potential for hybridization with vapor compression refrigeration for performance/sustainability enhancement. Additional approaches of enhancing the performance of vapor compression refrigeration through condenser cooling and/or operational control in DC are also discussed in subsequent sections. Concepts currently investigated to enhance the performance of vapor compression refrigeration are outlined in more detail in [44].

Key characteristics (i.e., working fluids, driving heat source temperature ranges, typical COP values) of heat-driven air-conditioning and refrigeration technologies (i.e., sorption- and ejector-based) are listed in Table 2. Although sorption cooling is widely employed in small- to large-scale air-conditioning and refrigeration applications [45–47], single/double-effect lithium bromide-water absorption is essentially the only type of sorption technology either integrated in installed DC central chiller plants or reported in DC research studies to date. Thus, most sustainable DC system analyses in Table 1 have incorporated a combination of both vapor compression and either single-effect [23,26–28,36,40] or double-effect absorption cooling [6,31,37,41,43]. DC investigations in which vapor compression was the sole cooling technology [21,24,25,34,35] are generally earlier efforts [21,24,25], while DC studies focusing on absorption cooling are recent efforts [29,30,33,42].

Table 2. Typical characteristics of heat-driven cooling technologies [46,48–51].

Cooling Technology	Working Pair	Chilled Fluid Temperature (°C)	Heat Source Temperature (°C)	COP (-)	DC Analysis
SE absorption chiller	H_2O/LiBr	5–10	80–120	0.5–0.8	[26,27,29,33,36,40,42]
	NH_3/H_2O	<0	80–200	~0.5	—
DE absorption chiller	H_2O/LiBr	5–10	120–170	1.1–1.51	[6,27,29–31,37,41–43]
	NH_3/H_2O		170–220	0.8–1.2	—
TE absorption chiller	H_2O/LiBr	5–10	200–250	1.4–1.8	—
GAX chiller	NH_3/H_2O	<0	160–200	0.7–0.9	—
HE absorption chiller	H_2O/LiBr	5–10	50–70	0.3–0.35	—
Adsorption chiller	H_2O/Silica gel	7–15	60–85	0.3–0.7	—
	Methanol/activated carbon	<0	80–120	0.1–0.4	
Liquid desiccant cooler	N/A	Dehumidified cold air 18–26	Hot water 60–90, hot air 80–110	0.5–1.2	—
Solid desiccant cooler	N/A	Dehumidified cold air 18–26	60–150	0.3–1.0	—
Ejector cooler	H_2O	5–15	60–140	<0.8	[52]

Note: SE—single-effect; DE—double-effect; HE—half effect; TE—triple-effect; GAX—generator absorber heat eXchange. —cooling technology not applied in the published DC literature to date.

Absorption-based and alternative developing air-conditioning technologies with potential for renewable or waste heat use, and/or operation in high ambient temperature/humidity conditions, are identified and discussed in the following paragraphs.

Absorption-Based Cooling

Absorption chillers are a mature and well-established sorption cooling technology employed since the 1960's in various air-conditioning and refrigeration applications [53,54]. Single-effect water-lithium bromide-absorption chillers are suitable for low-grade heat use (e.g., non-concentrating solar collector, geothermal, or waste heat), whereas double-effect water-lithium bromide-absorption chillers could be driven by higher-temperature sources (e.g., medium-concentration solar collectors, higher-temperature geothermal or waste heat). Triple-effect chillers would permit higher grade sustainable heat sources to be exploited for DC cooling, such as from concentrated solar power (CSP) plants as envisaged by [30], with improved COPs compared with lower-effect chillers [55]. However, the availability of triple-effect chillers at medium–large capacity is currently limited and would lead to higher investment cost than for lower-effect chillers [30,55]. An up-to-date list of commercially available, medium–large capacity, single- and higher-effect absorption chillers applicable in DC central chiller plants is compiled in Table A1, along with their key characteristics. (Absorption chiller capacities between 100 to 1000 kW and those exceeding 1000 kW are considered here medium- and large-scale capacities, respectively.) Alternative advanced absorption cycle configurations with potential for low-grade renewable/waste heat use, low energy consumption, and air-conditioning in cooling-dominated climates are identified below. These include bifunctional absorption, generator absorber heat exchange (GAX)-based absorption, and hybrid absorption-compression (also referred to as compression-assisted absorption).

Basic absorption cycles with insufficient external driving heat may be modified in their operation with internal recycling of absorber and condenser heat rejection to supplement the external heat input, as proposed by Arabkhoosar and Andresen [56]. The authors numerically evaluated this concept for evacuated-tube solar collector driven single-effect water-lithium bromide-absorption chillers to provide DC in summer in moderate/cool climates. The heat recuperated from the absorber and condenser served to compensate for the lack of excess heat diverted from a district heating system, which is generally insufficient in the hot season to cover absorption chillers' heat demand. The heat recuperated could also be employed for heating during other periods of the year. This absorption chiller configuration could be extended to cooling-dominated regions when insufficient low-carbon heat is available to drive absorption chillers.

GAX-based absorption and hybrid absorption-compression are currently attracting significant research attention and have technical features with potential for sustainable air-conditioning in cooling-dominated climates. GAX cycles make internal use of part of absorber heat rejection to supply heat to the desorber [57–59], while absorption-compression cycles incorporate a compression booster (i.e., mechanical compressor or ejector) [53,57]. Both types of cycles can achieve higher COPs than basic absorption cycles, and allow operation at higher ambient temperatures [53,55,57,58]. Hybrid absorption-compression systems can also operate at lower driving heat source temperatures than basic absorption cycles, which may be useful for the exploitation of low-grade renewable/waste heat [53,57]. Compression-assisted absorption is however considered more complex to operate than basic absorption [53]. GAX-based chillers are already applied to air-conditioning, but the capacities of commercially available systems [46] are presently insufficient for DC central chiller plant applications. Therefore, although the above advanced absorption cycle concepts hold potential for energy savings and performance improvements in hot climates, they require further development.

Alternative Thermally Activated Cooling

Additional heat-activated cooling technologies amenable to low-grade renewable/waste heat use, and/or operation in high ambient temperature/humidity conditions include adsorption refrigeration [60–62], desiccant-based sorption cooling [46,63–65] and ejector-based refrigeration [49,66]. Adsorption chillers are employed in for example industrial and data server air-conditioning applications, driven by on-site low-grade waste heat such as the server cooling fluid stream [62,67]. Solid desiccants can offer significant energy savings through separate air dehumidification via adsorption in high ambient air humidity conditions and can operate either as standalone or coupled with other air-conditioning technologies (e.g., compression refrigeration) [46,63–65]. Although of value for sustainable air-conditioning, the capacities of either adsorption chillers or desiccant systems are presently insufficient for DC applications, with both types of technologies currently undergoing further developments. Thus, their capacity limitations would restrain their potential application to air-conditioning at end-user sites' where a low-cost heat source may be available, thereby eliminating thermal energy distribution losses. If economically viable, such an approach could contribute to reduce the load of a DC central chiller in the district buildings considered.

Ejector-based refrigeration systems became widely applied in the air-conditioning of large buildings and railroad cars after the introduction of a vapor ejector-based refrigeration cycle in 1910, facilitated by the abundant availability of steam at the time [68,69]. The use of ejector-based air-conditioning systems ceased to expand with the introduction of higher performance, chlorofluorocarbon (CFC) refrigerant-based vapor compression chillers in the 1930's. However, ejector-based refrigeration systems are presently receiving considerable attention for space cooling due to their construction simplicity with no moving parts or seals, low maintenance requirements, low-grade renewable/waste heat use and energy consumption reduction potential, and compatibility with environmentally friendly working fluids [48,66,69,70]. Ejectors have been rarely applied in either existing DC systems (e.g., [52,71]) or researched DC systems to date. Although implementable at multi-MW scale, their main current limitations are the low performance (i.e., COP) and limited operational flexibility of the basic cycle with respect to generator (i.e., driving heat source) temperature and backpressure (which is related to ambient temperature). An actual DC system using a central chiller plant incorporating a 0.6 MW capacity steam jet ejector chiller and 0.6 MW compression chiller has however been in operation since 1998 in Gera, Germany [52,71]. The DC system uses both electricity and waste heat from a district heating network and CHP plants as energy sources to produce driving hot water for the steam ejector chiller. A two-stage triple ejector cycle configuration and steam pressure regulator are employed to provide flexibility in terms of cooling capacity and stability with respect to external conditions. The use of ejector refrigeration was estimated to reduce operating costs by up to 30% relative to an absorption chiller, due to reduced driving heat requirements [52,71]. Multi-ejector and hybrid ejector cycles (e.g., absorption-ejector, compression-ejector) are currently being researched for improved operating flexibility and performance (e.g., COP), as well as combined ejector-power cycles for co/tri-generation, in conjunction with control and storage systems [49,59,66].

In summary, multi-effect, GAX-based, and compression-assisted absorption, as well as desiccant-based sorption and ejector-based cooling, have promising technical features for sustainable air-conditioning, including in cooling-dominated regions. Of these technologies, ejector-based cooling has been occasionally implemented in DC, while multi-effect absorption may be realistically implementable in DC systems in the near future. Therefore, based on the current technical and commercial availability/capacity limitations of the above-listed technologies, the remainder of this article focuses on DC based on vapor compression and single/double-effect absorption refrigeration.

2.1.2. Heat Sources for Thermally Activated Cooling

DC system studies incorporating waste heat- and renewable heat-driven chillers are reviewed in in this section, in terms of system configuration, central chiller cooling technology, operating conditions, and key findings.

Heat Sources from Fossil-Fuel Conversion Systems

CHP technology has been widely used in district systems in heating-dominated regions [12]. Ondeck et al. [28] also highlighted that tri-generation plants driven by both fossil and renewable energy sources could also be a suitable choice to supply integrated utilities (i.e., cooling, heating, and electricity) to future districts in hot-climate regions. Their proposed tri-generation system (Figure 3), intended for application at the University of Texas, Austin, USA, included a natural gas-fueled CHP unit driving absorption chillers, while compression chillers were driven by a solar PV plant. In the CHP unit, superheated steam was generated from GT exhaust gas waste heat in a HRSG. When GT exhaust waste heat was not sufficient to produce the required amount of steam at a sufficient steam temperature, supplemental NG was fired in an in-duct burner to supply extra heat for steam production. Using utilities demand data obtained from a residential neighborhood in Austin, an optimum operating strategy was determined for the integrated CHP/PV system, with the effect of PV generation on system operation and operating profit evaluated. The 137 MW capacity plant could produce 581,505 kg/h of steam, 48,000 tons of chilled water, with a thermal storage capacity of 39,000 ton-hours of chilled water. Running in island mode (i.e., in isolation from the local/national electricity distribution network), the system could provide all the university's electricity, heat, and cooling requirements throughout the year, at an energy efficiency of over 80%.

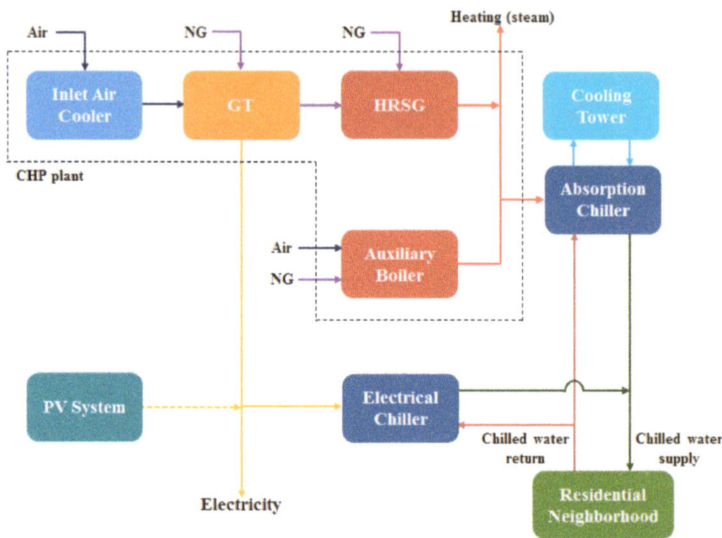

Figure 3. NG- and solar PV-powered tri-generation system proposed by Ondeck et al. for future residential neighborhoods in hot-climate regions, adapted from [28]. Note: CHP—combined heat and power; GT—gas turbine; HRSG—heat recovery steam generator; NG—natural gas; PV—photovoltaic.

Al-Qattan et al. [6] evaluated the performance of a modeled natural gas-fueled hybrid pressurized solid oxide fuel cell (SOFC)-gas turbine (GT) power plant integrated with a DC system (Figure 4) in Kuwait climatic conditions. Providing an estimated total cooling load of 96 MW, the cooling system included absorption chillers driven by GT exhaust heat, and electrical compression chillers driven by SOFC and GT electricity. Surplus SOFC-GT electricity was exported to the grid during periods of low cooling demand. A thermal cooling water storage tank was used to reduce system capacity. The combined system was operated at relatively high cooling tower inlet/outlet temperatures (i.e., 40 °C/35 °C) representative of the GCC's climatic conditions. To reduce the chilled water pumping power consumption, the temperature differential between the chilled water supply and return temperatures (i.e., 6.1 °C and 12.8 °C, respectively) was set marginally higher than typical values. The system was estimated to both improve the fuel-based efficiency of chilled water production by 346% and decrease annual fuel consumption by 750 TJ (54%) and peak power requirement by 24 MW (57%), relative to conventional packed air-conditioning units (PACs), which are commonly used for cooling homes. The DC system was estimated to contribute 53% reduction in capital and operating costs per ton-hour of cooling over PAC units.

As part of projected sustainable energy scenarios for Singapore in year 2050, Dominković et al. [40] identified waste heat sources from NG-fired and waste incineration power plants to supply absorption cooling to a future DC system, as well as waste cold from an LNG regasification terminal. Given its inclusion of renewable waste heat and cold energy use for DC, as well as renewable power generation, this work is discussed in more detail in sub-sections "Renewable Heat Sources" and "Artificial Cold".

Other studies in heating-dominated regions have also investigated the performance of cooling production at district level (together with heating production), by using waste heat from either existing thermal power plants (e.g., coal-fired power plants [29]) or coal-/oil-fired co-generation plants or boilers [23,26] during the hot season. Finally, Colmenar-Santos et al. [72] assessed the cost of converting European thermal electricity production plants to co-generation plants for electricity/heat-driven district heating and cooling. Their analysis demonstrated the cost and environmental benefits of using district heating networks for absorption cooling.

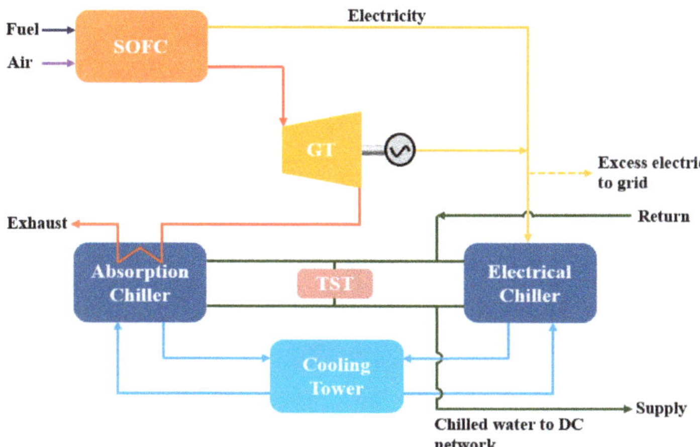

Figure 4. Integrated SOFC-GT and DC system proposed by Al-Qattan et al. for a new city in Kuwait, adapted from [6]. Note: GT—gas turbine; SOFC—solid oxide fuel cell; TST—thermal storage tank.

Renewable Heat Sources

Renewable thermal energy (e.g., solar, geothermal, biomass waste) can be transformed into cooling energy using heat-driven chillers, or into electrical/mechanical energy using thermal power plants to drive vapor compression chillers. DC systems based on renewable-heat-driven central chiller plants are discussed in this section, focusing on cooling-dominated regions.

Marugán-Cruz et al. [30] examined the technical and economic feasibility of using excess heat generated by the heliostats of a solar power tower for DC. Such heliostats require to be defocused from the central tower receiver placed when the maximum allowed thermal power is reached [30]. As proposed in Figure 5, the exceeding heliostats would be focused to an additional thermal power steam receiver located below the primary molten salt receiver. The steam produced from excess heat was used to drive double-effect $H_2O/LiBr$ absorption chillers. Cooling water at inlet/outlet temperatures of 30 °C/37 °C was used as a heat sink for these absorption chillers. The system could supply chilled water at 7 °C to cover 47% of the cooling demand of 30,027 Spanish dwellings. The system was also equipped with backup natural gas burners activated when solar cooling was not sufficient to meet the cooling demand. This hybrid system was found to be cost-effective, with end-users estimated to save approximately 75% of their electricity bills. The steam was available at a temperature sufficient to activate triple-effect absorption chiller technologies (i.e., >250 °C, Table 2), suggesting that system performance could be further improved. However, no commercially available indirect-fired triple-effect chillers were found at the time of the study [30].

Perdichizzi et al. [31] proposed and simulated a solar combined cooling and power (SCCP) plant (Figure 6) for latitude and weather conditions similar to those of Abu Dhabi, UAE, representing an isolated location (i.e., island) in a hot-climate region. The SCCP plant comprised a combined gas turbine-steam turbine cycle, parabolic trough solar collectors (PTCs), and steam-driven double-effect $H_2O/LiBr$ absorption chillers. Saturated steam extracted upstream of the low-pressure steam turbine was fed to the absorption chillers. Two alternative solutions to address peak power demand were evaluated: (i) the integration of CSP with the combined cycle to match the peak power demand of the electric grid; and (ii) the implementation of a DC system incorporating absorption chillers rather than mechanical chillers to reduce peak electricity demand and thus smoothen the electricity load profile. The performance of the proposed SCCP plant was compared with that of a conventional combined cycle driving mechanical chillers to meet similar power and cooling loads. The SCCP plant was found to offer the following benefits compared to the conventional fossil-fuel-based combined cycle: smaller gas turbine capacity, higher total energy efficiency, and 26% to 33% fossil-fuel savings in winter and summer, respectively. These significant fuel savings were mainly attributable to the integration of solar power and the use of absorption chillers.

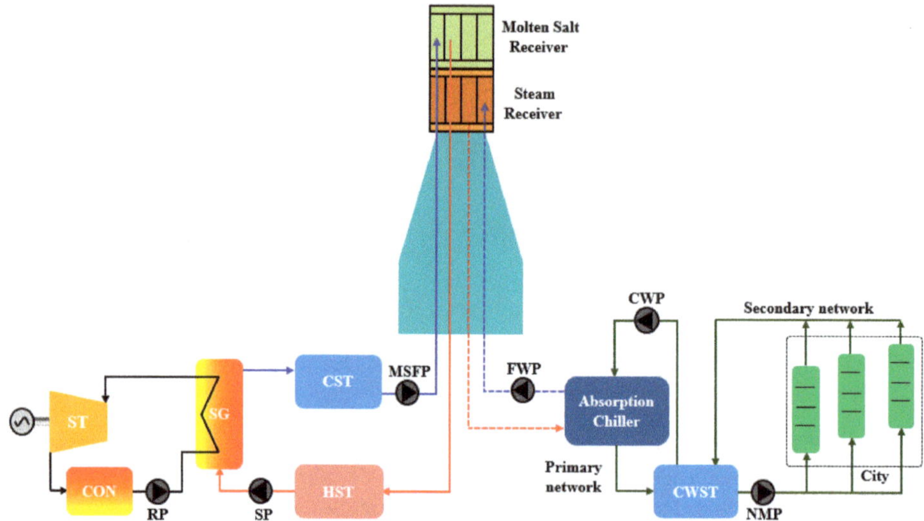

Figure 5. Solar power tower and DC system proposed by Marugán-Cruz et al., adapted from [30]. Note: CON—condenser; CST—cold storage tank; CWP—chilled water pump; CWST—chilled water storage tank; FWP—feed water pump; HST—hot storage tank; MSFP—molten salt feed pump; NMP—network main pump; RP—Rankine pump; SG—steam generator; SP—storage pump; ST—steam turbine.

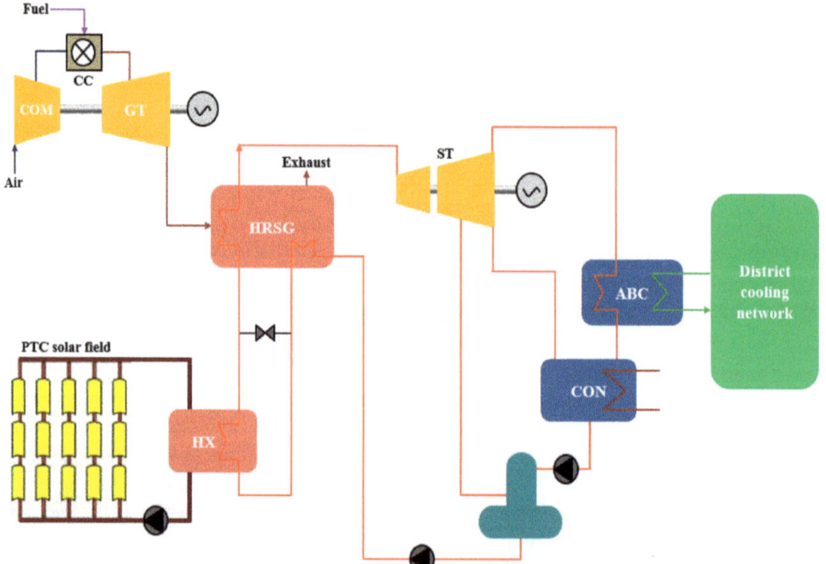

Figure 6. Solar combined cooling and power (SCCP) plant proposed by Perdichizzi et al., adapted from [31]. Note: ABC—absorption chiller; CC—combustion chamber; COM—compressor; CON—condenser; GT—gas turbine; HRSG—heat recovery steam generator; HX—heat exchanger, PTC—parabolic trough solar collector; ST—steam turbine.

Ravelli et al. [43] also proposed a CSP plant for the co-production of power and cooling to serve a Saudi Arabian community. The CSP plant included two different solar fields (i.e., PTCs, and central receiver system (CRS) with heliostats). The power unit consisted of a steam Rankine cycle including low- and high-pressure turbines. The CSP plant also included a thermal energy storage system that permitted to dispatch electricity in periods of low or no solar irradiation. Double-effect absorption chillers driven by steam extracted from the low-pressure steam turbine served to produce chilled water for distribution through a DC network. Supplemental electric chillers were required to meet peak summer cooling loads. The absorption chillers reduced DC electricity consumption by up to 5 MW and 30 MW in winter and summer, respectively, and contributed to peak shaving in summer.

Franchini et al. [42] investigated a solar DC system (i.e., solar field, central chiller plant, DC network and building load) incorporating two different central chiller plant technologies, consisting of a double-effect H_2O/LiBr chiller driven by a PTC, and a single-effect H_2O/LiBr chiller driven by evacuated-tube collectors (ETCs). The cooling load of a residential compound was simulated in the climatic conditions of Riyadh, Saudi Arabia. The sizes of all main DC system components were optimized by minimizing cost and maintaining an annual solar faction of 0.7. Annual transient simulations of the system configurations under variable operating conditions were undertaken to assess the annual energy performance of the solar DC system with a higher level of accuracy than for steady-state analysis-only. The solar DC system based on PTCs and double-effect H_2O/LiBr chillers was found to be more cost-effective than with single-effect H_2O/LiBr absorption chilling and led to ~30% reduction in primary costs including those of the DC network infrastructure. This was attributed to the high efficiency of the double-effect absorption chillers and the high level of local direct normal irradiation available. In comparison with conventional DC systems based on compression chillers, the solar DC system was anticipated to allow approximately 70% reduction in both annual primary energy consumption and CO_2 emissions.

Karlsson and Nilsson [33] investigated the benefits of using waste heat rejected by pyrolysis oil condensers in a biomass-based CHP plant to drive a DC system based on single-effect absorption cooling. Although for a heating-dominated region, the proposed DC concept based on the use of renewable-heat-driven thermal chillers is of interest to sustainable DC supply in cooling-dominated regions. The use of a single-effect absorption chiller driven by excess heat for DC cooling was investigated for three cooling demand scenarios, differing in both the magnitude and time duration of cold production. In a scenario maximizing pyrolysis oil output, 5 MW of DC could be provided using 6.4 MW of condenser heat rejected between June and August. The overall energy efficiency of the biomass CHP plant and DC system improved by 1.3%, and the efficiency of the pyrolysis process increased by 6%. In addition, mostly during summer months, the amount of electricity generated increased by approximately 9% to 19% depending on the cooling scenario considered.

Dominković et al. [40] investigated four different 2050 energy scenarios for Singapore (inclusive of electricity, heating, and transport), in terms of primary energy consumption, CO_2 emissions, and socio-economic expenses (i.e., capital and operating). The scenarios were based on either business-as-usual energy practices, the development of a DC system, increased share of solar PV generation as a replacement of natural gas power, or a combination of DC and solar PV deployment. In the two DC-based energy scenarios, waste heat was recovered from natural gas-fired and waste incineration power plants to supply absorption cooling to the DC system. In addition, cold energy was extracted from an LNG regasification terminal, and ice storage incorporated. Conventional electric chillers were used to provide cooling in periods of low waste heat availability, which overlapped high PV production periods, and served to absorb excess fluctuating PV electricity. In parallel, fossil transport fuels were partly displaced with electrically powered mobility that also contributed to reduce surplus electricity. Projected energy efficiency measures in the residential, industrial, and power sectors were accounted for. In comparison with the business-as-usual scenario, the best scenario (i.e., DC and 70.6% share of PV electricity generation) reduced primary energy consumption by 19.5%, CO_2 emissions by 41.5%, and socio-economic expenses by 38.4%, despite the significant capital investment incurred with DC. This study exemplifies a combination of waste heat (both from fossil and renewable waste sources) and waste cold recovery for DC, and synergy between DC and power generation for minimization of primary energy consumption, environmental emissions and excess electricity that could otherwise destabilize the electrical grid.

2.1.3. Cold Energy Sources

Cold sources that can be used by DC systems to increase heat rejection, hence overall cooling capacity and efficiency, may be categorized as either natural renewable sources or synthetic/artificial sources.

Natural Cold

Natural cold sources can be used for direct cooling either alone or in combination with active cooling technologies (e.g., mechanical compression and heat-driven technologies) to reduce active cooling energy consumption. Cold sources in actual DC systems include deep-seawater (e.g., [73] in China, [74] in Sweden), deep-lake water (e.g., [75], N.Y., USA and [76] in Toronto, Canada), aquifer (ground) water (e.g., [77] in London, UK), and lagoon and river water. A water temperature below either 5 °C (for 100% natural cooling) or 10 °C (for partial natural cooling) is typically required to achieve a DC supply temperature of 6–7 °C. After water filtration, the cold water is pumped to a heat exchanger station, where heat from a closed loop DC system is transferred to the cold water. Physical and chemical natural water properties (including harness, salt/contaminant concentrations) require monitoring to avoid damaging network components [8]. Discharge of the network water into natural water systems also requires temperature control to avoid undesired chemical/biological processes [8].

Given the lack of natural cold sinks in hot-climate regions, the use of natural cold sources in DC research studies [23,25,26] has been mainly for cooling in the summer season in heating-dominated regions. Svensson and Moshfegh [26] used deep-lake water for direct cooling in parallel with absorption chillers driven by heat from biomass-fired boilers and CHP units, to partly substitute vapor compression chillers and reduce electricity consumption in a Swedish DC system. Trygg and Amiri [23] supplied approximately 8% of their DC load using river water in Sweden, without the need for a cooling tower. Feng and Long [25] used river water at 6 °C to supply direct DC in summer in China.

A variant approach of using natural cold water is found in the cold district heating and cooling network concept defined by Pellegrini and Bianchini [8], which combines centralized energy supply with minimized heat losses in delivery. In this concept, low-temperature water (i.e., 10–25 °C, from e.g., surface, deep aquifer, sea, waste, and aqueduct water) is delivered to the network and serves as cooling fluid for either direct or active cooling when required. The supplied low-temperature water is chilled or heated by decentralized chillers/pumps for district cooling/heating, which reduces the thermal distribution losses and insulation costs that would arise in centralized water chilling/heating. The use of low-temperature water can also enable the use of renewable energy conversion technologies (e.g., solar, geothermal heat pumps) for heating, and facilitate the integration of district heating and cooling networks. Cold district heating/cooling networks can offer reductions in primary energy consumption through renewable energy use and reduced thermal losses, depending on the temperature-dependent COP of cooling equipment.

Artificial Cold

Given the lack of natural cold sinks in cooling-dominated regions, artificial cold from cryogens such as liquefied natural gas (LNG) for direct DC is a potential option that may take greater importance than in moderate climates. The cold exergy that can be practically recovered during LNG regasification at import terminals before gas distribution is of approximately 350–370 kJ/kg/s of LNG [78,79]. (Thermal exergy recovery, assuming an initial LNG state at ~70–80 bar for NG distribution and (-160)–(-165) °C.) With a global trade of 293 Mt in 2017, LNG currently represents 9.8% of the global gas supply [80]. LNG supply volumes are anticipated to double through 2016–2040, overtaking inter-regional pipeline supplies in the early 2020's [2]. The current global LNG regasification capacity (i.e., 851 Mt/annum (MPTA)) is distributed between 39 countries and 121 terminals, including in several hot-climate regions in Asia, the Middle East, and South Europe [80]. A further 87.7 MTPA is under construction including in China and hot-climate regions (i.e., India, Turkey, Bahrain, Kuwait, Bangladesh, Panama, Philippines, Brazil). Among the GCC countries that face domestic gas shortages (i.e., Bahrain, Kuwait, UAE), the UAE was the first to build an LNG receiving terminal.

The majority of LNG importing terminals however reject their cold energy to the environment (i.e., air or seawater) in either open rack, submerged combustion, or ambient air vaporizers, without cold energy recovery [78]. This is attributable to the following technical and non-technical barriers: (i) communities and cold applications being generally remotely located from regasification terminals, for safety and practical reasons [81], coupled with the lack of suitable fluid carriers for cold transportation; (ii) a lack of perceived incentives for regasification facilities to diversify their products beyond NG distribution, combined with a reluctance to cooperate with potential cold users (e.g., district energy companies), and unsuitable business models; (iii) end-user variable cooling demands and/or resistance to new cold provision technologies [81–83].

Regarding technical barrier (i), a suitable, secondary cold carrier fluid is required to transport cold energy recovered from LNG over long distances for off-site DC [81]. Water/glycol mixtures have been proposed but require large mass flows and exhibit high viscosities at very low temperatures, resulting in elevated pumping power. Nanofluids and ice slurries have also been considered but are not suitable for long distance transportation due to high pumping power consumption compared with conventional secondary fluids [81]. Liquid CO_2 has lower viscosity than water/glycol, is non-flammable and considered environmentally benign and low cost [83,84]. CO_2 has been suggested to transport cold energy from LNG regasification terminals [81,83,84] to end-users (e.g., agro-food industries, supermarkets, hypermarkets located 2 km away from an LNG regasification terminal [81,84]).

As a consequence of the above barriers, the implementation of LNG cold energy recovery technologies has been limited to a few countries to date, including Japan and Spain [78]. When recovered, LNG cold energy is mainly exploited to either enhance the efficiency of on-site thermal power generation plants (generally, Rankine and direct expansion-based cycles), and less frequently, for on-site direct cooling applications (e.g., cryogenic air separation for oxyfuel combustion, GT inlet air cooling) rather than off-site uses (e.g., air-conditioning in supermarkets and agro-food processing facilities) [78,81]. However, in support of its GHG emissions reduction goals, the EU Commission now recommends the exploitation of LNG cold energy and waste cold in general to reduce building energy demand [82].

The use of LNG cold energy for DC has hardly been considered in the literature, except in [40,83,85]. Jo et al. [85] investigated a Type 2 absorption system using an ammonia-water solution as working fluid to transport LNG cold energy over long distances. The maximum ammonia-water transportation distance in a 15 cm diameter pipe was estimated at 270 km. In Dominković et al. [40], the envisaged transportation of cold energy extracted from regasified LNG for DC was limited to 20 km within Singapore area. Based in Spain, Atienza-Márquez et al. [83] proposed the cascaded co-generation of DC and Rankine/direct expansion cycle-based power augmentation using cold energy recovered from LNG regasification. LNG cold energy was transported using CO_2. DC was provided at three different temperature levels, for low- and medium-temperature food refrigeration, and building air-conditioning. It was recommended to maintain LNG at a relatively low pressure (~8 MPa) for DC application, due to LNG's heat capacity reduction with increasing pressure. Electricity savings, and seawater savings for regasification, were estimated at 81 kWh/ton$_{LNG}$ and 68% respectively, relative to the standard power and cooling systems with no waste cold use.

Based on the existing and future LNG regasification capacity in cooling-dominated regions, LNG cold energy holds significant potential for DC, as well as for enhancing the efficiency of power generation and refrigeration cycles (e.g., compressor inlet air cooling in gas turbine power plants, heat sinking for heat-driven cooling and power generation technologies). The cold energy from LNG regasification terminals could be used for on-site space air conditioning or process-cooling applications, as well as space cooling in nearby buildings and industrial complexes.

2.2. Thermal Energy Storage

The concept of cool thermal energy storage (CTS) for DC applications was introduced in the early 1980s in the USA. Approximately 2000 CTS systems were installed in the USA by the 1990s, most of them (i.e., 80–85%) consisting of ice thermal storage (in the form of either crystals or slurries), then chilled water storage (i.e., 10–15%) and eutectic salt systems (i.e., 5%) [86,87]. CTS phase change materials (PCMs) include inorganic salt hydrates, organic chemicals (e.g., paraffins) and eutectic salt mixtures. Desirable CTS characteristics include low thermal losses during storage, high release efficiency of stored cold energy, low environmental impact, commercial availability, and cost effectiveness [10]. The characteristics of typical CTS systems are summarized in Table A2. Water has a high specific heat capacity, high availability, is non-toxic and low cost. Chilled water thermal storage is compatible with the evaporation temperature range of conventional chillers, while ice storage is more compact due to its higher volumetric energy density [87].

Centralized chiller plants facilitate the efficient and reliable integration of thermal energy storage, compared to conventional individual cooling systems [88]. The integration of CTS in building air-conditioning and DC has the following benefits [87,89–91]: improved cooling load management; increased cooling generation capacity by shifting operation from peak (i.e., daytime) to off-peak (i.e., nighttime) periods; reduced energy consumption and cost through load shifting for both the DC supplier and consumer; reduced installed cooling capacity, investment and operating cost; improved renewable (e.g., solar energy) integration by reducing energy production-demand mismatch, such as through excess solar energy storage for cooling production in non-sunshine periods; improved chiller efficiency by avoiding part load operation and transient/intermittent operation; and improved system reliability by using CTS as a backup [21,90]. To capitalize these benefits, the local electricity demand profiles and tariffs, and country's energy policy, are the most critical factors and should guide the selection of an operational configuration (e.g., series versus parallel CTS-chiller arrangement) and strategy (e.g., full versus partial storage) [21,90]. CTS is particularly beneficial with large day-night outdoor temperature swings [86,90], such as in for example the GCC. Hasnain [86] studied applications of CTS systems, including building air-conditioning and GT inlet air cooling, and their economic effects in hot-climate regions including the Kingdom of Saudi Arabia (KSA), at daily ambient temperature variations of ~18 °C. The use of CTS was found to reduce peak cooling and peak electrical demands in commercial buildings by approximately 30–40% and 10–20%, respectively. However, certain economies in cooling-dominated regions such as the GCC apply energy subsidies and flat-like tariff profiles that do not enable the potential of CTS to be fully exploited. Few DC design and analysis studies have reported details of CTS selection, design and/or operation/control (Table 1). Chan et al. [21] compared the energy savings achieved for different ice storage operational strategies, namely chiller-priority and storage-priority control, as a function of storage fraction in a Hong Kong-based DC system. The energy/economic savings were found to be highly dependent upon the local cooling demand profile, CTS capital cost and electricity tariff rates, with a prohibitive CTS payback period for application in Hong Kong at the time, and further research in these above areas was recommended. For another DC system in Malaysia, the electricity cost reductions obtained with ice thermal storage were also found to be marginal for the prevailing electricity tariff and limited day-night outdoor temperature variation (i.e., 8 °C) [90].

2.3. Distribution Network Infrastructure

A DC network infrastructure consists of pumping systems and distribution pipelines through which the cooling energy carrier is transferred and distributed from the central chiller plant (or cold energy source) to end-users via ETSs. The cooling energy carrier (i.e., cooling medium), could be chilled water, ice-water, ice slurry, water/glycol mixture or other secondary fluids. The DC network infrastructure requires careful design and optimum operation, as it often represents the largest initial investment in a DC system (i.e., typically 50–75% of total investment) [9].

The pumping system includes pumps operating at either variable or constant flow, that are located at the central chiller plant, within the distribution network infrastructure, and/or end-user substations. In practice, one of three types of pumping arrangements are employed in DC distribution networks: centralized (primary) pumping, primary–secondary pumping, and distributed pumping [92]. In the centralized pumping scheme, a single pumping system is employed to deliver chilled water to the entire system using two-way control valves for controlling the flow of chilled water supply to each end-user. In the case of a primary–secondary pumping arrangement, two separate pumping loops are involved: one at the central chiller plant, and one at the distribution network or circuit, which incorporates secondary variable-speed pumps. In distributed pumping, a distribution pump system is used at each distribution branch [92].

The pipe network structure and configuration are significant design aspects of DC distribution networks, as they directly relate to their operational and structural features. Three types of pipeline network layouts are encountered in DC systems, namely tree-shaped, radial, and looped networks [93]. These layouts are either used separately or more commonly, in combination within the same DC system. The selection of a piping material depends on both material cost and compatibility with the cooling medium being transported [94]. The most commonly used piping materials are welded steel, soldered copper, ductile iron, cement pipe, fiberglass-reinforced plastic, polyvinylchloride (PVC) and polyethylene (HDPE) [10,94]. Similarly to hot water distribution in district heating networks, chilled water distribution in DC networks also requires pipeline insulation, particularly in hot climates, to avoid heat gains that would adversely impact on overall system performance and translate into an economic loss. In addition, occurrences of high heat gains generally coincide with peak space cooling load periods, leading to additional plant capacity requirement. As the supply/return temperature difference in DC systems (i.e., ~10 °C) is lower than in district heating networks (i.e., ~40–50 °C), larger pipe dimensions are required, which tends to increase network cost relative to district heating networks [24]. However, unlike in district heating systems, no high-temperature resistant materials are required for DC pipelines, heat exchangers, valves, and instrumentation, and instead low-cost materials such as plastics can be employed [8]. Above-ground pipe networks are more easily accessible for maintenance, but are more prone to damage and heat gain, and should be protected from vapor condensation on their surface. Hence, buried insulated distribution pipeline networks are usually preferred in hot-climate regions.

2.4. Operational Conditions and Strategies

Key aspects related to the impact of DC operation and flexibility on performance and sustainability are discussed in this section.

The main factors affecting the operation of DC systems are the operating conditions of the sub-systems (i.e., central chiller plant, distribution network, CTS), usage and environmental conditions and economic factors. End-user behavior can vary significantly depending on user and time, strongly affects DC operating conditions and is a major aspect in determining operational strategies. Such strategies should aim at efficient chiller plant operation and cold distribution, reduction control of GHG emissions, and water conservation. The feasibility of operating the DC system at a high efficiency depends on the feasibility of operating the central chiller plant at optimum efficiency and maintaining the design chilled water temperature difference between the supply and return pipelines. Improvements in the efficiency of cold production lead to annual reductions in GHG emissions.

Typically, the central chiller plant and the cooling sub-systems located in end-user buildings are operated separately, which is the consequence of separate management. However, the coordinated operation of central and end-use site systems is of advantage to both the end-user and DC provider. The typical approach of designing DC systems is to size the system to meet the maximum peak cooling demand; however, the period of peak cooling load is significantly shorter than that of the base load. Consequently, this design approach contributes to large capital investments, energy losses, and distribution costs. Alternative DC design and operation approaches involve intelligent control and communication, such as through innovative demand-response management. Rifai [95] presented an example of DC demand-response model. The model involved forecasting the schedules of energy production and consumption, and resource planning to balance supply and demand. Energy demand can be forecasted based on historical and climatic data.

DC system flexibility is defined by Vandermeulen et al. [20] as an ability to adjust the rate of energy addition/removal to or from the system, respectively. The authors [20] identified thermal energy storage, the DC network, and buildings as flexibility providers. They highlighted the need for improved control strategies to enhance the distribution of DC thermal energy, operate efficiently and stabilize energy systems. Advanced hybrid control strategies combining attributes of centralized and distributed control were proposed as a future research direction.

Flexibility within energy systems, including through synergies between districts and other sectors, is advocated by the EU Commission to facilitate the use of waste heat/cold energy in buildings towards reduced GHG emissions [82]. DC system flexibility is of critical importance not only to the efficient operation of the DC system itself, but also in supporting a renewable-based energy system in which the DC system is integrated. Energy systems with high fluctuating renewable electricity/heat generation shares are prone to destabilization and inefficient use of resources unless excess electricity/heat could be absorbed or stored, or other system stabilization measures are applied [96]. DC systems can be used to absorb excess/waste electricity and thermal energy from other sectors to provide space cooling.

Gao et al. [41] analyzed the energetic performance of modeled integrated distributed energy systems and DC systems in subtropical conditions in Hong Kong, in comparison with standalone DC systems and individual cooling systems (ICSs) that only depended on the electrical network. The impact of four different control approaches on the performance of the coupled distributed energy and DC systems was assessed seasonally. These four control strategies were classified as following either the electric load, the cooling load, or the higher or smaller consuming load of primary energy (i.e., electric or cooling load). The analysis was performed using historical 2015 hourly cooling and electricity load data for Hong Kong Polytechnic University campus. The cooling provided by the coupled DC and distributed energy systems was found to be highly sensitive to the control strategy. The interaction between the DC system and the electrical network was also analyzed in terms of the amount of imported or exported electricity to the network. Integration of the DC and distributed energy systems was found to lead to 10–19% reduction in energy consumption relative to a DC system solely dependent on the grid. Under cooling load control or higher consumer load control, the coupled DC and distributed energy systems could fulfill 81.6% and 82.9% of the annual cooling loads, respectively, compared with only 64.1% and 62.6% for the electrical load or lower consumer load controls, respectively. Thus, under a regulation allowing surplus electricity to be exported to the grid, the control strategy following the higher primary energy consuming load led to the lowest energy consumption, was more self-sufficient in terms of electrical grid independence, and supported the electrical network in meeting peak demands. If surplus electricity cannot be exported to the grid, then the electric load following control strategy, which is presently advised in Hong Kong, would be preferable. The payback time of the coupled DC and distributed energy systems was estimated to range from 6.4 to 10.4 years, depending on the control strategy. The cooling load-based and higher consumer load-based control approaches had payback periods of less than 6.6 years.

The extension of DC to non-space cooling applications is another form of flexibility option discussed in Section 2.5.

2.5. Extension of DC to Non-Space Cooling End-Uses

Extensions of DC to non-space cooling end-uses could enable sharing of capital and energy/material resources with other systems/applications towards smart energy systems, leading to thermodynamically more efficient systems and improved sustainability. Conceivable extensions of DC to non-space cooling end-uses of interest to cooling-dominated regions include electrical power augmentation, hot water production [10,12] and water desalination/treatment [12,97].

Regarding power augmentation, although not previously reported in the DC literature but proposed here, the efficiency of GT power plants in high ambient temperatures could be improved through either direct or indirect compressor inlet air cooling provided by the DC system. Compressor inlet air cooling using waste heat-driven absorption refrigeration has been shown to yield significant enhancements in GT power generation and efficiency in harsh climatic conditions in for example the GCC [98], where traditional evaporative coolers perform poorly in high ambient humidity and consume large deionized water volumes that is not recycled. The GT plant could be located remotely from the DC central chiller plant, as long as the cooling energy distribution system has sufficient capacity.

DC central chiller heat rejection, which is at a temperature (~45 °C) insufficient for direct exploitation in most potential non-DC heat use applications, could be recycled for either hot water generation or (sea)water thermal desalination/treatment (e.g., multi-stage flash (MSF), multi-effect distillation (MED)) after upgrading the heat rejected to a suitable temperature (~70–110 °C). This could be achieved by incorporating additional units such as absorption heat transformers, the application of which to desalination and other uses is reported in [99,100]. This could contribute to reduce the energy consumption of thermal desalination/treatment, which is energy-intensive and is a major mean of fresh water provision in hot-climate regions such as the GCC [101]. Hugues at al. [97] found that the cumulative capital and operating cost of MED driven by DC waste heat would break even with that of reverse osmosis (RO) desalination after 4 to 6.5 years of operation in the UAE (depending on utility prices), mainly due to a 25% reduction in MED annual operating cost relative to RO.

Applications of DC to GT power augmentation and recycling of DC chiller heat rejection are scarce or lacking in DC design/analysis studies to date, leaving opportunities for future investigations.

3. Analysis, Modeling and Optimization

Although DC systems can offer higher energy efficiency, and economic and environmental benefits over on-site cooling systems, they represent substantial capital investment and operational costs. Consequently, the performance evaluation and optimization of DC systems at the planning and design stages are key to capitalize their potential benefits. This section firstly reviews district-level cooling load evaluation and analysis methodologies, which is a critical input for DC system design/analysis (Section 3.1), followed by DC thermodynamic performance, economic and environmental impact evaluation methodologies (Section 3.2) and optimization (Section 3.3).

3.1. Cooling Load Estimation and Analysis

DC systems can provide both space cooling, and as discussed in Section 2.5, could also support non-space cooling applications of interest to cooling-dominated regions, such as power augmentation, hot water generation and water desalination/treatment.

With regard to space cooling and dehumidification, the cooling capacities and usage patterns of different end-users (i.e., residential, commercial, institutional, transport, and industrial buildings) can differ significantly and require detailed analysis, as cooling load is often considered the most critical input for the design, performance, and economic viability analysis of DC systems. For example, commercial buildings (e.g., office type, shopping malls, supermarkets) have high cooling loads for regular air-conditioning on weekdays but also to cool server rooms. On the other hand, shopping malls require a significant cooling energy throughout the week, with peak loads during weekends and evenings. Industrial complexes have high cooling demands for both air-conditioning and process applications.

DC systems can be most effectively used in districts where the cooling loads and number of full-load operating hours are high. A high cooling load density is required to recover the distribution network infrastructure investment cost, which represents approximately 60% of the total overall system capital cost [10]. This makes DC systems most attractive in densely populated urban areas and high-density building groups requiring high cooling loads. New urban developments, either at the planning or design stages, are good candidate DC applications, as they allow building owners to make maximum use of building footprint by placing most of the cooling and heat-rejection equipment off-site. Since DC systems are capital intensive, the equivalent full-load hours of cooling (i.e., annual cooling energy demand in MWh per installed capacity in MW) are important to maximize benefits.

Cooling loads need be quantified at an early stage of the DC design and analysis process. Three types of cooling loads are typically sought in standard DC design/analysis practices: (i) peak cooling load data for system capacity sizing, (ii) annual average hourly cooling load data for economic (i.e., cost-benefit) analysis, and (iii) hourly daily cooling demand data for operational and control design/analysis [10]. The most straightforward district-level cooling load estimation approach uses cooling load density data per unit area (m^2/kW) from engineering standards such as ASHRAE [10]. These data are applicable with caution at the preliminary DC planning stage, given their accuracy limitations.

A limited number of research studies have discussed more advanced cooling load estimation [6,21,22,30,31,34,36] and analysis [38] methodologies at district level. This is in contrast with individual buildings, for which significant attention has been devoted to cooling load analysis and simulation [102–104]. District loads are however the largest source of uncertainty in DC system design [34]. District-level cooling load calculation/analysis approaches employed or developed in DC research studies may be categorized under the following six types (i)–(vi), described below. Their applications to DC design/analysis is also summarized in Table 3, with emphasis on studies for cooling-dominated regions, in terms of input data, DC system characteristics, modeling software, output data and key findings as applicable.

In the first approach (i), applied in for example [6,31,105], actual measured end-user data for existing buildings is employed, such as chilled water flow rate combined with the temperature difference between the supply and return chilled water [105], electricity demand data [31], historical and present district cooling load data [6], energy metering data from an energy monitoring system, utility bills, and installed equipment capacities [10]. However, end-user data are often incomplete or not available [10].

In the second approach (ii), the cooling load a standard building is approximated using the overall heat transfer conductance (UA) for the building envelope (accounting for both external and internal loads), and an indoor-outdoor dry bulb air temperature difference derived from measured meteorological data, with dynamic heat transfer effects neglected [30]. Alternatively, the cooling load temperature difference (CLTD)/solar cooling load factors (SCLs)/cooling load factors (CLFs) method may be applied to account for dynamic effects, in conjunction with hourly weather data for a typical day in the cooling period of the year considered [106]. The cooling load of the standard building is multiplied by the number of buildings in the district. This approach is applicable to districts made of buildings with similar heat transfer characteristics.

In the third approach (iii), presented in for example [34,35], the cooling load of each building in the district is calculated using either commercially available or proprietary dynamic building simulation software, a review of which is provided in [104], in conjunction with building construction data and measured weather data. The detailed simulation of each building to derive individual building cooling loads is however unlikely to be practically feasible in large districts, with reduced-order models employed in [34,35].

Gang et al. [34] extended approach (iii) by quantifying uncertainties in district cooling loads due to uncertainties in weather data, indoor data and building design/construction data. Historical weather data for three decades were employed to estimate uncertainties in outdoor conditions, rather than typical metrological year (TMY) data. For other sources of cooling uncertainty, normal and uniform distributions were assumed. DC system size was identified based on the peak cooling load distribution, while DC performance distribution was evaluated based on the annual cooling load for several alternative DC configurations/technologies. For a planned DC system in Hong Kong, uncertainties in indoor conditions (i.e., occupant density, lighting density, plug-in load density, and particularly ventilation rate) were found to have the largest impact on cooling load estimation, whereas uncertainties in building construction/design had the smallest impact. The DC system annual average and peak cooling loads were found to have 80% and 90% chance, respectively, of being lower than those estimated using a conventional design method (i.e., without uncertainties), which would tend to overestimate such loads and lead to oversized DC system capacity, particularly when a safety margin is applied. The impacts of uncertainties on DC system technology selection, configuration, sizing, and performance prediction were examined using sensitivity analyses. In an accompanying optimization study [35] (discussed in terms of optimization aspects in Section 3.3), DC system design and total annual cost were found to be less sensitive to uncertainties in cooling loads compared with individual cooling systems.

The fourth approach (iv), adopted in [21,22,92,107,108] extends approach (iii) through the development of a database of space cooling loads per unit floor area (i.e., cooling load intensity, in W/m^2, which represents the peak cooling load of a building normalized by its total gross floor area (GFA).) for each type of building in the district. This approach may particularly be applicable to large districts, either existing or future ones, during their planning and design phase. The approach involves categorizing end-user buildings into groups according to their type of operation and energy consumption patterns/magnitude [107]. Typical categories include office buildings, dwelling houses, hotels, and retail buildings. Representative buildings are then selected from each building category, with their detailed description and characteristics (e.g., floor plan, construction materials, occupancy, operating schedule, lighting, and equipment specifications). The total GFA that requires air-conditioning in each building category is then established. The cooling load intensity for each building category is determined using local building energy codes, local weather data, and dynamic building simulation software. From the cooling load data obtained from simulation of each building category, the total district peak load, and total district hourly, monthly, and yearly cooling loads, can be estimated. This method avoids the detailed modeling of all individual buildings in an entire district, which would be practically infeasible for large districts. However, although energy simulation tools have the potential to provide more accurate cooling load data than cooling load density data from engineering standards, numerical prediction of cooling loads is prone to errors in absence of model validation using reference data, particularly in large districts.

Focusing on district cooling loads spatial/temporal distribution, Yan et al. [38] proposed a new methodology (referred to as approach v) to analyze the effects of (i) discrepancies in buildings' hydraulic pressure requirements within a given group of buildings on secondary network variable-speed pumps' energy consumption and overall DC system energy consumption, (ii) spatial/temporal inequalities in normalized DC system buildings' load distribution (i.e., normalized to buildings' rated loads), and deviations in chillers' load from full load, on both DC chiller plant energy consumption and overall DC system efficiency savings relative to an ICS. Discrepancies in buildings' hydraulic pressure requirements were quantified using a newly introduced grouping coefficient. This grouping coefficient was used to identify buildings groups than can be served by a common branch of the DC water network, to minimize secondary pump energy consumption. Lorenz curve and the Gini coefficient were proposed as criteria to capture spatial/temporal differences in district building cooling load distribution and deviation in chillers' load from full load. These metrics were applied to evaluate the energy performance analysis of a modeled DC system under different cooling load profiles in a developing area of Hong Kong. Using the grouping coefficient to optimize the layout of secondary chilled water pumps, secondary pumping power and total DC energy consumption could be reduced by up to 50% and 4%, respectively. A low Gini coefficient, reflective of a more homogenous cooling load distribution, was found to correlate with higher annual chiller plant COP and higher annual DC system energy efficiency, and to generally lead to higher annual DC system energy savings relative to an ICS. The grouping coefficient had a more significant influence on energy consumption than the Gini coefficient, suggesting the greater importance of the chilled water network in terms of DC energy performance, than building cooling loads. At low grouping and Gini coefficients, overall DC annual energy consumption could be reduced by up to 14%.

Table 3. Overview of district-level cooling load evaluation and analysis methodologies in DC design/analysis studies.

Source	Approach Type and Implementation Specifics	Simulation Tool	End-User Type and Location	Remarks
Perdichizzi et al. (2015) [31]	Type (i), with hourly electrical consumption of compression chillers on a typical summer day and winter day estimated from total electricity demand	N/A	Mid-size community (~50,000 inhabitants) assumed in Abu Dhabi (UAE)	• Total peak electricity demands of 83 MW_e and 42 MW_e in summer and winter, respectively • Daily total electricity demand varies by 40% and 50% between day and night in winter and summer, respectively • Electric chillers represent 15–20% and 45% of total electricity demand in winter and summer, respectively
Al-Qattan et al. (2014) [6]	Type (i), with DC hourly cooling load profiles for a selected April day and August day predicted based on historical and present weather data	Unspecified numerical solver	Block of 805 residential villas (each with 800 m^2 air-conditioned space area), 4 schools, 2 community shopping centers and 4 houses of worship in a district of a new city in Kuwait	• Peak DC load in summer estimated at 96 MW • Chilled water supply/return temperature of 6.1/12.8 °C
Marugán-Cruz et al. (2015) [30]	Type (ii), with standard building (UA) value taken as 0.38 kW/°C, outdoor dry bulb air temperature estimated using local weather station temperature data, and assumed indoor dry bulb temperature (i.e., 25 °C) based on the Spanish Regulation on Thermal Facilities in Buildings, and external and internal loads	N/A	30,027 dwellings in Ciudad Real (Spain)	• Daily cooling demand profile of an average dwelling from mid-May to early October 2012 and hourly cooling data profile for the hottest day (August 10) reported
Ameri and Besharati (2016) [36]	Type (iii), with building energy simulation based on building architectural drawings and data for complete DC project	Dynamic building simulation software EnergyPlus	137 buildings in residential complex (~500,000 m^2) in Eastern Tehran (Iran)	• Hourly building cooling loads for a set of typical days used to derive reported annual hourly cooling demand profile • Weather data and indoor conditions data N/R
Gang et al. (2016) [34,35]	Type (iii), using historical weather data, both sample building design/construction data, and sample indoor conditions for each building, with uncertainties in these three input data types accounted for	Reduced-order building energy model implemented in EPC software	Planned district in Hong Kong, including 37 buildings (i.e., office, government, hospitals, hotels, metro stations, etc.)	• Uncertainties in DC loads estimated at the DC system design stage for each building, based on three sources of uncertainties, namely outdoor weather data, indoor conditions and building design/construction • Uncertainties in indoor conditions (particularly ventilation rate) found to have the largest impact on cooling loads • DC system annual average and peak cooling loads found to have 80% and 90% chance, respectively, of being lower than those estimated using a conventional design method, which would overestimate such loads; Uncertainties have less impact on district- than individual cooling systems' loads

Table 3. *Cont.*

Source	Approach Type and Implementation Specifics	Simulation Tool	End-User Type and Location	Remarks
Chow et al. (2004) [107]	Type (iv), with cooling load profile of each building category for 12 months represented by 2016 h (i.e., a typical week in each month)	Dynamic building simulation software DOE-2	Five building categories in Hong Kong, including offices/residential buildings, shopping mall, hotel, and mass-transit railway station	• Cooling load profile of each building category reported for 12 months • Daily cooling load profiles reported for three types of days (i.e., weekday, Saturday, Sunday)in a summer month
Chan et al. (2006) [21]	Type (iv), with typical weather year data for Hong Kong used as input	Dynamic building simulation software DOE-2	Planned building mix in West side of Hong Kong City, including offices, hotels, retails, government depots, schools, indoor recreational centers, and magistracy	• District annual hourly cooling demand profile determined (peak load, 116 MW and minimum load, 1 MW)
Chan et al. (2007) [22]	Type (iv), with hourly weather data of a typical metrological year (TMY) for Hong Kong developed in [109] used as input	Dynamic building simulation software EnergyPlus	Hypothetical site in Hong Kong with different building mixes, including offices, hotels, retails, schools, hospitals, and mass-transit railway station	• Annual hourly cooling load and hourly chilled water demand for each building category determined (data N/R)
Yan et al. (2017) [38]	Type (v), with Lorenz curve and Gini coefficient employed to quantitatively describe building cooling load characteristics and their effects on DC system energetic performance	TRNSYS software applied to predict DC system energy consumption	Three planned development areas including different type of buildings (office, hotel, commercial mall, residential, school and hospital) with a total GFA of 80,000 m^2	• Impact of building load characteristics on DC system energy performance quantitatively evaluated
Gros et al. (2016) [110]	Type (vi), with 2008 meteorological data, and indoor setpoints of 19 °C and 26 °C in non-occupation and occupation periods, respectively	• Outdoor direct and diffuse solar irradiance calculated using SOLENE software • Outdoor reflected solar irradiance and longwave interchanges calculated using EnviBatE software • Outdoor wind fields computed using QUIC software • Building heat transfer modeling software N/R	• 10 nine story building blocks (70,000 m^2) in Buire district, Part-Dieu, Lyon (France) • Mix of historical and 1960's 7-storey buildings, plus one 51 m and one planned 153 m high towers (60,000 m^2), in Money district, Part-Dieu, Lyon (France)	• District cooling energy demand (MWh/yr) reported for several alternative urban plans (green versus albedo) in summer period (i.e., May 1–September 30) • DC load reduced by 3%, 35% and 76% with tree growth, green roofs, and increased used of surface albedo

Note: Cooling load evaluation/analysis approaches (i)–(v) described in the text body of Section 3.1. EPC—Engineering, Procurement & Construction (EPC). N/A—not applicable. N/R—not reported.

In approach (vi), presented in [110], district-level cooling loads are evaluated using coupled building energy and outdoor microclimatic simulations. Microclimatic simulations of hourly atmospheric flow and temperature fields around buildings are employed to yield more detailed outdoor environmental data than single-site, urban-level data from a near weather station. Such simulations can capture heat island effects on district outdoor air temperatures, district cooling loads and pollutant concentrations in dense urban areas, and enable the evaluation of alternative urban plans (e.g., landscaping, vegetation, surface albedo) to reduce such effects. The additional computational expense associated with atmospheric modeling would depend on the fluid flow and heat/mass transport calculation methods used, and may currently restrain the application of this method to a selected annual period.

Combinations of the above approaches (e.g., accounting for the effects of both uncertainties and district microclimate on cooling loads) could potentially yield further accuracy improvements in cooling load prediction, at the expense of increased analysis complexity and time.

3.2. Performance Evaluation

Energy performance, and economic and environmental impact evaluation methodologies employed in published DC system analyses are reviewed in this section.

3.2.1. Energy Performance

The DC system design and analysis studies in Table 1 have focused on energy analyses, except for Coz et al. [39], who evaluated DC system performance based on exergy and exergoeconomic analyses. Although still limited in number, exergy-based analyses of combined district heating and cooling systems in heating-dominated regions have received more attention [39,111]. Energy-related DC operating and performance parameters typically include annual cooling energy supplied, maximum cooling load supplied over the operational life consumed, central chiller plant COP, annual energy consumption, chilled water supply temperature, design chilled water temperature difference, average chilled water temperature difference achieved, and distribution system makeup water rate (% of circulation) [10]. Based on the studies in Table 1, the annual energy savings achieved using a DC system relative to a conventional cooling system is the most widely employed energy consumption-related metric.

In addition to the above widely used performance metrics, Yan et al. [38] proposed a water transport factor (WTF),

$$WTF = \frac{Q_c}{E_d} \tag{1}$$

where Q_c is the cooling energy produced by the central chiller plant, and E_d is the energy expended by the chilled water distribution system. An accompanying system coefficient of performance (SCOP) was also proposed [38]:

$$SCOP = \frac{Q_{DC}}{E_c + E_d} \tag{2}$$

where Q_{DC} is the effective cooling energy delivered to end-users after subtracting distribution losses, and E_c is the energy consumed by the central chiller plant. Distribution losses were assumed to represent approximately 80% of chilled water pump energy consumption [38].

In addition, the non-renewable primary energy factor (PEF) may be useful in comparing the fuel use of different sustainable DC design options or technologies. The non-renewable PEF measures the combined effects of efficiency and use of renewable and waste energy sources. The lower the PEF value, the more fossil energy is being saved. The non-renewable PEF accounts for the process of extraction, processing/refining, storage, and transportation of the fuel considered [112,113]. The non-renewable PEF for DC applications is a dimensionless variable defined as [112,113]:

$$PEF_{DC} = \frac{\sum_{i=1}^{n} E_{F(i)} \cdot f_{F(i)}}{\sum_{j=1}^{n} Q_{DC(j)}} \tag{3}$$

where PEF_{DC} is the non-renewable PEF for delivered cooling by a DC system to the end-user within a considered period (in kWh/kWh), $E_{F(i)}$ is the net energy content of fuel (i) (in kWh) delivered to the system where it is finally converted to cooling within the considered period, $f_{F(i)}$ is the non-renewable PEF factor for fuel i (values of which are provided in Table A3), and $Q_{DC(j)}$ is the cooling energy (in kWh) delivered to end-user substation (j) within the considered period.

Gros et al. [110] proposed an energy performance index (EPI) to compare the effects of alternative urban modeling strategies (e.g., landscaping, vegetation, surface albedo) on district cooling energy consumption:

$$EPI_i = \frac{E_{ref} - E_{coolstrat(i)}}{E_{ref} - E_{ideal}} \tag{4}$$

where E_{ref}, $E_{coolstrat(i)}$ and E_{ideal} designate the district cooling energy consumption for the reference uncooled district, district cooling strategy (i) under evaluation, and ideal cooling strategy. In the latter strategy the district topology is unchanged, but the albedo and emittance of all opaque and non-vegetated surfaces are increased to 0.8 and 0.9, respectively. The EPI was applied by Gros et al. [110] to analyze the outputs of coupled building energy and microclimatic models in terms of the effects of alternative urban designs on district cooling energy consumption.

3.2.2. Environmental Impact

The higher efficiency of DC systems relative to on-site cooling, in conjunction with the use of low-carbon energy sources can contribute to significantly reduce environmental emissions. Emissions from biomass and waste in district heating/cooling systems have however prompted concerns [12,18].

Environmental impact assessments of DC systems are generally based on the operational phase CO_2 equivalent emissions avoided through the introduction of both centralized versus individual cooling systems, and sustainability enhancement options discussed in Section 2. Environmental impact can also be quantified using the amount of CO_2 emissions released relative to the amount of energy delivered to the end-user [112], which is a useful performance parameter to compare the fuel use of different DC design options (or technologies):

$$K_{DC} = \frac{\sum_{i=1}^{n} E_{F(i)} \cdot K_{F(i)}}{\sum_{j=1}^{n} Q_{DC(j)}} \tag{5}$$

where K_{DC} is the CO_2 factor for supplied cooling to the end-user (in kg of CO_2/MWh of cooling), $E_{F(i)}$ is the net energy content of fuel (i) delivered to the system where it is finally converted to cooling within the considered period, $K_{F(i)}$ is the specific CO_2 emission factor for energy source (i) (in kg of CO_2/MWh$_{fuel}$, Table A3), and $Q_{DC(j)}$ is the cooling energy delivered to end-user substation (j) within the considered period.

As an alternative CO_2 emission-related metric, the CO_2 payback time (CPT) (Equation (6)) was applied in [114] to quantify the environmental benefits obtained through replacement of air-source heat pumps by ground-source ones for a district heating and cooling system in Tokyo (Japan). The CPT can be a useful concept to evaluate and interpret net CO_2 emissions, and assess systems from a life-cycle CO_2 emissions point of view:

$$CPT = \frac{CO_2 \text{ emissions during the construction phase}}{annual\ CO_2\ emission\ reduction\ by\ the\ introduction\ of\ new\ systems} \tag{6}$$

Genchi et al. [114] found that 87% of life-cycle emissions resulted from the ground-source heat pumps digging process. Their district heating and cooling system contributed annual CO_2 emission reductions of 54%, with a CPT at 1.7 years.

Gros et al. [110] proposed an ambient temperature mitigation index (ATMI) to compare the effects of alternative urban modeling strategies (e.g., landscaping, vegetation, surface albedo) on outdoor temperature:

$$ATMI_i = \frac{DH_{ref} - DH_{coolstrat(i)}}{DH_{ref} - DH_{ideal}} \tag{7}$$

where DH_{ref}, $DH_{coolstrat(i)}$ and DH_{ideal} are the number of degree-hours higher than the cooling temperature setpoint in the district under analysis (e.g., 26 °C in French regulation [110]) for the reference uncooled district, district cooling strategy (*i*) under evaluation, and ideal cooling strategy. In the latter strategy the district topology is unchanged, but the albedo and emittance of all opaque and non-vegetated surfaces are increased to 0.8 and 0.9, respectively. The ATMI was employed by Gros et al. [110] in conjunction with the EPI (Equation (4)) to analyze the outputs of coupled building energy and microclimatic models.

Although not suggested in [110], district outdoor air quality indicators could also be employed, since atmospheric modeling software can yield predictions of pollutant concentrations.

The ozone depletion potentials (ODPs) and global warming potentials (GWPs) of commonly used refrigerants in both conventional air-conditioning and DC systems are compiled in Table A4. Widely employed refrigerants in compression refrigeration cycles include hydrochlorofluorocarbons (HCFCs), hydrofluorocarbons (HFCs), ammonia, CO_2, sulfur dioxide and hydrocarbons, as alternatives to highly ozone-depleting chlorofluorocarbons (CFCs) [115]. HCFC-22 is used in most air-cooled chillers in buildings in developing countries, and although less ozone-depleting than CFCs, unrepaired refrigerant leaks have an adverse impact on the environment as well as on chiller efficiency [116]. Certain DC systems in developing regions including the GCC still use CFC-11 and CFC-12 refrigerants [117]. DC systems can facilitate the phase-out of ozone-depleting refrigerants through the use of HFCs and ammonia, which are not restricted by international protocols (e.g., Montreal Protocol) [118]. DC systems also offer opportunities to use cooling technologies with near-zero ozone depletion and global warming, such as $H_2O/LiBr$ and NH_3/H_2O absorption chillers.

Another environmental benefit of DC systems is that they are less likely to cause adverse noise and vibration in a densely populated urban environment, than conventional on-site air-conditioning systems. The latter systems generate approximately 74 dB sound level, which is equivalent to heavy traffic.

3.2.3. Economics

The total investment cost of a DC system comprises the costs of the central chiller plant, distribution network infrastructure, ETSs, and project development costs.

The investment and production costs of cooling energy using the central chiller plant depend on several factors, including plant size, type and characteristics of energy sources, system diversity factor (i.e., required cooling capacity divided by total market demand), ambient air conditions, supply/return temperatures, local legislation (e.g., for refrigerants) and other conditions [119]. The operation and maintenance costs include the administrative cost for operating and maintenance staffs, maintenance contracts, and materials for preventive maintenance. The space savings associated with replacement of on-site cooling equipment by an off-site, centralized DC chiller plant should also be taken into consideration in an economic analysis, as this space becomes available for other purposes at the end-user site. This can result in economic benefits over the system lifetime.

The cost of the distribution network infrastructure generally represents 50–75% of the initial investment [9]. This cost is influenced by several factors including pipeline density, which is defined as cooling energy or cooling capacity provided by the DC system per unit length of pipeline network [119]. A pipeline cooling energy density greater than 4 MWh/m and cooling capacity greater than 3 kW/m indicate that the cost of the distribution network infrastructure is within a reasonable range [119].

The capital cost of end-user equipment (e.g., heat exchangers, distribution/regulation valves) is considered the least significant [9]. Standardized prefabricated ETSs have relatively well known and stable capital cost. ETS installation cost in end-user buildings depends mainly on local labor costs.

Operational costs (i.e., utilities, service, maintenance) are influenced by the types of energy sources and their pricing, and the efficiency of cold production and delivery, which are affected by chiller/storage efficiency, system layout and control strategy, chilled water supply/return temperatures, thermal energy, and pumping losses, hence district density [12]. The overall cost of DC can be significantly influenced by regulations (e.g., utility pricing, carbon taxes/benefits, customer connection costs) [12].

The DC system economic assessment metrics employed in the studies in Table 1 include investment cost [26], annual operating cost [34], annual fuel/electricity cost savings [3,21], total production cost of cooling (including investment and operation costs) [23–25,27,32,36], specific production cost of cooling (i.e., per unit cooling energy) [6], profit from cooling and electricity sales [28] or annual net earnings [30], total net present value (TNPV) [30], payback period [37,41], and reduction in peak energy demand charges [31]. The impact of uncertainties and availability risk (i.e., reliability) were included as part of total annual cost in [35]. Rezaie et al. [12] presented a generic district enviro-economic function to extend investment and operating costs to business environmental costs, including carbon tax and carbon benefits. The exergoeconomic cost of cooling was only evaluated in one study [39].

Short-term district regulation models generally do not permit long-term technical, economic, and environmental sustainability goals to be attained [12]. A life-cycle cost analysis (LCCA) and the use of TNPVs are recommended to reach rational decisions on engineering projects involving district energy systems and are useful to compare DC systems with on-site cooling production [10]. Such analyses should include not only quantitative (quantifiable) but also qualitative (non-quantifiable) life-cycle economic parameters [120]. Quantifiable parameters include the capital costs of the central chiller plant and distribution network infrastructure, energy and utility costs, and operational and maintenance costs. Non-quantifiable parameters include: re-use of space for other valuable purposes (rentable area and roof garden); visible architectural and environmental impacts; lower GHGs emission; cost stability; reliable service; freeing up maintenance staff; makeup water; only pay for the energy used; refrigerant storage and management; and others.

3.3. Optimization

In general, the mathematical optimization of district heating and/or cooling systems has been less investigated and is thus less advanced than at building-scale [121], particularly in regions where district energy systems are not well developed, and/or where energy conservation incentives are limited. However, district systems are spatially and temporally large-scale systems with complex characteristics and that require dedicated optimization methodologies [32,111,121], including in terms of objectives and constraints. Their optimizations involve both continuous and discrete variables (e.g., existence of a power generator, chiller, pump, pipe or other network element, pipe insulation material and thickness), non-linear performance parameters, and a larger design space than at building-scale [121].

An overview of DC system optimizations [22,24,25,32,35,37,39,121,122], with emphasis on studies for cooling-dominated regions, is presented in Table 4. Most of these studies [22,24,25,32,122] have focused on the optimization of the DC distribution network infrastructure, which represents substantial investment, to investigate possible distribution pipeline designs and network configuration alternatives that would reduce piping capital and installation costs, and pumping energy cost. Coz et al. [39] extended these efforts to exergetic and exergoeconomic optimizations of a DC distribution pipeline network.

Table 4. Overview of DC system optimization studies.

Source (Country)	Item Optimized	Model	Optimization Objective(s)	Optimization Decision Variable(s)	Optimization Constraint(s)	Optimization Algorithm(s)
Chow et al. (2004) [107] (Hong Kong)	Mix of overall GFA of building types in a district	N/A	Minimize central chiller plant cooling load fluctuations	Percentage shares of GFA of each building types to the total GFA served by the DC system	Sum of percentage share of GFA of each building type equal to 100%	GA
Chan et al. (2007) [22] (Hong Kong and UK)	Configuration of distribution piping network	Graphical-based, with undirected links and integer string representation in encoding method	Minimize piping cost and pumping energy cost	Combination of piping configurations to connect central chiller plant and end-users	Structural	GA with local and looped local search methods; stochastic uniform selection function
Söderman (2007) [24] (Finland)	Central cold production unit, distribution pipe lines, cold medium storage	MILP	Minimize overall DCS annualized investment and operating costs	Operational and structural variables	Material and energy balances; components' existence	IBM CPLEX 9.0 solver
Feng and Long (2008) [122] (location-generic)	Pipe network layout design	Mathematical, with constrained conditions	Minimize network annual equivalent cost, consisting of total investment, annual operating, maintenance and amortization cost, and annual cooling loss costs	Discrete standard pipe diameters	Flow equilibrium$^{(1)}$; water velocity ($V \leq V_{max} = 3.5$ m/s); pipe diameter ($d_{min} \leq d \leq d_{max}$) where d_{min} is DN15 and d_{max} is DN1400	Single-parent GA (SPGA)
Feng and Long (2010) [25] (China)	Pipe network layout design	Mathematical, with constrained conditions	Minimize network total annual cost, including initial investment, and operation, depreciation, and maintenance costs	Discrete standard pipe diameters	Pipe diameter; flow equilibrium; pressure equilibrium$^{(2)}$; users' flow requirements; water velocity; DCS hydraulic stability	GA
Khir and Haouari (2015) [32] (Qatar)	Capacity of central chiller plant; storage tank capacity; layout and size of main distribution piping network	MILP	Minimize DCS total investment and operational costs	Temperature and pressure of supplied chilled water; chilled water flow rate and stock level in storage tank at the end of each period; existence of sub-systems (central chiller, storage, type, and layout of pipe)	Structural, chilled fluid temperature and pressure	IBM ILOG CPLEX 12.5 solver
Gang et al. (2016) [35] (Hong Kong)	DCS central chiller	Cooling loads calculated as in [34]; Cooling load uncertainties propagated using Monte Carlo approach; Sub-system reliability modeled using Markov method	Minimize DCS total annual capital (i.e., DCS chiller), operation (i.e., electricity) and availability risk cost (i.e., non-fulfilled cooling demand)	DCS central chiller plant capacity	Chiller capacity identified by restricting non-fulfilled hours to < 35 h with 100% probability	N/R
Kang et al. (2017) [37] (Hong Kong)	Sizes of distributed power generators, and both DCS absorption and electric chillers	N/R	Minimize payback time of distributed power generators and DCS, i.e., ratio of additional capital cost to annual operating cost savings relative to centralized power plant and individual cooling systems	Capacities of distributed power generators and DC chiller	• Electricity and cooling outputs $\leq$ equipment capacities • Primary energy consumption savings maximized in each annual hour	N/R

Table 4. *Cont.*

Source (Country)	Item Optimized	Model	Optimization Objective(s)	Optimization Decision Variable(s)	Optimization Constraint(s)	Optimization Algorithm(s)
Coz et al. (2017) [39] (Slovenia)	Pipe network material, diameter, insulation thickness	Exergetic and exergoeconomic	• Maximize exergy efficiency of cold exergy transportation, or • Minimize total exergoeconomic product (i.e., cold) cost	Pipe diameter, insulation thickness	N/R	N/R
Al Noaimi (2018) [88] (hypothetical site with operational and economic parameters representative of Qatar DC systems)	Tree-like network configuration, including chiller plants' technology, locations and sizes, pipe network layout, distribution strategy	MILP	Minimize DCS total annual capital and operating cost for electricity and cold production, storage, and distribution	Chiller/storage system location and capacity, pipe location, chiller water production/delivery flow, chilled water temperature/pressure	• Chiller output cannot exceed chiller capacity • Network edges should route chilled water to all end-users simultaneously • No redundant network edges • Maximum chilled water temperature rise in network connection • Maximum pressure drop	IBM ILOG CPLEX 12.6.1.0 solver
Perez et al. (2018) [121] (France)	Building materials, energy (heating, cooling, hot water) production and distribution systems	Energy, economic and environmental	• Energy-based: minimize total annual primary energy consumption (electricity/gas for heating, cooling, hot water, lighting, auxiliaries, electrical equipment) • Economic-based: minimize total investment (building structural/insulation materials, fenestrations, heating/cooling/hot water production systems), installation, maintenance, and energy (electricity, gas) costs • Environmental-based: minimize district life-cycle equivalent CO_2 emissions	• Building (structural, insulation, window) material type, insulation location and thickness • Building configuration • Energy production (boilers, CHP, heat pumps, electric/solar heaters, solar PV roof area coverage), ventilation and storage (hot water tank) systems • Energy distribution network	• Building energy demand < 45 kWh/m^2/year • Total primary energy consumption for electricity and gas < 120 kWh/m^2/year • Solar irradiation (for selection of roof surfaces for solar PV installation) > 900 kWh/m^2/year	GA; brute force search; shortest path (Dijkstra)

Note: DCS—district cooling system; GA—genetic algorithm; GFA—gross floor area; MILP—mixed integer linear programing; MTR—mass-transit railway; (1) Algebraic sum of water flow of all branches connected to one nodal point is zero; (2) denotes that algebraic sum of the pressure differences of all branches included in any circuit of pipe network is zero. N/A—not applicable. N/R—not reported.

By contrast Chow et al. [107] focused on identifying an optimal mix of DC end-user building categories and cooling load patterns to minimize fluctuations in cooling demand, which can improve DC chiller use and reduce payback time. Al Noaimi et al. [88] optimized chiller plant/storage system parameters (chiller technology selection, location, size), network pipe layout, chiller water flow, and temperature/pressure, to minimize investment and operating cost. Kang et al. [37] determined the optimum size of distributed power generators, absorption, and electric chillers to minimize the payback time of an energy system including distributed power generators and DC, relative to centralized power generators and individual cooling systems. Gang et al. [35] presented a DC cost minimization methodology focusing on central chiller sizing, by accounting for the impact of both uncertainties in cooling loads and availability risk (i.e., reliability). Their work extends the cooling load uncertainty-based DC design methodology of Gang et al. [34], which was discussed in Section 3.1, and is examined here in terms of optimization aspects.

Unlike for building-scale optimizations, most DC optimizations [22,24,25,32,35,37,122] have been single-objective. The above optimization efforts are reviewed in this section, focusing on cooling-dominated regions. Additional region-generic district heating/cooling optimization studies are summarized in [88].

Chan et al. [22] used a modified version of the genetic algorithm (GA) to find optimum or near-optimum piping network configurations for a DC system in a hypothetical site in Hong Kong. A modified GA method using a local search and looped local search techniques was employed to improve performance in the search of a near-optimal piping network configuration. The optimization objectives were to minimize both the investment (i.e., capital) and operational (i.e., pumping energy consumption) costs associated with the piping configuration. Flexible pipe connection options were applied using different types of network layouts (e.g., radial, tree-shaped or combinations of both). Accordingly, structural constraints were applied to include restrictions on the number of links in relation to a given number of nodes (e.g., n nodes must have n-1 links), and to prevent flow recirculation (i.e., cycles between nodes). Using a series of parametric simulations, the mathematical model developed was tested for two cases, namely eight and 16 consumer buildings, with different GA population sizes and mutation rates investigated. Incorporating local search led to improved search performance (i.e., reduced convergence time) and better optima (i.e., improved objective functions values).

For a new urban area requiring a DC system, or either the expansion or retrofitting of existing DC systems, Söderman [24] employed a mixed integer linear programming (MILP) model to optimize both DC structural and operational aspects. The model was tested using actual 2006 cooling demand data for an existing DC system, as well as using predicted demand data for the year 2020, representing an expansion of the existing 2006 DC system. The single optimization objective was to minimize total annual cost, which included annualized investment cost and running costs of the central chiller plants with auxiliary units and main distribution pipelines. The optimization solution would assist in decision-making on how the existing DC system should be expanded in the future, in terms of the number, capacity and spatial arrangement of DC sub-systems (i.e., central chiller plants, storage tanks, and pipeline networks). Furthermore, the optimization solution contained information on how the DC system should be operated over different time periods, for example in terms of the chilled water flow rates in different sections of the pipeline network during different periods of the year (day, night, seasons), and the charging/discharging of storage tanks, and operation of the chiller plants. The author considered constraints related to supply and demand, enthalpy flow balance with heat gains/losses, and power consumption constraints. Technical hydraulic and temperature constraints were however omitted in the model.

Feng and Long [122] developed a mathematical model to optimize the pipe network layout of a DC system by minimizing the system's total annual cost. The total cost comprised annualized overall investment, operating, maintenance and amortization costs, as well as the cost of piping energy losses. An improved form of GA (i.e., single-parent GA (SPGA)), was applied. The optimization problem was constrained by flow equilibrium constraints at each nodal point, an upper bound chilled water flow velocity (i.e., 3.5 m/s), and limits on the pipe diameter range, which is considered to be an important DC system design factor. The solution obtained using the SPGA was compared with Dijkstra's algorithm, and the results showed that SPGA was useful in network piping scheme assessment and decision-making. It was also useful to select the optimal location of the central chiller plant(s). According to the authors, SPGA demonstrated high search efficiency, rapid convergence, and stability. It was also suggested that road and building distribution in the district should be considered before a final optimum DC design is selected.

Feng and Long [25] subsequently applied a standard GA to the optimization of a pipe network layout in the same DC system as in [122], with additional optimization constraints to obtain a more realistic model than in [122]. The additional constraints related to pressure balance (equilibrium), hydraulic stability, and end-user cooling demand. The objective function incorporated initial investment, operational, depreciation and maintenance costs.

The above optimizations [22,24,25,122] focused on economic objectives, by imposing structural optimization constraints, without temperature and pressure constraints. Khir and Haouari [32] considered pressure and temperature-related technical constraints, which are critical for DC system functionality and integration with both the central chiller plant and end-user ETS. Such constraints were used to minimize the sum of the fixed costs of a building chiller plant and storage tank, the variable production and storage costs, and the fixed costs related to the purchase and installation of the distribution pipelines. A MILP model of the DC network was constructed using a reformulation-linearization technique (RTL). All relevant data (i.e., cooling demand, costs, capacities, and other parameters) were collected from appropriate sources (i.e., market, services) and prescribed as inputs to the MILP model. Optimal and near-optimal solutions could be obtained within a reasonable computational time for a distribution network up to 60 nodes.

Coz et al. [39] performed an exergoeconomic optimization of a DC distribution pipeline network, by modeling exergy transfer due to heat gains and viscous (pumping) losses (i.e., pressure drop) associated with cold distribution in the network. The optimization sought to minimize the cost of cold (i.e., final product) for the main consumer, and the exergetic efficiency of cold transportation, through determination of the optimum pipe diameter and pipe insulation thickness. This optimization was performed for polyurethane-pre-insulated steel pipes and non-insulated polyethylene pipes, at cooling capacities of 50 to 1500 kW. Slovenian unit electricity and input heat prices were applied. Higher cooling capacities were found to reduce the exergetic cost of cold (i.e., full cost of cold production). In addition, polyethylene pipes had smaller exergetically optimum diameters than pre-insulated steel ones, due to the lower surface roughness of the former material, hence lower frictional losses and flow rate requirement to carry the same quantity of cold. However, polyethylene pipes exhibited lower exergetic efficiency than insulated steel pipes due to larger parasitic heat gains. By contrast the exergoeconomically optimum polyethylene pipes were found to be larger in diameter than those of the insulated steel pipes. The minimized cost of transporting cold was lower for insulated steel than polyethylene pipes. For steel pipes, the price of the inlet cold exergy was found to have the largest impact on the total exergoeconomic product (i.e., cold) cost, when compared with the impacts of the specific costs of insulated steel pipes and construction, and electricity price to drive pumping. For steel pipes, higher polyurethane insulation thicknesses improved exergetic efficiency because of reduced heat gains, while increasing pipe diameters resulted in lower frictional losses but higher heat gains. Consequently, an exergoeconomically optimal pipe diameter existed at which the total product cost was minimum. In summary, pre-insulated steel pipes were somewhat advantageous in terms of exergy efficiency and exergoeconomic product cost of cold. Such results however depend on the input electricity and inlet heat prices. The analysis methodology could be extended in future work to a complete DC system and undertaken for several cooling technologies.

The existence of a large variety of end-user buildings and different cooling load profiles in a DC system can result in a relatively steady cooling demand profile [107]. Thus, a design approach could seek to identify and select the optimum mix of end-user building categories and cooling load patterns that could minimize the cooling energy production cost to maximize effective system use and ultimately lead to a shorter payback period. Chow et al. [107] proposed an approach to determine the optimal mix of building types using a GA. The buildings mix was described by the percentage share of the overall GFA of the different types of end-user buildings that are served by the DC system. The optimization and cooling load estimation approaches were illustrated for three case studies in Hong Kong and reported to be effective to produce optimal or near-optimal solutions. In the first case study, five building types were considered which included hotels, residences, offices, shops, and mass-transit railway (MTR) station. The optimum GFA shares of the building types were found to be 67.0%, 13.8%, 14.1%, 3.4% and 1.6%, respectively. As the MTR station had the least effect, it was disregarded in the second case study, leading to an optimized end-user building mix shares of 81.4%, 12.0%, 5.1% and 1.5% for the hotels, residence, office, and shops, respectively. In practice, it would be difficult to implement a high share of hotel buildings (such as 81.4% as in the second case study) within a district [107]. Therefore, in the third case study, the GFA share of the hotel category was set to a constant 5% value, with the shares of the remaining three other end-user building types optimized. This case led to more stable solutions than the other two case studies and the fitness function converged to two close specific fitness values (i.e., cooling demand fluctuation index of 0.60415 and 0.60425). The slightly better fitness was obtained for an office GFA share of 2% to 7.5%, residential building share of 30% and 40%, and shop share of 40% to 80%.

Kang et al. [37] investigated the energetic performance and economics of a modeled distributed energy system comprising distributed power generators (i.e., gas engines), DC chillers (i.e., absorption and electrical), electrical and chilled water distribution networks, and end-users in a group of 12 buildings in Hong Kong Polytechnic University campus. The power generators provided both electricity and waste heat to the electrical and absorption chillers respectively, for district cooling. The optimum size of the distributed generators, absorption and electric chillers were determined to minimize the payback time of the overall energy system, destined to substitute a centralized energy system, in which cooling was provided by individual water-cooled centrifugal chillers powered by the grid (i.e., coal-fired centralized power plant). The absorption chillers provided most of the cold supply in winter, but the opposite occurred in summer. From April to November, the DC system chillers achieved a higher COP than those of the centralized cooling system. Electricity consumption from the grid was reduced through the substitution of grid power by distributed generators and of electric chillers by absorption systems, and effectively shaved peak electrical loads. Such loads were reduced by up to 72% in summer. The distributed energy system led to 9.6% reduction in primary energy consumption relative to the centralized energy system, at the expense of three times the capital cost, but 45% lower operating expenditure. The payback time of the distributed energy system investment was estimated at 1.9 years.

Gang et al. [35] extended their cooling load uncertainty-based DC design approach [34] (Section 3.1) to a cost minimization methodology, by accounting for the economic impact of uncertainties in both DC cooling loads and operational reliability (i.e., DC sub-system availability risk). The DC system total annual cost included capital cost (which depended on DC chiller capacity), operational cost (i.e., electricity cost, which depended on both cooling loads and DC chiller capacity), and availability risk cost (i.e., costs associated with non-fulfilled cooling demand due to underestimation of DC system capacity, and/or DC sub-system performance loss or malfunction; such costs depended on both cooling loads and DC chiller capacity). Uncertainties in cooling loads were propagated using a Monte Carlo approach, while sub-system reliability was quantified using Markov method. The impact of incorporating uncertainties on DC chiller sizing and annual cost were evaluated relative to standard DC capacity sizing (i.e., without uncertainties and with a safety margin). The optimal DC chiller capacity obtained by accounting for all uncertainties (i.e., in cooling loads and operational reliability) was lower than that obtained using either standard capacity sizing or by only accounting for operational risk, but higher than by only accounting for uncertainties in cooling loads. The minimized total annual cost of a DC system sized by including either all uncertainties or using a standard approach were similar (as the contribution of DC chiller capital cost to total annual cost was not significant), but significantly lower than by accounting for operational risk only, particularly at assumed high availability risk prices. Finally, uncertainties in both DC cooling loads and operational reliability were found to have larger impacts on the design of individual on-site cooling systems than on the design of DC systems.

Perez et al. [121] highlighted the need for multi-objective optimizations of district heating/cooling systems, as single-objective optimizations generally do not effectively incorporate holistic sustainability criteria. The authors presented a five-level, multi-objective (energy-, environmental- and economic-based) district heating and cooling system optimization procedure. Although the methodology was demonstrated for a heating-dominated region, it would be applicable to cooling-dominated environments and is of interest in terms of handing building to district-scale features and incorporating multiple sustainability-related performance criteria. The optimization procedure steps involved concurrent building envelope optimization and rooftop solar PV exploitation potential evaluation, followed by distribution networks optimization, then complete building optimization and identification of optimal combinations of complete building configurations in the district (at local decentralized branch heat, cold or hot water production scale), with in parallel combinations of building envelope configurations and complete district optimization (at centralized production scale). The procedure closed with an aggregation of all results at local (branch) and central (urban) production scales. Several optimum search methods were employed including the non-dominated sorting GA. The number of possible solutions could be effectively reduced using sustainable energy-related constraints. The optimization procedure was applied to a future industrial estate planned in Grenoble (France) for the year 2030, including 48 residential, office, and retail buildings and parking lots, combined in 12 urban islands. Cooling was delivered to retail and office buildings using absorption heat pumps in summer. Comparing the results obtained in Grenoble (France) and Stockholm (Sweden) climatic conditions, it was found that for the hotter climate, the optimization led to solutions that focused on reducing solar-induced overheating, whereas in the colder climate, insulation from the outdoor environment was emphasized. This methodology has potential for DC systems in cooling-dominated climates.

4. GCC Regional Air-Conditioning Challenges, DC Status and Future Opportunities

A considerable potential exists for DC in the Middle East, in particular in the GCC, due to its harsh climatic conditions that make air-conditioning a necessity almost all year round, high-pace and dense urbanization, and popular building architectures with extensive glass exteriors. In this section, space cooling challenges specific to the GCC region are summarized, before providing an overview of the current and forecasted development of the DC market in the region, including economic and environmental opportunities. Potential technical and policy-related solutions to address regional space cooling challenges are then identified.

4.1. Air-Conditioning Challenges

Air-conditioning in the GCC faces the following challenges: a lack of natural cold sinks (i.e., air, water) for either direct cooling or heat rejection from air-conditioning systems; natural water scarcity; a growing cooling energy demand driven by population and economic developments; elevated energy use per capita and high reliance on fossil fuels including for cooling production with an associated environmental impact; high utility tariff subsidies paid by governments; and a lack of air-conditioning and urbanization legislation [123,124]. These aspects are developed in this section.

General population, economic, energy, water and CO_2-equivalent GHG emissions indicators for GCC countries are tabulated in Table 5 and compared with the world average and other major energy users. From approximately 52.7 M inhabitants in 2015, the region's population will grow to 66.5 M (+26%) and 76.7 M (+46%) by 2030 and 2050, respectively, at annual growth rates above the world average through most of the 2015–2050 period [4]. As highlighted by the data in Table 5, GCC countries have had among the highest annual energy, electric power consumption, and CO_2 emission rates per capita in the world. Natural gas and oil are the dominant primary energy resources used in the region. The UAE, Qatar and Oman cover above 60% of their total domestic energy needs, particularly electricity production, using natural gas. Bahrain, KSA, and Kuwait mostly use oil, which fulfills 53.8%, 71.2%, and 77.9% of their total energy demands, respectively, and a substantial portion of their electricity generation.

Table 5. General population, economic, energy, water, and emissions indicators for GCC countries, compared with the World average and other major energy users [4,125,126].

	Population [a]	GDP [b]	Energy Use [c]	Electricity Use [d]	Energy Intensity [e]	Renewable Water [f]	Water Consumption [g]	CO$_2$ Emissions [h]
Bahrain	1.4 (2.01%)	23.7	10,158	18,415	244	84.2	348 (2003)	22.7
Kuwait	3.9 (5.44%)	35.4	10,093	15,689	134	5.1	447.2 (2002)	28.2
Oman	4.2 (6.45%)	17.8	6502	6097	153	311.7	509.3 (2003)	16.3
Qatar	2.5 (4.93%)	72.9	17,221	14,960	138	26.0	376.6 (2005)	41.9
Saudi Arabia	31.6 (6.65%)	22.2	6663	8556	137	76.1	907.5 (2006)	18.7
UAE	9.2 (2.03%)	41.2	7619	10,899	127	16.4	665.2 (2005)	20.1
World	7383.0 (1.19%)	10.5	1890	3050	132	6064	506 (2002)	4.9
USA	319.9 (0.72%)	56.2	6983	13,116	138	9538	1864 (2005) 1543 (2010)	16.7
EU	508.2 (0.10%)	34.1	3251	6108	94	2961	593 (2002)	6.9
China	1397 (0.54%)	8.0	2129	3483	192	2018	435.2 (2007) 425 (2015)	7.3
India	1309 (1.23%)	1.7	597	0.727	123	1458	559.9 (2000) 602.3 (2010)	1.6
Japan	127.975 (−0.09%)	38.1	3615	8100	99	3397	652.1 (2007) 640.6 (2009)	9.5

[a] 2015 population in millions, with percentage annual average rate of population change (2010–2015) in parentheses () [4]. [b] Annual GDP per capita (2013–2017), in current kUSD/capita [125]. [c] Annual energy use per capita (2010–2014), in kgoe/capita [125]. [d] Annual electricity use per capita (2010–2014), in kWh/capita [125]. [e] Annual energy intensity (2010–2014), in kgoe per USD 1000 [125]. [f] Total renewable water (internal and external surface/groundwater) availability, in m^3/capita/year. Data for the year 2014, except for world average and EU (both for the year 2012) [126]. [g] Water consumption (year), in m^3/capita/year [126]. [h] Annual equivalent CO$_2$ emissions per capita (2010–2014), in Mt/capita [125].

Space cooling in the GCC is essentially provided using fossil electricity-driven window units, split systems, central air-cooled or water-cooled chillers, and to a limited extent by DC systems (i.e., ~5% of space cooling capacity) [5]. Whereas 10% of electricity consumption is expended globally for space cooling, in the GCC this application represents approximately 50% of total electricity use [5] and up to 70% of peak-period electricity use in the UAE and Kuwait [3,6]. This is contributed by climatic conditions, high living standards with insufficient emphasis on energy conservation, including in the use of low-efficiency air-conditioning systems and building designs/constructions [123]. The annual energy use per capita for space cooling is of approximately 590 kWh in the Middle East (which extends beyond the GCC), compared with the following figures in other regions, including in predominantly hot climates: 35 kWh in Africa, 60 kWh in India, 70 kWh in Indonesia, 320 kWh in the EU and China, 760 kWh in Japan and above 1800 kWh in the USA [3]. By 2050 these figures will more than double in the Middle East and China, and be multiplied by over 12 in India and Indonesia, with significantly less pronounced rises in the EU, USA, and Japan [3]. The GCC is known for its significant variations in building cooling loads both on a daily basis between day and night (i.e., 40% and 50% in winter and summer, respectively, in for example the UAE [31]), and on a seasonal basis between winter and summer, with peak winter loads of order 50% lower than summer loads in the UAE [31]. Peak cooling demand periods (i.e., afternoon/evening) also coincide with peak ambient temperatures and humidity, during which the efficiency of thermal power generation significantly decreases [3]. The resulting electricity demand fluctuations for vapor compression cooling adversely impact power plant efficiency as they operate at off-design conditions for a major part of the year and induce stress on electrical networks [31]. In for example the UAE, the minimum total annual electricity demand (i.e., in a winter night) is approximately 40% lower than the annual peak demand (i.e., in a winter day) [127].

The GCC peak cooling demand is expected to triple between 2010 and 2030 (Figure 7) as a consequence of population growth, rapid urbanization and the sought for improved comfort. KSA and the UAE will continue to have the highest peak cooling demands among the GCC members at 19–52 MRT (2010–2030) and 8–21 MRT (2010–2030), respectively. It has been estimated that if GCC countries were to continue using conventional air-conditioning technology and keep their cooling energy consumption patterns, based on projected population and economic developments, it would cost approximately $100 billion to acquire the projected new cooling hardware by 2030, and over $120 billion to provide the associated power supply capacity [5].

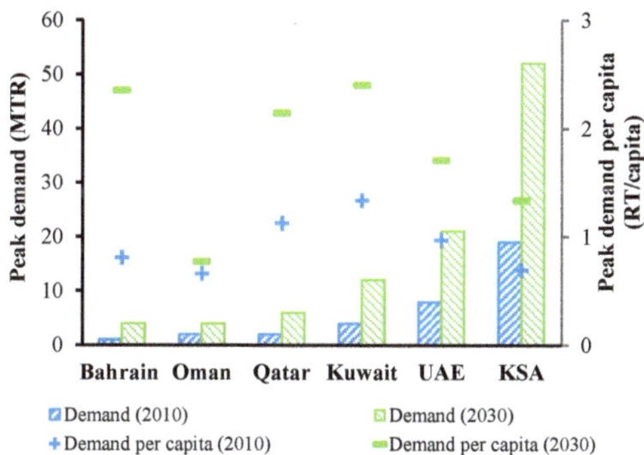

Figure 7. Peak cooling demand (2010–2030) of GCC countries, based on [4,5,128].

Both electricity generation for space cooling (currently essentially thermal, with increasing shares of solar PV and CSP generation in the future [127]), and the operation of cooling towers in air-conditioning systems require water [129,130]. Water requirements for power generation vary widely depending on technology and climate, and increase in arid and semi-arid (i.e., desert) conditions such as in the GCC [129]. Solar electricity installations require regular surface cleaning in dusty (e.g., sandy) environments [129]. Air-conditioning cooling towers require makeup water to compensate for water losses through evaporation or blow-down effect (to maintain water quality, within specified limits) [131]. Due to a lack of rainfall and groundwater, the available renewable water volume in the GCC ranges from 5 to 312 m^3/capita (in Kuwait and Oman, respectively), with an average at approximately 77 m^3/capita, which is far below the global water scarcity limit (i.e., 1000 m^3/capita) [129]. Conversely GCC regional average water consumption (i.e., 760 m^3/capita [123]) significantly exceeds the corresponding worldwide average (i.e., 506 m^3/capita [126]). The resulting water stress indexes in the region range from 106.2% in Oman to 2610% in Kuwait, compared with a European average of 8% (with values of less 70% in any European country) and world average of 13% [126]. (SDG 6.4.2 water stress, defined as 100* (total freshwater consumption)/(total renewable water resources-environmental flow requirements) [126]. Kuwait, Saudi Arabia, and the UAE have three of the top four highest water stress levels in the world [132].) Water scarcity in the GCC is accompanied by a degradation in water quality and extensive reliance on seawater desalination [129,133]. The GCC desalination sector accounts for approximately 40% of the global desalination capacity, with approximately two thirds of the GCC's capacity met by thermal desalination (essentially MSF), which is energy-intensive [123]. In addition, 66% of surface water in Arab countries is from outside the region, which has been and remains a potential target and source of political instabilities [129]. Finally, space requirements for the installation of cooling towers make it increasingly difficult for individual end-users in densely populated areas to install individual water-cooled chiller technologies [116].

Despite the abundance and affordability of fossil fuels in the GCC, the above context and need to grow regional fossil-fuel exports rather than consume fuel production domestically, will require local energy/water conservation and efficiency improvement efforts, and increased penetration of low-carbon energy resources. The deployment of more sustainable air-conditioning is a key opportunity in the GCC to improve the efficiency of energy use and address climate change.

4.2. Existing and Developing District Cooling Market

Interest in DC in the GCC began in the late 1990s, prompted by its higher efficiency and lower environmental impact than conventional on-site cooling systems. The first commercially successful DC system in the region was introduced in 1999 at Zayed Military City in Abu Dhabi [134]. This plant was a significant technical success which led to further progress in the design and implementation of DC systems in the GCC. DC systems in the GCC range from small-scale (i.e., capacity < 10,000 RT) to large-scale (i.e., capacity > 10,000 RT) and are operated by several private and government-owned DC companies. In 2010, the largest DC system in the world was built at The Pearl Island in Qatar, with a capacity of 130,000 Ton of Refrigeration (RT) [134]. Other large-scale DC systems deployed or under construction in the GCC are listed in Table A5.

As of 2014, the GCC DC market (4.1 MRT or 44.3 GW$_{th}$, with 10,000 km distribution network) represented 32% of the worldwide installed DC capacity (i.e., 12.6 MRT), with a market size of approximately 5.5 billion USD [17,135]. 65% of the current GCC DC capacity is installed in the UAE, followed by 22% in KSA and 7% in Qatar [17]. DC end-users in the GCC are mostly residential (56.2%), followed by commercial (39.1%) and industrial (3%) users [17].

Figure 8 contrasts the breakdown of installed GCC DC capacity by country in 2014, with its forecasted potential by 2019 and 2030. Despite progress in DC implementation in the region, the current penetration of DC in the GCC cooling market remains limited to approximately 5% (i.e., 4.1 MRT) of total space cooling capacity, with the largest share in the UAE (i.e., 11%) [5]), highlighting significant potential for further growth. In addition, the currently installed GCC DC capacity is estimated to be used at only 50% of its capacity [135]. 60% of the DC capacity is powered by DC utilities, and the rest by government- and large single-owned utility companies [135]. The GCC DC market is forecasted to rise at an average 16% cumulative annual growth rate through 2014–2019 [17]. These forecasts are based on large-scale development projects in real estate and the commercial sectors, population growth, and increased awareness of and governmental focus on energy saving measures. KSA is estimated to be the fastest growing user of DC systems in the GCC, with a projected growth rate of 34% between 2014 and 2019, followed by Qatar at 18% growth rate over the same period [17]. DC growth in Qatar is mainly associated with increased construction activity driven by the 2022 FIFA World Cup, and government initiatives for the promotion of DC systems. In Bahrain and Oman, real estate developments are the key drivers for the growth of DC capacity, whereas in Kuwait, infrastructure and industrial development are the main drivers. From cooling market shares of <1% and 11% in KSA and the UAE in 2014, by 2030, the share potential of DC in these countries is estimated at 26% and 50% of national cooling capacity in KSA and the UAE, respectively (Figure 8).

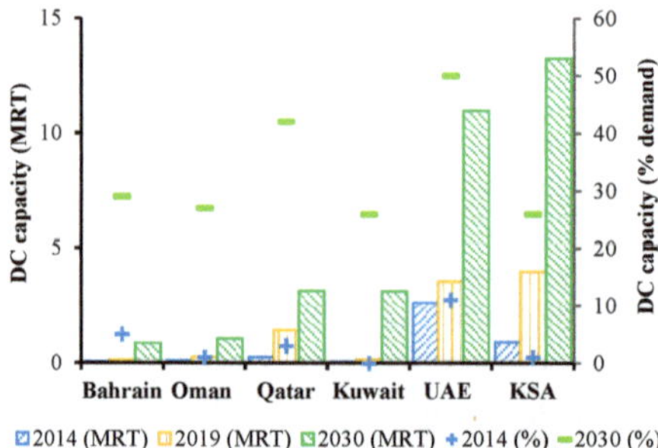

Figure 8. Installed (2014) and projected (2019, 2030) DC capacity potential in GCC countries, based on [5,17]. Percentage DC capacity is relative to total cooling requirement.

Average energy savings of 0.78 kWh (46%) per ton-hour of cooling are reported by a leading DC provider for 43 monitored actual water-cooled DC systems implemented in the GCC [116] relative to real-world data for monitored air-cooled building chillers. Based on data for approximately 72 existing regional DC plants (63 of which are located in the UAE) operated by the same provider (i.e., TABREED [136]) and its associate companies and joint ventures, and delivering approximately 1.1 MRT of combined cooling annually, these plants are estimated to reduce the electricity consumption of the region by approximately 1.53 TWh annually, with an accompanying 768,000 tons of avoided CO_2 emissions annually.

4.3. Future Opportunities and Directions

The forecasted energy, economic and environmental benefits anticipated through the expansion of DC capacity and space cooling share in the GCC, assuming continued use of existing DC technologies, are summarized in Section 4.3.1. Further technical and policy-related opportunities that remain to be tackled and potential solutions to fully exploit DC to reduce the energy, economic and environmental impact of space cooling in the region, are identified in Sections 4.3.2 and 4.3.3, respectively.

4.3.1. Forecasted Energy, Economic and Environmental Benefits

Based on the existing and forecasted 2012–2019 GCC DC capacity (i.e., 3.2–8.7 kTR [14,17]), and annual average specific energy consumptions for monitored DC systems and on-site air-cooled chillers in the UAE (i.e., 0.92 and 1.7 kWh/ton-hour, respectively [10,116]) and other sources, GCC regional space cooling electricity savings of approximately 46% could be achieved annually between 2012 and 2019 using DC instead of conventional on-site air-cooled chillers. The corresponding annual CO_2 equivalent emissions avoided, based on 2008 annual average specific GHGs (i.e., CO_2, CH_4 and N_2O) electricity emission factors in the GCC [137], assuming a reliability of 99.96% for the DC system [17], would be in the range of 19.2–51.3 Mt per annum between 2012 and 2019. These energy and emissions savings would even be more substantial considering that the installed capacity of existing DC systems is typically 15% lower than that of conventional on-site cooling systems. Energy and emissions savings estimated by comparison of DC with window/split air-conditioning units, which have annual average specific energy consumptions of approximately 2.0 kWh/ton-hour [116], rather than air-cooled chillers, would also be more significant.

By 2030, it has been estimated that the regular use of DC in the GCC region could result in 20 GW reduction of in new power capacity requirements, thereby saving USD 120 billion in power capital expenditure, and the equivalent of 200,000 oil barrels per day in fuel consumption [5]. The avoided CO_2 emissions in for example the UAE are estimated at 31 Mt/year [5].

The above energy, economic and environmental estimates are based on existing current DC system design and operational practices in the GCC. Potential technical and non-technical solutions to improve the sustainability of DC systems in the region are outlined in Sections 4.3.1 and 4.3.2, that could enable higher energy, economic and environmental benefits.

4.3.2. Technical Opportunities

Due to a history of abundant and low-cost fossil fuels, compounded by the lack of natural cooling potential in the region, current GCC energy systems, including DC systems, rely almost solely on fossil primary energy supplies. DC systems in the GCC typically use fossil electricity-powered, water-cooled fixed-speed compression chillers [138], that are more efficient than their air-cooled counterparts [10]. Forced-draft cooling towers are employed to reject heat from chiller condensers [5,131,138]. Elevated ambient temperatures in the region have an adverse impact on both chiller and cooling tower efficiencies. Cooling tower efficiency also degrades at elevated inlet air humidity, depending on geographical location and daily/seasonal period [138]. DC capacity sizing and distribution planning, operational strategies, and performance evaluation methodologies (e.g., without life-cycle considerations) also require improvements [138]. Technical opportunities and potential solutions to improve DC system sustainability in the region are identified below in terms of DC energy sources, cooling/storage technologies, operation, and performance evaluation.

A diversification of the energy mix (e.g., including renewable and nuclear energy) and energy efficiency improvements, including through the replacement of conventional cooling with DC, and use of renewable and waste energy sources, is feasible and already under way in the region to reduce fossil-fuel consumption and its environmental impact, with the additional benefit of more diversified economies. Such a shift is expected to increase the availability of crude oil for export for KSA and Kuwait, and reduce the dependence on costly natural gas imports for Bahrain, Kuwait, and the UAE. Considering electricity generation, renewable shares of up to approximately 15%, 20–25%, 25% and 30% are envisaged in Kuwait, the UAE, Saudi Arabia, and Qatar by year 2030, with emphasis on solar power [127]. In addition, Saudi Arabia aims at 30% renewable electricity by 2040, while the UAE targets 44% and 75% renewable and clean electricity generation (including nuclear and clean coal power), respectively, by 2050 [127]. Progress in more sustainable electricity generation will be complemented by solar thermal energy use, less energy-intensive water desalination/treatment and conservation, and developments in other sectors [124]. Therefore, there is scope to prepare for a shift from fossil-dominated space cooling to more sustainable cooling in the region. However, the appropriate mix of solar electrical and thermal power conversion technologies (in particular PV and CSP) should be carefully planned in the region to provide both electricity and cold in future GCC energy systems.

Potential pathways of sustainable DC production to assist in meeting the significant year-round cooling demands of the GCC region using locally available energy sources include indirect cooling via low-carbon electricity-driven and low-carbon heat-driven chiller technologies, direct cooling using waste cold energy recovered from LNG regasification terminals, as well as optimized use of thermal storage systems [138].

Based on future energy plans in the regions, low-carbon electricity for vapor compression chilling could be provided by solar PV/CSP, and to some extent by wind installations, the economic potential of which is less evident than that of solar energy in the region, as well as waste-to-energy, and nuclear power in the UAE's case [127]. As daily periods of high cooling demand coincide with sunshine hours, electrically driven space cooling could also contribute to reduce future excess electricity levels at growing solar/wind power shares in future GCC energy systems [127] and its adverse effects on grid stability and the efficiency of resource use. In parallel, heat-driven chillers could reduce electricity demand in non-sunshine (i.e., night time) periods. The efficiencies of solar PV and CSP technologies will however be affected by high ambient air temperature and humidity, respectively, in afternoon/evening peak cooling demand periods [127]. The cooling demand-provision match could be improved using short-term weather forecast in conjunction with historical energy demand data, as well as smart energy monitoring and storage management [138].

Sustainable energy sources for heat-driven cooling include waste heat from regional energy-intensive industrial facilities (e.g., aluminum, cement, and hydrocarbon production facilities), and thermal energy from solar and biomass waste plants. In addition to single/double-effect absorption chillers, triple-effect chillers, subject to increased commercial availability at large capacity, could be considered to exploit higher-temperature CSP/industrial heat with improved efficiency, as well as reversible heat pumps/chillers, as recently envisaged in [138]. In parallel, the deployment of tri-generation, waste heat recovery technologies and a thermal grid and heat trading system have also been recognized as potential solutions to reduce the energy consumption of GCC DC systems [138]. Gas engine CHP systems could provide both electricity to vapor compression chillers and heat to absorption chillers for both space and non-space cooling applications, as well as heat to other applications (e.g., hot water generation). Flows of hot/cold thermal energy (including waste heat/cold) from a variety of sources between the DC and other systems would be enabled by an intelligent thermal network, with reduced energy distribution losses.

The extension of DC end-uses to non-space cooling applications, including augmentation of power generation, recycling of DC chiller heat rejection for either hot water production or water treatment/desalination, are also envisageable.

Improving the efficiency of electricity/heat-driven DC central chiller and cooling towers in the regions' harsh climatic conditions requires customized technical solutions. For example, in moderate climates, the heat source and chilled water inlet temperatures of single-effect H_2O/LiBr absorption chillers may typically be set at 100 °C and 10 °C, respectively [139]. However, the COP of a basic H_2O/LiBr absorption chiller reduces from 0.75 to 0.65 when the heat-rejection medium temperature of the absorber and condenser, connected in parallel, increases from 15 °C to 35 °C, while its cooling capacity drops by 55% [139]. Potential concepts that could be investigated to improve chiller efficiency would involve reducing the heat-rejection temperature of the DC chiller plant, and could include:

- Incorporation of innovative thermodynamic cycles with operational flexibility (e.g., hybrid absorption-compression cycles and combined power-refrigeration absorption cycles) to improve chiller thermodynamic performance by extending the operating conditions of its absorber and/or condenser (e.g., operating the absorber at a relatively higher pressure for the same heat rejection, chilled water, and heat source temperatures). Such cycles could provide flexibility to cope with both seasonal and daily fluctuations in ambient conditions and thus cooling loads.

- Exploitation of artificial cold sources, such as LNG from regasification terminals, as a heat sink for the condenser (and/or absorber) of the chiller plant, to reduce its heat-rejection temperature. Among the GCC countries that face domestic gas shortages (i.e., Bahrain, Kuwait, UAE), the UAE was the first to build an LNG receiving terminal in Dubai. Kuwait and the UAE (in Abu Dhabi and Dubai) currently have 5.8 and 9.8 MTPA regasification capacities, respectively [80]. Kuwait and Bahrain will have additional 11.3 and 6 MTPA regasification capacities by 2021 and 2019, respectively [80]. Even if partly exploited, the above regasification capacities could enhance the thermodynamic, economic, and environmental performance of DC and other direct cooling applications, and power generation in the region, subject to adequate cold transportation. CO_2 has been proposed as a potential cold transportation medium, which could offer additional opportunities of CO_2 sinking and valorization, which in the GCC, have essentially been limited to CO_2 injection into hydrocarbon reservoirs for enhanced recovery to date.

Customized solutions to improve cooling tower efficiency could involve development and integration of innovative dehumidification technologies to reduce the moisture content of cooling tower inlet air, as suggested in [138]. Approaches to improve the sustainability of water provision/use for DC should also be pursued. The water required by cooling towers is either scarce or difficult to access in the region and requires treatment before being used. Although seawater can be employed in cooling towers, it requires more costly corrosion- and fouling-resistant equipment, with higher maintenance costs. Similarly, alternative low-quality water, such as treated sewage effluent (TSE), brackish groundwater and partially desalinated water [131], could be used but also with increased capital and maintenance costs. TSE is used as makeup water in for example the cooling tower of Sheikh Zayed Road DC plant in Dubai [131]. Challenges are encountered with the use of TSE in DC systems, including fluctuating availability (i.e., quantity, quality, pressure) which may not match demand, and management of discharge products. Alternatively, seawater, TSE [138] or other water sources could be employed directly as heat-rejection medium–cool chiller condensers, rather than as working fluids for a cooling tower. This approach has been employed in a DC system in Bahrain [140].

In parallel, cold distribution losses from the DC piping network and pumping stations, which are significant in hot-climate regions, should also be addressed. As the chilled water temperature supplied to end-users increases, cooling capacity reduces. To compensate for increased water supply temperature, larger chilled water volumes need to be supplied to meet the demand, that require additional pumping energy. Alternatively, additional cooling energy needs to be supplied to maintain the desired system temperatures, resulting in increased primary energy consumption. Therefore, the distribution network pipe materials and their geometry, and their installation environment, along with external factors that affect cold distribution losses, should be carefully designed/planned. The need for novel pipe designs with special insulation materials to reduce cold distribution losses, and their life-cycle assessment in GCC DC plants, has been recently suggested [138].

The need for improved engineering models, operational strategies, and life-cycle assessments to reduce the energy consumption, capital/operational cost, and environmental impact of DC systems in the GCC has also been raised [138]. Suggested development areas towards more efficient DC design and operation include (i) identification and prioritization of energy saving opportunities; (ii) improved modeling of electric centrifugal compressors, accounting for equipment environment- and operating condition-dependent performance; (iii) improved control strategies, including variable-speed rather than fixed-speed operation for electric centrifugal compressors, thermal energy storage and water conservation/recycling. Predictive chiller/storage control and management strategies with the use of short-term weather forecast, historical and real-time energy consumption monitoring should optimize cooling production, storage, and consumption. Ultimately, holistic life-cycle DC performance and impact assessment approaches will be required to evaluate the sustainability of DC systems more meaningfully.

Finally, urban outdoor ambient air temperatures and pollutant concentration, and DC cooling energy consumption can be reduced through thermal design of urban areas (e.g., building construction/design, landscaping, vegetation, surface albedo), which can be optimized using dedicated simulation methodologies, such as through coupled building-atmospheric flow, heat, and mass transport models.

4.3.3. Policy-Related Opportunities

The importance of regulations in enabling technically, economically, and environmentally efficient district heating/cooling systems was outlined in [12,18]. Despite initiating an energy transition, GCC economies still have indulgent energy and environmental regulations that do not effectively support the introduction of energy/water conservation measures, clean energy sources, efficient energy conversion technologies, and GHG emissions reduction actions [141]. It is widely recognized that GCC governments need to bring energy and environmental policy reforms and regulations [124,141]. Regarding air-conditioning, its supply is not considered by most GCC governments as a utility service that requires public policy and planning, which has resulted in an unregulated environment in the deployment and operation of DC systems, which are sub-optimal from energy use, environmental and economic perspectives. Based on critical observations relating to the implementation of DC in the GCC and international DC guidelines [5,10,138], five key policy-related areas that need to be tackled are identified below in terms of challenges and potential solutions, namely reforming of utility tariffs, support of DC front-load investments, integration of planning/decision-making between building, DC and urban development stakeholders, improvement of DC cost recovery practices, and establishment of local DC service standards and technical codes.

Low, subsidized electricity and water tariffs in GCC countries, rather than tariffs based on real costs, mask the true economic advantages of DC systems over conventional on-site cooling systems. This is compounded by flat tariffs rates (i.e., with no or limited peak hour tariffs), that reduce the economic gains that could be realized with CTS systems, and the application of different tariff rates between regions of the same country (e.g., UAE emirates). Electricity and water tariffs require conservation-driven reforms across the region to stimulate sustainability improvements in space cooling provision and use.

Rapid large-scale urbanization developments and their subsequent deceleration in the GCC have revealed the risks associated with real estate investments. DC systems, particularly their cold distribution network infrastructure, require substantial early front-load investment. Although chiller plants are more suited to modular investment than distribution infrastructure, they still involve some degree of front-load investment. These front-load investments have been exaggeratedly high due to overestimation of cooling load intensities at the design stage, and under-realization of DC development goals due to excess capacity. Government backing for front-loaded distribution network infrastructure investments can mitigate DC system investment risks and improve DC competitiveness relative to conventional space cooling.

Building development and associated engineering decisions are made independently. Hence, different decision makers and timelines are involved as well as different design practices. Therefore, developers mostly follow non-aggregated development decisions, that form an easier, quicker, and cheaper process, rather than aligning plans with other developers. Instead, the implementation of DC systems requires integrated planning and design practices to achieve optimal energy use. Governments can assist in such planning including the designation of appropriate zones for DC deployment as part of future urban development.

The cost recovery models typically employed by DC providers are complex, diverse, and inappropriate, resulting in discrepancies in the distribution of costs (i.e., connection, capacity, use) among involved stakeholders (i.e., DC developers, owners, tenants) from project to project and buildings to building. In general, developers include the capital and fixed operating costs of conventional on-site air-conditioning systems in the purchase price or building rent. By contrast, DC providers typically require a fixed charge to recover capital costs. In addition, the operating costs of conventional on-site air-conditioning systems are also often hidden. For instance, in buildings with central chilled water supply, there is generally no metering for in-building cooling systems—instead, the usage costs are recovered indirectly via tenant rent or management fees. Greater consistency should be adopted in the allocation of DC costs between projects and between buildings [5].

As with other utilities (e.g., electricity), DC could be better benefited from using a proper regulatory framework to protect all involved stakeholders (e.g., developers, providers, and end-users). Guidelines should be set to define areas for DC implementation where cooling load density is appropriate. Guidelines should also include integration of DC planning with urban and infrastructure planning, including power and water. Governments can support the DC market by establishing appropriate national tariff frameworks for DC. They should also define the minimum level of requirements for DC providers for reliability and performance. Such requirements should be accompanied by technical codes to ensure quality in design, installation, and operation of DC systems.

In summary, the GCC DC market requires regulation to address the above issues, to better benefit from the economic advantages of DC systems, ensure end-user protection regarding pricing and service quality, and increase energy and water use efficiency [5]. GCC governments can play a key role in developing and enforcing such regulations. Even if the form of government intervention varies from country to country, it should be focused on the above five areas. Finally, as for any energy/water system in the GCC, the development of sustainable DC systems will be highly dependent on the future availability and affordability of local fossil and renewable resources, economic market forces, energy-environmental policies/legislation, social acceptance, and education.

5. Research Trends and Future Outlook

Sustainable DC technologies, operational aspects, and analysis, modeling, and optimization methodologies, were reviewed based on published DC design and analysis studies in Sections 2 and 3, focusing on the needs of cooling-dominated regions. The status and challenges of DC implementation in the GCC region were examined in Section 4, with potential solutions and opportunities identified. In this section, key collective research trends are compiled from the works reviewed, leading to suggested future research directions that could facilitate the performance improvement, and wider deployment of sustainable DC in cooling-dominated regions. DC trends and future opportunities are categorized in this section under the following thematic areas:

- Geographical deployment location
- Sustainable energy sources and cross-sectorial DC integrations
- Sustainable cooling and storage technologies
- Thermodynamic, environmental, and economic analysis and optimization methodologies

- Energy, environmental and economic benefits findings.

5.1. Geographical Deployment Location

In terms of DC geographical deployment location, the largest single group of DC studies in cooling-dominated regions has focused on South Asian regions, followed by another group of studies for the Persian Gulf. Most of DC systems analyzed in South Asia were based in Hong Kong. These works are complemented by studies in non-cooling-dominated regions, essentially in Europe including in Scandinavia. Despite the size of the existing DC market in the USA, published research focusing on DC appears to be limited. There has been little if no published DC research activity to date in the MENA other than in the Persian Gulf. These observations suggest that there is significant scope for extending sustainable DC research and deployment to other cooling-dominated regions, including in the Americas, Australia, Malaysia, and MENA.

Despite progress in DC deployment in the GCC, only a limited fraction of regional cooling loads is yet fulfilled by this technology. Based on this and considering the regions' extreme climatic conditions, natural cooling water scarcity, dense and expanding urbanization, renewable resource potential, and availability of industrial waste heat and waste cold, the GCC therefore represents a major opportunity for further DC research and implementation. The energy demand and cost of space cooling in the region is currently compounded by elevated energy intensity- and emissions per capita, domestic natural gas shortages, volatile revenues from hydrocarbon production, and indulgent energy and environmental regulations. GCC-specific technical and non-technical solutions identified to improve the sustainability of DC cooling while addressing harsh climatic conditions and renewable water scarcity were discussed in Sections 4.3.2 and 4.3.3. If effectively implemented, DC could play a key role in advancing sustainable space cooling in the GCC residential, commercial, and industrial sectors, and contribute to decarbonizing this historically fossil-fuel-reliant region. Such progress could serve as model to DC development in other cooling-dominated regions.

5.2. Sustainable Energy Sources and Cross-Sectorial DC Integrations

Renewable energy sources pursued in DC research studies to date have essentially consisted of solar thermal power (as reported in Saudi Arabia, Spain, and the UAE) to drive thermally activated cooling, solar electricity (as reported in Iran, Singapore, and USA) to drive compression chillers, and biomass electricity/heat (i.e., waste/biofuels, as reported in Singapore, Sweden, and Thailand). In comparison with district heating studies, the use of CHP technology for DC in cooling-dominated regions has been less common, partly due to a lack of heating demand in the residential sector. Most investigations have focused on a single renewable energy conversion technology, with few studies comparing the performance of alternatives, such as different CSP options. Mixes of renewable energy conversion technologies, and 100% renewable energy-driven DCs (such as in a Saudi Arabian investigation) were rarely envisaged. Non-solar low-carbon energy sources, such as wind, hydroelectricity, geothermal, marine, and nuclear power, have rarely or not been reported in DC systems for cooling-dominated climates to date, suggesting further opportunities to integrate renewables in DC applications. Low- to ultra-low-grade renewable heat such as produced by solar collectors, or extracted from geothermal or sea water, as well as ultra-low-grade waste heat sources, have received little attention. Their potential upgrade and exploitation to drive sorption chillers could be investigated in future work. Although integrating DC with intermittent renewable utilities (e.g., solar, wind) could provide means of absorbing excess electricity in future energy systems, this aspect has been rarely discussed, unlike for district heating. This may be related to less progress being achieved in the integration of high shares of fluctuating renewables in energy systems in cooling-dominated regions, compared with for example European regions.

Low–medium-grade heat use from fossil-fired utility production installations (i.e., gas-fired combined cycles, gas engines, coal-fired plants, coal/oil-fired CHP plants, hybrid solid oxide fuel cell-gas turbine plants) to drive thermally activated cooling was considered in both hydrocarbon producing regions (Iran, Kuwait, USA) and others (Hong Kong, Singapore, Sweden, Turkey).

The use of free cold sources (i.e., river water in China/Sweden, lake water in Sweden) and artificial cold sources (i.e., LNG in Singapore) has been limited in DC studies to date, reflecting a lack of natural cold sinks in hot climates, and lack of cryogenic cold energy recovery implementation. LNG cold energy was only considered in two instances for direct district cooling including in parallel with power augmentation, while its use for enhancing the efficiency of thermodynamic cooling cycles has not been reported in previous DC studies. Therefore, given the projected growth of the LNG market, the exploitation of cryogenic cold for either direct cooling provision or enhancing the efficiency of thermodynamic power/cooling cycles serving DC applications deserves greater attention, in conjunction with cold transportation. Potential cold transportation media require further investigation, such as CO_2, which could offer opportunities of CO_2 sinking/valorization.

In terms of cross-sectorial DC integrations, which can enable material/energy recycling, hence efficient capital use, energy conservation and flexibility, the following synergies between DC systems and energy-intensive industrial heating and power production processes, have been considered. DC systems analyzed in Sweden, USA, Turkey, and Iran were integrated with district heating and CHP units, understandably because of climatic conditions. In such regions, excess waste heat generated from thermal power, co-generation/CHP, or biomass-fired power plants during the hot season, was exploited for cooling production at district level in the hot season. In cooling-dominated regions, cross-sectorial integrations have essentially been limited to centralized or distributed fossil/renewable-fueled power generators, and in one instance, an LNG regasification terminal. Few studies have analyzed DC systems as part of broad or complete energy systems, which may have limited the consideration of cross-sectorial integration options in the literature. Examples of potential DC cross-sectional integrations of interest to cooling-dominated regions and that have not yet been considered include conventional thermal power augmentation, fresh water production and synthetic gas production from fluctuating excess electricity/heat in future renewable-based energy systems. Thus, hot-climate regions generally experience significant losses in gas turbine power generation capacity and efficiency, and suffer from a scarcity of natural water sources, resulting in an extensive reliance on seawater desalination. Two possible concepts of DC cross-sectorial integration or extension of DC to non-space cooling end-uses in such regions would be the supply of (i) compressor inlet air cooling by the DC system to maintain power generation performance in yearly elevated high ambient temperature conditions, as well as (ii) recycling DC chiller plant waste heat to reduce the energy consumption of thermal desalination technologies, which currently dominate the desalination market in for example the GCC, as well as waste water treatment. In future renewable-based energy systems with excess electricity, the recovery of power-to-gas parasitic heat losses from either electrolysis or catalytic methanation, which has already been proposed in the power-to-gas literature for district heating, could also potentially serve to drive thermally activated DC cooling.

To conclude, based on published DC analyses to date, there is scope for the more systematic identification and exploitation of a broader mix of sustainable electricity/heat sources and natural/artificial cold sources available locally, to drive DC systems based on an optimized combination of electrical and thermally activated chillers. In hot-climate, densely populated regions that are also geo-politically sensitive, the exploitation of local energy/material sources rather than imported ones to drive DC systems would play a strategic role towards energy self-sufficiency. The integration of DC systems in future 100% renewable energy scenarios, and their role in absorbing excess fluctuating electricity/heat to facilitate renewable penetration, hence contribute to energy security and climate change mitigation, are also suggested areas of further research.

5.3. Sustainable Cooling and Storage Technologies

Based on published DC research studies to date, DC design/operation sustainability improvement approaches in cooling-dominated regions have involved one or more of the following approaches: (i) replacement of fossil electricity-driven compression cooling with either renewable electricity-driven compression cooling, and/or with waste/renewable-heat-driven absorption cooling, in conjunction with capacity optimization; (ii) introduction of cold thermal energy storage; (iii) artificial cold energy recovery; (iv) reducing DC distribution energy consumption (i.e., pumping power, thermal energy losses); (v) improved chiller/storage control strategies; and (vi) improved modeling/optimization approaches. However, when considered individually, DC studies have generally exploited a restricted range of technology alternatives in a given application environment.

The focus on renewably powered electrical compression cooling and low-carbon heat-driven absorption cooling is mainly attributable to technology maturity, availability at MW-scale equipment capacity, and affordability. Thus, at present, commercially available thermally activated cooling technologies with sufficient capacity and reliability for DC applications are essentially limited to single/double-effect water-lithium bromide-absorption refrigeration.

In the future, triple-effect absorption chillers, subject to commercial availability at sufficient capacity, may potentially be integrated in DC chiller plants to exploit sustainable heat sources with higher temperatures, such as from high concentration solar power plants or industry, with improved COPs compared with lower-effect chillers, albeit at higher equipment cost. Based on an existing DC plant incorporating an ejector chiller, with further technology developments, ejector-assisted refrigeration could also potentially become an option in parallel with other chiller technologies, with benefits including ejector construction simplicity, amenability to low-grade renewable/waste heat conversion, low energy consumption, and low maintenance requirements, providing that sufficient performance and operational flexibility can be achieved through ejector cycle enhancements (e.g., layout, controls, storage). Other identified sustainable air-conditioning technologies having promising technical features for cooling-dominated regions include GAX-based absorption, compression-assisted absorption, and desiccant-based cooling. However, these concepts are presently not available in commercial systems with sufficient capacity and require further developments. In addition, the need for innovative dehumidification technologies for DC cooling tower inlet air has been suggested to improve tower efficiency in high ambient humidity conditions. Potential future research topics also include hybrid refrigeration cycle concepts.

Technical concepts that could be investigated to enhance heat rejection from DC central chiller plants in high ambient temperature conditions include the use of cold energy as heat-rejection medium if available (e.g., LNG from regasification terminals), in conjunction with suitable cold energy transportation options, as well as incorporation of innovative thermodynamic concepts in the design of the central chiller plant, as mentioned above.

Regarding cold thermal energy storage, chilled water thermal storage is compatible with the evaporation temperature range of conventional chillers, while ice storage can offer compactness in dense districts. Local ambient conditions, combined with consumer and renewable generation profiles, result in unique electrical/thermal load and energy supply profiles. The potential of cold thermal energy storage to reduce cooling energy consumption is known to be more significant where large daily ambient temperature variations exist between day and night, such as in arid regions. However, based on the present review, the incorporation of thermal energy storage technologies in DC systems or the specific storage technology employed are however frequently not reported in DC design/analysis studies. When reported, chilled water and ice storage received similar levels of attention. Either hot water or molten salts were employed for concentrated solar thermal energy storage. However, the justification for selecting a given storage technology and consideration of alternative storage options is generally lacking, with limited efforts devoted to storage design and optimization. This suggests that greater attention could be given to the design and evaluation of thermal storage in DC research studies for cooling-dominated regions. In parallel, electricity demand profiles and tariffs, and local energy policies, may require adjustments or reforms in certain regions, as these are critical factors in the selection of a storage technology and in capitalizing its benefits.

5.4. Thermodynamic, Environmental and Economic Analysis and Optimization Methodologies

Cooling loads (i.e., peak loads—typically for system capacity sizing, and annual hourly cooling loads—for operational, control and economic analyses) are an essential if not the most essential input data to the design, simulation, and optimization of DC systems. Therefore, a high-priority research task should be the collection and analysis of measured cold deliveries in existing DC systems. Such data could not only directly guide potential DC design retrofits and/or operational improvements, but also provide supporting reference information for the development and validation of predictive district cooling load and performance models. Among published DC-level cooling load estimation procedures, a trade-off between analysis accuracy and computational expense/time may be achieved using the combined use of local building energy codes, local weather data, and dynamic building simulation applied to identified building categories in the district, to develop a database of space cooling loads per unit floor area for each building category in a given district. Such a procedure should however be combined with uncertainties in district cooling loads (as well as sub-systems reliability/availability risk), which have been rarely incorporated in DC systems design and analysis studies to date. Based on numerical investigations, uncertainties in cooling loads have been found to essentially arise from indoor conditions (i.e., mainly ventilation rate, but also occupant, lighting, and plug-in load densities), rather than outdoor weather or building design/construction. These uncertainties have been shown to likely significantly impact technology selection, capacity sizing and performance evaluation. Incorporating these uncertainties in DC system design could avoid under- or over-designs, and more accurately estimate performance, cost, and environmental impact, with quantified confidence levels. The evaluation of district cooling loads could be refined further using coupled building energy and microclimatic simulations to capture heat island effects on outdoor temperatures and district cooling energy consumption, and evaluate alternative urban designs for planned districts.

Most DC performance evaluations in cooling-dominated environments to date have been steady-state, energy analysis-based, and have focused on the operational phase. In addition to common performance criteria generically applicable to space cooling systems, the non-renewable PEF, EPI, CO_2 factor for supplied cooling to the end-user (in kg of CO_2/MWh of cooling), CO_2 payback time (CPT), and ATMI have been proposed as additional DC system sustainability-related metrics. However, the application of such criteria to DC systems has been limited to date. Similarly, holistic sustainability metrics have rarely been employed, with hardly any exergy, exergoeconomic and life-cycle analyses of DC systems reported. Collectively, analyses of combined district heating and cooling systems in heating-dominated regions appear to be more advanced from these perspectives.

In terms of DC system optimization, published works tend to focus on the optimization of the distribution pipeline network layout design and end-user's facilities (e.g., mix of buildings, cooling load patterns), rather than either the central chiller plant or complete DC system including demand and supply sides. This is related to the significant investment cost of the distribution network, as well as prohibitive optimization complexity and computational cost. Most optimizations have been single-objective, focusing on an economic criterion, typically the sum of annualized investment and operational costs. In terms of software tools, mixed integer linear programing and genetic algorithms have been the most commonly applied.

Based on these observations, there is scope to further extend the range of DC performance metrics and analysis approaches to (i) dynamic analyses incorporating dynamic system operation characteristics (e.g., capacity-, load-, and temperature-dependent COPs of cooling equipment), and dynamic utility prices, with sensitivity assessments to projected fuel prices (i.e., fossil, biomass, synthetic); (ii) design methods that incorporate uncertainties in cooling loads and sub-system reliability; (iii) design methods that account for the effects of outdoor microclimate on district outdoor temperature and cooling loads; (iv) more comprehensive environmental impact assessments, rather than solely operational CO_2 emissions-based; (v) concurrent demand- and supply-side optimizations, with linkage with other sectors; (vi) use of holistic sustainability metrics and their incorporation in multi-objective optimizations, including exergy, exergoeconomic, and reliability-based criteria, with account made of DC energy/material recycling (e.g., waste heat/cold sources, waste water, emissions), as well as quantitative and qualitative (non-quantifiable) life-cycle economic parameters. Regarding item (vi), additional sustainability criteria proposed for districts but not previously reported in DC studies, could include for example exergy-based COPs for DC chiller plants, primary exergy ratios, compound CO_2 emissions, composite rationality indicators [142] and emergy-based indicators [143] to contribute to the analysis of district metabolism, including in terms of energy, waste, and material (e.g., waste/fresh water, emissions) flows (i.e., including production, use and re-use/recycling), intensity and efficiency. Aspects (i)–(vi) could contribute to the better design and life-cycle management of DC systems as parts of smart energy hubs.

5.5. Energy, Environmental and Economic Benefits Findings

The energy savings contributed by implemented DC systems compared with on-sight cooling systems are in large part due to the higher efficiency of large-scale central water-cooled chiller plants compared with on-site small-capacity cooling systems. In addition, the use of a CTS system shifts electricity/thermal energy consumption from peak to off-peak periods, which can significantly contribute to more effective energy use. As a relevant example, average energy savings of 0.78 kWh (46%) per ton-hour of cooling have been reported by a leading DC provider for several conventional water-cooled DC systems implemented in the GCC relative to air-cooled building chillers. Such DC systems were based on fossil electricity vapor compression cooling and standard design/operational practices, rather than energy conversation-driven, state-of-the-art designs/operation.

The studies reviewed in this article highlight further energy savings that could be contributed by the use of sustainable energy sources and cooling, storage, and cold distribution technologies, in conjunction with appropriate control strategies. The energy benefits associated with a given type of sustainability enhancement are generally not isolated in published studies when several sustainability enhancement options (discussed in Sections 5.2 and 5.3) are applied in the same DC system. However overall, the replacement of fossil electricity-driven compression cooling with either renewable electricity-driven compression cooling, and/or with waste/renewable-heat-driven absorption cooling, which have been the most widely investigated options, have led to reported reductions in DC system energy consumption spanning a wide range (i.e., 10–70%), depending on DC system design/operating characteristics and modeling methodologies. As part of such enhancement options, DC system cross-sectional integration can enable the exploitation of low-cost, low-emission excess electrical/thermal energy generated in the power and industrial sectors, to drive DC. Benefits are also obtained for the excess energy provider, in terms of avoided electricity curtailment and/or thermal energy losses, which may be sold at low cost to the DC system operator to generate revenues.

The benefits of cold thermal storage are known to be more significant in regions with large daily ambient temperature variations between day and night, such as in for example the GCC climate. The benefits of thermal energy storage include shifting cooling operation from peak to off-peak demand periods, which can (a) reduce peak electricity demand; (b) reduce energy cost for both the cooling supplier and consumer; (c) enable the more efficient use of solar thermal energy through absorption of excess heat/electricity generated during sunshine hours, the use of which is shifted to non-sunshine demand periods; (d) reduce installed cooling capacity requirements; and (e) improve system reliability by using stored thermal energy as a backup. However, the energy, environmental and economic benefits of a given thermal energy storage technology strongly depend on electricity demand profiles and tariffs, and local energy policies. In conjunction with technical sustainability improvement efforts, DC-related energy policies require greater attention in cooling-dominated regions, with examples of thematic regulatory areas outlined in this article for the case of the GCC region.

Published DC system analysis studies indicate that annual CO_2 emissions in sustainable DC systems can be reduced in relative magnitudes comparable to those of energy consumption reductions (i.e., up to 80%), relative to conventional cooling systems. Heat-activated sorption cooling can also contribute to reduce not only fossil energy consumption, but also the use of conventional ozone-depleting and/or global warming refrigerants, as well as cooling equipment noise and vibration.

In terms of economic benefits, implemented DC systems have been shown to offer several cost reduction advantages over conventional on-site cooling systems including lower energy-, maintenance- and construction costs, with payback periods of a few to ten years. In addition, the expected life of DC systems is at least 25–30 years, compared with 10–15 years for conventional on-site air-conditioners. The findings from the reviewed DC literature are consistent with those from larger energy system studies (i.e., including but not limited to districts) [144], that increased investment costs associated with sustainable energy technologies are often offset by reduced operating costs (i.e., including fuel costs, maintenance/operating costs, peak energy demand charges). However, the estimated economic benefits are also sensitive to the economic assumptions and modeling/optimization methodology applied. In particular, future commodity price projections, constant versus variable cooling equipment performance parameters, (non)incorporation of the impacts of uncertainties (particularly in indoor conditions) or reliability (i.e., availability risk), and analyzing the DC system as part of a broader energy system rather than in isolation or with limited cross-sectorial interactions, can have significant impacts on the estimated economic benefits. Economic benefits are also obtained for excess energy providers, in terms of avoided electricity curtailment and/or thermal energy losses, which may be sold at low cost to the DC system operator to generate revenues.

6. Concluding Remarks

Due to economic and population growth driving energy demand in hot/humid climate regions, compounded by climate change and its effect on cooling loads, sustainable district cooling (DC) systems will gain increasing importance over the coming decades. Most previous publications related to district energy systems have focused on the needs of heating-dominated regions, thereby lacking specificity to cooling-dominated regions. In the present article, available and developing sustainable space cooling technologies driven by low-carbon energy sources and with present or potential future applicability to DC were reviewed, as well as DC analysis, modeling, and optimization methodologies. The challenges, status and future energy, environmental and economic potential of DC specifically in the GCC region were discussed, with opportunities for DC technology customization and market regulation highlighted.

Based on the present review, the potential of DC systems to contribute significantly to energy conservation, improvements in operational cooling capability, efficiency, flexibility, and reliability, as well as reductions in the environmental impact and cost of building air-conditioning has been well established. However, based on the collective research trends identified in this article, several research gaps remain to be addressed to enable the potential of this technology to be more effectively exploited in cooling-dominated regions. Key directions proposed include increased DC cross-sectorial integrations and synergies to enable exploitation of sustainable and low-cost energy/material flows (including recycled waste/excess flows), more systematic exploitation of locally available renewable electricity/heat supply options and natural/artificial cold sources for direct cooling, and application of more holistic thermodynamic, environmental and economic performance evaluation methods, under an appropriate regulatory framework.

Author Contributions: V.E. proposed the topical area. D.S.A. drafted an initial manuscript under V.E.'s direction and editing, who subsequently added contents to all sections and re-wrote the manuscript.

Funding: This research was funded by The Petroleum Institute grant number LTR 14502 and the APC was funded by the lead author.

Conflicts of Interest: The authors declare no conflict of interest.

Appendix A

Table A1. Medium–large-capacity, commercially available absorption chillers from leading manufacturers and suitable for DC applications.

Manufacturer (Country)	Capacity (kW)	Working Fluid Pair	Technology
AGO (Germany) [145]	50–1000	NH_3/H_2O	SE (Indirect-fired)
Broad (China) [146]	105–5272	$H_2O/LiBr$	SE (Indirect-fired) DE (Direct/Indirect-fired)
Carrier (USA) [147]	387–3516 359–5803	$H_2O/LiBr$	SE (Indirect-fired) DE (Direct/Indirect-fired)
Hitachi (Japan) [148]	106–19,690	$H_2O/LiBr$	DE (Direct/Indirect-fired)
Kawasaki (Japan) [149]	281–1758 281–3517 563–1196	$H_2O/LiBr$	SE (Indirect-fired) DE (Direct/Indirect-fired) TE (Direct-fired)
LG (South Korea) [150]	98–3587 176–5275 258–3427	$H_2O/LiBr$	SE (Indirect-fired) DE (Direct/Indirect-fired) DE Double-lift (Indirect-fired)
Sakura (Japan) [151]	105–5274 176–5274 264–4571	$H_2O/LiBr$	SE (Indirect-fired) DE (Direct/Indirect-fired) SE Double-lift (Indirect-fired)
Shuangliang (China) [152]	2901–1630	$H_2O/LiBr$	SE, DE (Direct/Indirect-fired)
Thermax (India) [153]	175–35,001 35–140,001 350–3500	$H_2O/LiBr$	SE (Indirect-fired) SE, DE (Direct/Indirect-fired) TE (Indirect-fired)
Trane Company (USA) [154]	392–4725 350–5775	$H_2O/LiBr$	SE (Indirect-fired) DE (Direct/Indirect-fired)
Johnson Controls (USA) [155]	420–4850 700–2460	$H_2O/LiBr$	SE (Indirect-fired) DE (Direct/Indirect-fired)
Fischer Eco Solutions (Germany) [156]	15–5000	$H_2O/LiBr$ Methanol/LiBr	SE (Indirect-fired)

Table A2. Characteristics of typical CTS systems suitable for building air-conditioning [10,86,87].

Characteristic	Chilled Water System	Ice Thermal Storage System	Eutectic Salt
Chiller fluid	Standard water	Low-temperature secondary fluid	Standard water
Latent heat of fusion (kJ/kg)	N/A	334	80–250
Specific heat (kJ/kg·K)	4.19	2.04	N/R
Tank volume (m^3/kWh)	0.089–0.169	0.019–0.023	0.048
Charging temperature (°C)	4–6	(-6)–(-3)	4–6
Discharge temperature (°C)	1–4 (above charging temperature)	1–3	9–10
Chiller charging (kW/ton)	0.6–0.7	0.85–1.4	0.6–0.7 (PCM)
Chiller charging efficiency (COP)	5.0–5.9	2.9–4.1	5.0–5.9
Footprint (plant area/ton-h)	Fair	Good	Good
Modularity	Poor	Excellent	Good
Ease of retrofit	Excellent	Fair	Good
Simplicity and reliability	Excellent	Fair	Good
Economy-of-scale	Excellent	Poor	Good
Dual-use as fire protection	Excellent	Poor	Poor

Note: N/A = not applicable; N/R = not reported.

Table A3. Non-renewable PEF and CO_2 emission factors for fuels/energy carriers [113].

Fuel/Energy Carrier	Non-Renewable PEF, $f_{F(i)}$ (-)	CO_2 Emission Factor, $K_{F(i)}$ (kg CO_2/MWh$_{fuel}$)
Lignite	1.02	369
Hard coal	1.19	369
Heavy fuel oil	1.35	296
Light fuel oil	1.35	283
Natural gas	1.36	222
Peat	1.02	417
Bioenergy (primary)	0.1	7
Bioenergy (refined)	0.2	12
Bioenergy (secondary)	0.06	3
Residual fuel	0.05	88
Waste as fuel	0	94
Electricity (input and output)	2.6	420
Industrial waste heat	0	0
Geothermal heat	0	0
Solar heat	0	0

Table A4. GWP and ODP values of commonly used refrigerants in conventional air-conditioning and DC systems.

Refrigerant	Application	GWP [1]	ODP [2] (-)
HCFC-123	DC system/Conventional	77	0.06
HCFC-22	Conventional	1810	0.055
HFC-134a	DC system/Conventional	1430	0
HFC-245fa	Conventional	1030	0
Propane (R-290)	Conventional	3	0
HFO-1234yf	Conventional	4	0
Ammonia (R-717)	DC system/Conventional	0	0

[1] GWP (100 year), GWP for CO_2 = 1 [157]. [2] ODP, CFC-11 = 1 [158].

Table A5. Large-scale (i.e., capacity > 10,000 RT) DC projects in operation or under development in the GCC [17,136,159–162].

Project Name (Country)	Company	Capacity (TR)	Chiller Technology	Starting Year	End-Users
West Bay (Qatar)	Qatar Cool	Plant 1: 25,000 and 25,000 RT-hour CTS Tank	10 Mechanical chillers	2006	Residential and commercial buildings
		Plant 2: 32,000 and 20,000 RT-hour CTS Tank	12 Mechanical chillers	2009	
		Plant 3: 35,000 and 25,000 RT-hour CTS Tank	14 Mechanical chillers	2016 (third quarter)	
The Pearl Qatar (Qatar)	Qatar Cool	130,000	52 Mechanical chillers	2010	Residential accommodations and mixed-use developments
Barwa City (Qatar)	Marafeq Qatar	37,000	Multiple mechanical chiller plants	2014	6000 apartments, schools, nurseries, shopping centers, a bank, health club, mosques, and restaurants
Lusail City (Qatar)	Marafeq Qatar	Up to 500,000	Multiple mechanical chiller plants	Under construction	Residential and commercial buildings, schools, mosques, medical facilities, sport, entertainment, and shopping centers.
Dubai Motor City (UAE)	Emicool	80,000 and 77,600 RT-hour CTS Tank	Two mechanical chiller plants (each 40,000 RT)	2010 / 2012	Residential buildings, business towers, motor-sports complexes, retail, and theme park
Dubai Investment Park (UAE)	Emicool	Plant 3: 60,000 and 15,000 RT-hour CTS Tank	Centrifugal water-cooled chillers	2012	Residential buildings, commercial towers, motor-sports facilities, shops, and theme park
		Plant 4: 20,000 and 5000 RT-hour CTS Tank		2007	
		Plant 6: 30,000 and 30,000 RT-hour CTS Tank		2010	
Dubai Metro (UAE)	Tabreed	Total: 36,400 Al Rigga (10,000); Al Barsha (7500); Jumeirah Island (7000); Jebel Ali Industrial (4400); Al Rashidiya (7500)	Multiple mechanical chiller plants	2010	Dubai metro stations (47 stations)
Business Bay Executive Towers Dubai (UAE)	Empower	35,200	16 Centrifugal chillers (each 2200 RT)	2009	Mixed-use community development within Business Bay
Saudi Aramco Dhahran complex (KSA)	Saudi Tabreed	Total: 32,000 (20,000 RT and 7000 RT-hour CTS Tank)	8 York Centrifugal chillers (each 2500 RT)	2013	Office complex, communities, hospital, and data centers
King Abdullah Financial District (KSA)	Saudi Tabreed	100,000 (two phases each 50,000 RT) and 15,000 RT-hour CTS Tank	Multiple mechanical chiller plants	2013	Office towers
Jabal Omar Development Project (Makkah, KSA)	Saudi Tabreed	55,000	Multiple mechanical chiller plants	2014	13 twin towers catering for hotels, malls, commercial outlets, and residential buildings
Knowledge Oasis Muscat (Oman)	Tabreed Oman SAOC	Total: 25,000 (planned) / 5000 (cooling load)	Multiple mechanical chiller plants	2010	Technology park
Bahrain Financial Harbor (Bahrain)	Tabreed Bahrain	22,000	Mechanical chiller plant using sea water as a heat-rejection medium	2009	Offices, luxury residential accommodation, shopping complex housing, retail outlets, and hotels
Reef Island (Bahrain)	Tabreed Bahrain			2010	
Bahrain World Trade Center (Bahrain)	Tabreed Bahrain			2006	

References

1. Intergovernmental Panel on Climate Change (IPCC). *Climate Change 2014: Mitigation of Climate Change. Contribution of Working Group III to the Fifth Assessment Report of the Intergovernmental Panel on Climate Change*; Edenhofer, O.R., Pichs-Madruga, Y., Sokona, E., Farahani, S., Kadner, K., Seyboth, A., Adler, I., Baum, S., Brunner, P., Eickemeier, B., et al., Eds.; Cambridge University Press: Cambridge, UK; New York, NY, USA, 2014.
2. BP. BP Energy Outlook. 2018 Edition. Available online: https://www.bp.com (accessed on 25 November 2018).
3. International Energy Agency (IEA). The Future of Cooling. Opportunities for Efficient Air Conditioning. 2018. Available online: https://www.iea.org/publications/freepublications/publication/The_Future_of_Cooling.pdf (accessed on 25 November 2018).
4. Strategy. Unlocking the Potential of District Cooling: The Need for GCC Governments to Take Action. 2012. Available online: https://www.strategyand.pwc.com/media/file/Unlocking-the-potential-of-district-cooling.pdf (accessed on 25 November 2018).
5. Al-Qattan, A.; Elsherbini, A.; Al-Ajmi, K. Solid oxide fuel cell application in district cooling. *J. Power Sources* **2014**, *257*, 21–26. [CrossRef]
6. United Nations. Population Division. Department of Economic and Social Affairs (DESA). Available online: https://population.un.org/wpp/Download/Standard/Population/ (accessed on 25 November 2018).
7. Liew, P.Y.; Walmsley, T.G.; Wan Alwi, S.R.; Abdul Manan, Z.; Klemeš, J.J.; Varbanov, P.S. Integrating district cooling systems in Locally Integrated Energy Sectors through Total Site Heat Integration. *Appl. Energy* **2016**, *184*, 1350–1363. [CrossRef]
8. Pellegrini, M.; Bianchini, A. The Innovative Concept of Cold District Heating Networks: A Literature Review. *Energies* **2018**, *11*, 236. [CrossRef]
9. Dinçer, İ.; Zamfirescu, C. *Sustainable Energy Systems and Applications*; Springer International Publishing: New York, NY, USA, 2011; ISBN 9780387958606.
10. American Society of Heating, Refrigerating and Air-Conditioning Engineers (ASHRAE). *ASHRAE District Cooling Guide*; ASHRAE: Atlanta, GA, USA, 2013; ISBN 9781936504428.
11. International District Energy Association (IDEA). *District Cooling Best Practices Guide*; IDEA: Westborough, MA, USA, 2008; ISBN 10 0615250718.
12. Rezaie, B.; Rosen, M.A. District heating and cooling: Review of technology and potential enhancements. *Appl. Energy* **2012**, *93*, 2–10. [CrossRef]
13. Seeley, R.S. District cooling gets hot. *Mech. Eng.* **1996**, *118*, 82–84.
14. MarketsandMarkets. District Cooling Market. Global Trend & Forecast to 2012–1019. 2014. Available online: https://www.marketresearch.com/product/sample-8361266.pdf (accessed on 25 November 2018).
15. Werner, S. International review of district heating and cooling. *Energy* **2017**, *137*, 617–631. [CrossRef]
16. Gang, W.; Wang, S.; Xiao, F.; Gao, D.C. District cooling systems: Technology integration, system optimization, challenges and opportunities for applications. *Renew. Sustain. Energy Rev.* **2016**, *53*, 253–264. [CrossRef]
17. Marafeq Qatar. District cooling GCC and Qatar. In Proceedings of the 37th Euroheat & Power Congress, Tallin, Estonia, 27–28 April 2015. Available online: http://www.ehpcongress.org/archive-2015/wp-content/uploads/2015/04/62_Eric_LINDSTR%C3%96M.pdf (accessed on 25 November 2018).
18. Lake, A.; Rezaie, B.; Beyerlein, S. Review of district heating and cooling systems for a sustainable future. *Renew. Sustain. Energy Rev.* **2017**, *67*, 417–425. [CrossRef]
19. Palm, J.; Gustafsson, S. Barriers to and enablers of district cooling expansion in Sweden. *J. Clean. Prod.* **2018**, *172*, 39–45. [CrossRef]
20. Vandermeulen, A.; van der Heijde, B.; Helsen, L. Controlling district heating and cooling networks to unlock flexibility: A review. *Energy* **2018**, *151*, 103–115. [CrossRef]
21. Chan, A.L.S.; Chow, T.T.; Fong, S.K.F.; Lin, J.Z. Performance evaluation of district cooling plant with ice storage. *Energy* **2006**, *31*, 2414–2426. [CrossRef]
22. Chan, A.L.S.; Hanby, V.I.; Chow, T.T. Optimization of distribution piping network in district cooling system using genetic algorithm with local search. *Energy Convers. Manag.* **2007**, *48*, 2622–2629. [CrossRef]
23. Trygg, L.; Amiri, S. European perspective on absorption cooling in a combined heat and power system—A case study of energy utility and industries in Sweden. *Appl. Energy* **2007**, *84*, 1319–1337. [CrossRef]
24. Söderman, J. Optimisation of structure and operation of district cooling networks in urban regions. *Appl. Therm. Eng.* **2007**, *27*, 2665–2676. [CrossRef]

25. Feng, X.; Long, W. Optimal design of pipe network of district cooling system based on genetic algorithm. In Proceedings of the 2010 6th International Conference on Natural Computation, Yantai, China, 10–12 August 2010; Volume 5, pp. 2415–2418. [CrossRef]
26. Svensson, I.L.; Moshfegh, B. System analysis in a European perspective of new industrial cooling supply in a CHP system. *Appl. Energy* **2011**, *88*, 5164–5172. [CrossRef]
27. Udomsri, S.; Martin, A.R.; Martin, V. Thermally driven cooling coupled with municipal solid waste-fired power plant: Application of combined heat, cooling and power in tropical urban areas. *Appl. Energy* **2011**, *88*, 1532–1542. [CrossRef]
28. Ondeck, A.D.; Edgar, T.F.; Baldea, M. Optimal operation of a residential district-level combined photovoltaic/natural gas power and cooling system. *Appl. Energy* **2015**, *156*, 1–14. [CrossRef]
29. Erdem, H.H.; Akkaya, A.V.; Dagdas, A.; Sevilgen, S.H.; Cetin, B.; Sahin, B.; Teke, I.; Gungor, C.; Atas, S.; Basak, M.Z. Renovating thermal power plant to trigeneration system for district heating/cooling: Evaluation of performance variation. *Appl. Therm. Eng.* **2015**, *86*, 35–42. [CrossRef]
30. Marugán-Cruz, C.; Sánchez-Delgado, S.; Rodríguez-Sánchez, M.R.; Venegas, M.; Santana, D. District cooling network connected to a solar power tower. *Appl. Therm. Eng.* **2015**, *79*, 174–183. [CrossRef]
31. Perdichizzi, A.; Barigozzi, G.; Franchini, G.; Ravelli, S. Peak shaving strategy through a solar combined cooling and power system in remote hot climate areas. *Appl. Energy* **2015**, *143*, 154–163. [CrossRef]
32. Khir, R.; Haouari, M. Optimization models for a single-plant District Cooling System. *Eur. J. Oper. Res.* **2015**, *247*, 648–658. [CrossRef]
33. Karlsson, V.; Nilsson, L. Co-production of pyrolysis oil and district cooling in biomass-based CHP plants: Utilizing sequential vapour condensation heat as driving force in an absorption cooling machine. *Appl. Therm. Eng.* **2015**, *79*, 9–16. [CrossRef]
34. Gang, W.; Augenbroe, G.; Wang, S.; Fan, C.; Xiao, F. An uncertainty-based design optimization method for district cooling systems. *Energy* **2016**, *102*, 516–527. [CrossRef]
35. Gang, W.; Wang, S.; Augenbroe, G.; Xiao, F. Robust optimal design of district cooling systems and the impacts of uncertainty and reliability. *Energy Build.* **2016**, *122*, 11–22. [CrossRef]
36. Ameri, M.; Besharati, Z. Optimal design and operation of district heating and cooling networks with CCHP systems in a residential complex. *Energy Build.* **2016**, *110*, 135–148. [CrossRef]
37. Kang, J.; Wang, S.; Gang, W. Performance and Benefits of Distributed Energy Systems in Cooling Dominated Regions: A Case Study. *Energy Procedia* **2017**, *142*, 1991–1996. [CrossRef]
38. Yan, C.; Gang, W.; Niu, X.; Peng, X.; Wang, S. Quantitative evaluation of the impact of building load characteristics on energy performance of district cooling systems. *Appl. Energy* **2017**, *205*, 635–643. [CrossRef]
39. Coz, T.D.; Kitanovski, A.; Poredo, A. Exergoeconomic optimization of a district cooling network. *Energy* **2017**, *135*, 342–351. [CrossRef]
40. Dominković, D.F.; Rashid, K.A.; Romagnoli, A.; Pedersen, A.S.; Leong, K.C.; Kraja, G.; Dui, N. Potential of district cooling in hot and humid climates. *Appl. Energy* **2017**, *208*, 49–61. [CrossRef]
41. Gao, J.; Kang, J.; Zhang, C.; Gang, W. Energy Performance and Operation Characteristics of Distributed Energy Systems with District Cooling Systems in Subtropical Areas Under Different Control Strategies. *Energy* **2018**, *153*, 849–860. [CrossRef]
42. Franchini, G.; Brumana, G.; Perdichizzi, A. Performance prediction of a solar district cooling system in Riyadh, Saudi Arabia—A case study. *Energy Convers. Manag.* **2018**, *166*, 372–384. [CrossRef]
43. Ravelli, S.; Franchini, G.; Perdichizzi, A. Comparison of different CSP technologies for combined power and cooling production. *Renew. Energy* **2018**, *121*, 712–721. [CrossRef]
44. Shea, X.; Cong, L.; Nie, B.; Leng, G.; Peng, H.; Chen, Y.; Zhang, X.; Wen, T.; Yang, H.; Luo, Y. Energy-efficient and -economic technologies for air conditioning with vapor compression refrigeration: A comprehensive review. *Appl. Energy* **2018**, *232*, 157–186. [CrossRef]
45. Ziegler, F. State of the art in sorption heat pumping and cooling technologies. *Int. J. Refrig.* **2002**, *25*, 450–459. [CrossRef]
46. Deng, J.; Wang, R.Z.; Han, G.Y. A review of thermally activated cooling technologies for combined cooling, heating and power systems. *Prog. Energy Combust. Sci.* **2011**, *37*, 172–203. [CrossRef]
47. Best, R.; Rivera, W. A review of thermal cooling systems. *Appl. Therm. Eng.* **2015**, *75*, 1162–1175. [CrossRef]
48. Jradi, M.; Riffat, S. Tri-generation systems: Energy policies, prime movers, cooling technologies, configurations and operation strategies. *Renew. Sustain. Energy Rev.* **2014**, *32*, 396–415. [CrossRef]

49. Besagni, G.; Mereu, R.; Inzoli, F. Ejector refrigeration: A comprehensive review. *Renew. Sustain. Energy Rev.* **2016**, *53*, 373–407. [CrossRef]

50. Thermax Global Triple Effect Absorption Chiller. Available online: http://www.thermaxglobal.com/thermax-absorption-cooling-systems/vapour-absorption-machines/triple-effect-chillers/ (accessed on 25 November 2018).

51. Kawasaki Thermal Engineering. Direct-Fired Double-Effect Absorption Chiller. Available online: http://www.khi.co.jp/corp/kte/EN/product/df-chiller1.html (accessed on 25 November 2018).

52. Schwerdt, P. Activities in Thermal Driven Cooling at Fraunhofer Umsicht. In Chapter 4: Thermally Driven Heat Pumps for Cooling. Available online: https://depositonce.tu-berlin.de/bitstream/11303/5164/1/117_schwerdt.pdf (accessed on 25 November 2018).

53. Srikhirin, P.; Aphornratana, S.; Chungpaibulpatana, S. A review of absorption refrigeration technologies. *Renew. Sustain. Energy Rev.* **2001**, *5*, 343–372. [CrossRef]

54. Wu, W.; Wang, B.; Shi, W.; Li, X. An overview of ammonia-based absorption chillers and heat pumps. *Renew. Sustain. Energy Rev.* **2014**, *31*, 681–707. [CrossRef]

55. Abed, A.M.; Alghoul, M.A.; Sopian, K.; Majdi, H.S.; Al-shamani, A.N.; Muftah, A.F. Enhancement aspects of single stage absorption cooling cycle: A detailed review. *Renew. Sustain. Energy Rev.* **2017**, *77*, 1010–1045. [CrossRef]

56. Arabkoohsar, A.; Andresen, G.B. Supporting district heating and cooling networks with a bifunctional solar assisted absorption chiller. *Energy Convers. Manag.* **2017**, *148*, 184–196. [CrossRef]

57. Alefeld, G.; Radermacher, R. *Heat Conversion Systems*; CRC Press, Inc.: Boca Raton, FL, USA, 1994; ISBN 9780849389283.

58. Jawahar, C.P.; Saravanan, R. Generator absorber heat exchange based absorption cycle—A review. *Renew. Sustain. Energy Rev.* **2010**, *14*, 2372–2382. [CrossRef]

59. Xu, Z.Y.; Wang, R.Z. Absorption refrigeration cycles: Categorized based on the cycle construction. *Int. J. Refrig.* **2016**, *62*, 114–136. [CrossRef]

60. Wang, L.W.; Wang, R.Z.; Oliveira, R.G. A review on adsorption working pairs for refrigeration. *Renew. Sustain. Energy Rev.* **2009**, *13*, 518–534. [CrossRef]

61. Shmroukh, A.N.; Ali, A.H.H.; Ookawara, S. Adsorption working pairs for adsorption cooling chillers: A review based on adsorption capacity and environmental impact. Renew. Sustain. *Energy Rev.* **2015**, *50*, 445–456. [CrossRef]

62. Goyal, P.; Baredar, P.; Mittal, A.; Siddiqui, A.R. Adsorption refrigeration technology—An overview of theory and its solar energy applications. *Renew. Sustain. Energy Rev.* **2016**, *53*, 1389–1410. [CrossRef]

63. Sahlot, M.; Riffat, S.B. Desiccant cooling systems: A review. *Int. J. Low-Carbon Technol.* **2016**, *11*, 489–505. [CrossRef]

64. Enteria, N.; Mizutani, K. The role of the thermally activated desiccant cooling technologies in the issue of energy and environment. *Renew. Sustain. Energy Rev.* **2011**, *15*, 2095–2122. [CrossRef]

65. Liu, X.H.; Yi, X.Q.; Jiang, Y. Mass transfer performance comparison of two commonly used liquid desiccants: LiBr and LiCl aqueous solutions. *Energy Convers. Manag.* **2011**, *52*, 180–190. [CrossRef]

66. Chen, J.; Jarall, S.; Havtun, H.; Palm, B. A review on versatile ejector applications in refrigeration systems. *Renew. Sustain. Energy Rev.* **2015**, *49*, 67–90. [CrossRef]

67. Ruch, P.; Saliba, S.; Ong, C.L.; Al-Shehri, Y.; Al-Rihaili, A.; Al-Mogbel, A.; Michel, B. Heat-driven adsorption chiller for sustainable cooling applications. In Proceedings of the 11th IEA Heat Pump Conference 2014, Montréal, QC, Canada, 12–16 May 2014. Available online: https://pdfs.semanticscholar.org/7db6/bc0436cd1db75103f7bffaf24a0fabc17e66.pdf (accessed on 25 November 2018).

68. Elbel, S. Historical and present developments of ejector refrigeration systems with emphasis on transcritical carbon dioxide air-conditioning. *Int. J. Refrig.* **2011**, *34*, 1545–1561. [CrossRef]

69. Chen, J.; Havtun, H.; Palm, B. Screening of working fluids for the ejector refrigeration system. *Int. J. Refrig.* **2014**, *47*, 1–14. [CrossRef]

70. Saleh, B. Performance analysis and working fluid selection for ejector refrigeration cycle. *Appl. Therm. Eng.* **2016**, *107*, 114–124. [CrossRef]

71. Energieversorgung Gera GmbH District Cooling System in Gera Cold Supply with Steam Jet Ejector Chilling Technology. Available online: http://dampfstrahlkaelte.de/links/downloads/fernkaelteversorgung_gera_e.pdf (accessed on 25 November 2018).

72. Colmenar-Santos, A.; Rosales-Asensio, E.; Borge-Diez, D.; Collado-Fernández, E. Evaluation of the cost of using power plant reject heat in low-temperature district heating and cooling networks. *Appl. Energy* **2016**, *162*, 892–907. [CrossRef]

73. Zhen, L.; Lin, D.M.; Shu, H.W.; Jiang, S.; Zhu, Y.X. District cooling and heating with seawater as heat source and sink in Dalian, China. *Renew. Energy* **2007**, *32*, 2603–2616. [CrossRef]

74. Lind, L.; Mroczek, S.; Bell, J. Seawater used for district cooling in Stockholm. GNS Science, 2014. Available online: https://www.yumpu.com/en/document/view/9050455/seawater-used-for-district-cooling-in-stockholm-gns-science (accessed on 25 November 2018).

75. Peer, T.; (Lanny) Joyce, W.S. Lake-Source Cooling. *ASHRAE J.* **2002**, *44*, 37–39.

76. Tredinnick, S.; Phetteplace, G. District cooling, current status and future trends. In *Advanced District Heating and Cooling (DHC) Systems*; Wiltshire, R., Ed.; Woodhead Publishing (Elsevier): Cambridge, UK, 2016; pp. 167–188.

77. Ampofo, F.; Maidment, G.G.; Missenden, J.F. Review of groundwater cooling systems in London. *Appl. Therm. Eng.* **2006**, *26*, 2055–2062. [CrossRef]

78. Gómez, M.R.; Garcia, R.F.; Gómez, J.R.; Carril, J.C. Review of thermal cycles exploiting the exergy of liquefied natural gas in the regasification process. *Renew. Sustain. Energy Rev.* **2014**, *38*, 781–795. [CrossRef]

79. Invernizzi, C.M.; Iora, P. The exploitation of the physical exergy of liquid natural gas by closed power thermodynamic cycles. An overview. *Energy* **2016**, *105*, 2–15. [CrossRef]

80. International Gas Union (IGU). 2018 IGU World LNG Report, 2018 Edition. 27th World Gas Conference & Exhibition Edition. Available online: https://www.igu.org/sites/default/files/node-document-field_file/IGU_LNG_2018_0.pdf (accessed on 25 November 2018).

81. La Rocca, V. Cold recovery during regasification of LNG part one: Cold utilization far from the regasification facility. *Energy* **2010**, *35*, 2049–2058. [CrossRef]

82. European Commission. An EU Strategy on Heating and Cooling, COM(2016) 51 Final, Brussels, 16.2.2016. Available online: http://ec.europa.eu/energy/sites/ener/files/documents/1_EN_ACT_part1_v14.pdf (accessed on 7 January 2019).

83. Atienza-Márquez, A.; Bruno, J.C.; Coronas, A. Cold recovery from LNG-regasification for polygeneration applications. *Appl. Therm. Eng.* **2018**, *132*, 463–478. [CrossRef]

84. La Rocca, V. Cold recovery during regasification of LNG part two: Applications in an Agro Food Industry and a Hypermarket. *Energy* **2011**, *36*, 4897–4908. [CrossRef]

85. Jo, Y.K.; Kim, J.K.; Lee, S.G.; Kang, Y.T. Development of type 2 solution transportation absorption system for utilizing LNG cold energy. *Int. J. Refrig.* **2007**, *30*, 978–985. [CrossRef]

86. Hasnain, S.M. Review on sustainable thermal energy storage technologies, Part II: Cool thermal storage. *Energy Convers. Manag.* **1998**, *39*, 1139–1153. [CrossRef]

87. Yau, Y.H.; Rismanchi, B. A review on cool thermal storage technologies and operating strategies. *Renew. Sustain. Energy Rev.* **2012**, *16*, 787–797. [CrossRef]

88. Al-Noaimi, F.; Khir, R.; Haouari, M. Optimal design of a district cooling grid: Structure. *Eng. Optim.* **2019**, *51*, 160–183. [CrossRef]

89. Oró, E.; de Gracia, A.; Castell, A.; Farid, M.M.; Cabeza, L.F. Review on phase change materials (PCMs) for cold thermal energy storage applications. *Appl. Energy* **2012**, *99*, 513–533. [CrossRef]

90. Abdullah, M.O.; Yii, L.P.; Junaidi, E.; Tambi, G.; Mustapha, M.A. Electricity cost saving comparison due to tariff change and ice thermal storage (ITS) usage based on a hybrid centrifugal-ITS system for buildings: A university district cooling perspective. *Energy Build.* **2013**, *67*, 70–78. [CrossRef]

91. Lizana, J.; Chacartegui, R.; Barrios-Padura, A.; Ortiz, O. Advanced low-carbon energy measures based on thermal energy storage in buildings: A review. *Renew. Sustain. Energy Rev.* **2018**, *82*, 3705–3749. [CrossRef]

92. Chow, T.T.; Au, W.H.; Yau, R.; Cheng, V.; Chan, A.; Fong, K.F. Applying district-cooling technology in Hong Kong. *Appl. Energy* **2004**, *79*, 275–289. [CrossRef]

93. Chow, T.T.; Fong, K.F.; Chan, A.L.S.; Yau, R.; Au, W.H.; Cheng, V. Energy modelling of district cooling system for new urban development. *Energy Build.* **2004**, *36*, 1153–1162. [CrossRef]

94. Skagestad, B.; Mildenstein, P. *District Heating and Cooling Connection Handbook—Programme of Research, Development and Demonstration on District Heating and Cooling*; International Energy Agency—IEA District Heating and Cooling: Paris, France, 2002; ISBN 9057480298.

95. Rifai, S.M.M. Demand-Response Management of a District Cooling Plant of a Mixed Use City Development. Master's Thesis, KTH—Royal Institute of Technology, Stockholm, Sweden, 2012.

96. Eveloy, V.; Gebreegziabher, T. A Review of Projected Power-to-gas Deployment Scenarios. *Energies* **2018**, *11*, 1824. [CrossRef]

97. Hughes, B.R.; Rezazadeh, F.; Chaudhry, H.N. Economic viability of incorporating multi-effect distillation with district cooling systems in the United Arab Emirates. Sustain. *Cities Soc.* **2013**, *7*, 37–43. [CrossRef]

98. Eveloy, V.; Rodgers, P.; Popli, S. Trigeneration scheme for a natural gas liquids extraction plant in the Middle East. *Energy Convers. Manag.* **2014**, *78*, 204–218. [CrossRef]

99. Parham, K.; Khamooshi, M.; Tematio, D.B.K.; Yari, M.; Atikol, U. Absorption heat transformers—A comprehensive review. *Renew. Sustain. Energy Rev.* **2014**, *34*, 430–452. [CrossRef]

100. Rivera, W.; Best, R.; Cardoso, M.J.; Romero, R.J. A review of absorption heat transformers. *Appl. Therm. Eng.* **2015**, *91*, 654–670. [CrossRef]

101. Eveloy, V.; Rodgers, P.; Qiu, L. Hybrid gas turbine-organic Rankine cycle for seawater desalination by reverse osmosis in a hydrocarbon production facility. *Energy Convers. Manag.* **2015**, *106*, 1134–1148. [CrossRef]

102. Harish, V.S.K.V.; Kumar, A. A review on modeling and simulation of building energy systems. *Renew. Sustain. Energy Rev.* **2016**, *56*, 1272–1292. [CrossRef]

103. Shi, X.; Tian, Z.; Chen, W.; Si, B.; Jin, X. A review on building energy efficient design optimization from the perspective of architects. *Renew. Sustain. Energy Rev.* **2016**, *65*, 872–884. [CrossRef]

104. Wang, H.; Zhai, Z.J. Advances in Building Simulation and Computational Techniques: A Review between 1987 and 2014. *Energy Build.* **2016**, *128*, 319–335. [CrossRef]

105. Shimoda, Y.; Nagota, T.; Isayama, N.; Mizuno, M. Verification of energy efficiency of district heating and cooling system by simulation considering design and operation parameters. *Build. Environ.* **2008**, *43*, 569–577. [CrossRef]

106. Jing, Z.X.; Jiang, X.S.; Wu, Q.H.; Tang, W.H.; Hua, B. Modelling and optimal operation of a small-scale integrated energy based district heating and cooling system. *Energy* **2014**, *73*, 399–415. [CrossRef]

107. Chow, T.T.; Chan, A.L.S.; Song, C.L. Building-mix optimization in district-cooling system implementation. *Appl. Energy* **2004**, *77*, 1–13. [CrossRef]

108. Gang, W.; Wang, S.; Gao, D.; Xiao, F. Performance assessment of district cooling systems for a new development district at planning stage. *Appl. Energy* **2015**, *140*, 33–43. [CrossRef]

109. Chan, A.L.S.; Chow, T.T.; Fong, S.K.F.; Lin, J.Z. Generation of a typical meteorological year for Hong Kong. *Energy Convers. Manag.* **2006**, *47*, 87–96. [CrossRef]

110. Gros, A.; Bozonnet, E.; Inard, C.; Musy, M. A New Performance Indicator to Assess Building and District Cooling Strategies. *Procedia Eng.* **2016**, *169*, 117–124. [CrossRef]

111. Sameti, M.; Haghighat, F. Optimization approaches in district heating and cooling thermal network. *Energy Build.* **2017**, *140*, 121–130. [CrossRef]

112. Renewable Smart Cooling for Urban Europe (RESCUE). Cool Conclusions How to Implement District Cooling in Europe. 2015. Available online: http://www.rescue-project.eu/fileadmin/user_files/FinalReport/Rescue_Cool_Conclusions_Final_Report_A4_EN_RZ.pdf (accessed on 25 November 2018).

113. Ecoheat4cities. Guidelines for Technical Assessment of District Heating Systems. Available online: https://www.euroheat.org/wp-content/uploads/2016/04/Ecoheat4cities_3.1_Labelling_Guidelines.pdf (accessed on 25 November 2018).

114. Genchi, Y.; Kikegawa, Y.; Inaba, A. CO_2 payback-time assessment of a regional-scale heating and cooling system using a ground source heat-pump in a high energy-consumption area in Tokyo. *Appl. Energy* **2002**, *71*, 147–160. [CrossRef]

115. Calm, J.M. The next generation of refrigerants—Historical review, considerations, and outlook. *Int. J. Refrig.* **2008**, *31*, 1123–1133. [CrossRef]

116. Kassim, J. Efficiency Gains and Emissions Reductions Achieved Through District Cooling. In Proceedings of the International District Energy Association Annual Conference & Trade Show 2014 (IDEA 2014), Seattle, WA, USA, 8–11 June 2014. Available online: https://www.districtenergy.org/HigherLogic/System/DownloadDocumentFile.ashx?DocumentFileKey=e8e778c9-452c-8b50-eccb-8850764965e8 (accessed on 25 November 2018).

117. Montreal Protocol on Substances that Deplete the Ozone Layer. Available online: http://conf.montreal-protocol. org/meeting/mop/mop-27/presession/Background%20Documents%20are%20available%20in%20English% 20only/RTOC-Assessment-Report-2014.pdf (accessed on 25 November 2018).

118. Bolaji, B.O.; Huan, Z. Ozone depletion and global warming: Case for the use of natural refrigerant—A review. *Renew. Sustain. Energy Rev.* **2013**, *18*, 49–54. [CrossRef]

119. Swedblom, M.; Peter, M.; Anders, T.; Frohm, H.; Rubenhag, A. District Cooling and the Customers' Alternative Cost (Renewable Smart Cooling for Urban Europe (RESCUE) WP2. 2014. Available online: http://www.rescue-project.eu/index.php?id=2 (accessed on 25 November 2018).

120. Tredinnick, S. Benefits of Economic Analyses (Part 2): Real World Examples. District Energy. 2011. International District Energy Association. pp. 66–69. Available online: https://www.eesi.org/files/ districtenergy2011Q3-dl.pdf (accessed on 25 November 2018).

121. Perez, N.; Riederer, P.; Inard, C. Development of a multiobjective optimization procedure dedicated to the design of district energy concept. *Energy Build.* **2018**, *178*, 11–25. [CrossRef]

122. Feng, X.; Long, W. Applying single parent genetic algorithm to optimize piping network layout of district cooling system. In Proceedings of the 4th International Conference on Natural Computation (ICNC 2008), Jinan, China, 18–20 October 2008; Volume 1, pp. 176–180. [CrossRef]

123. Amulla, Y. Gulf Cooperation Council (GCC) Countries 2040 Energy Scenario for Electricity Generation and Water Desalination. Master's Thesis, School of Industrial Engineering and Management, KTH Royal Institute of Technology in Stockholm, Stockholm, Sweden, 2014. Available online: https://www.diva-portal.org/ smash/get/diva2:839740/FULLTEXT01.pdf (accessed on 25 November 2018).

124. International Renewable Energy Agency (IRENA). Pan-Arab Renewable Energy Strategy 2030. 2014. Available online: https://www.irena.org/DocumentDownloads/Publications/IRENA_Pan-Arab_Strategy_ June%202014.pdf (accessed on 25 November 2018).

125. World Bank. Available online: http://databank.worldbank.org/data/indicator/ (accessed on 25 November 2018).

126. Aquastat. Food and Agriculture Organization of the United Nations. 2014. Data. Available online: http://www.fao.org/nr/water/aquastat/data/query/results.html (accessed on 25 November 2018).

127. Eveloy, V.; Gebreegziabher, T. Excess electricity and power-to-gas storage potential in the future renewable-based power generation sector in the United Arab Emirates. *Energy* **2018**, *166*, 426–450. [CrossRef]

128. Malit, F.; Naufral, G. Labour Migration, Skills Development and the Future of Work in the Gulf Cooperation Council (GCC) Countries. Working Paper, International Labour Organization (ILO), October 2017. Available online: https://www.ilo.org/newdelhi/whatwedo/publications/WCMS_634982/lang--en/index.htm (accessed on 25 November 2018).

129. United Nations-Water. Managing Water Report under Uncertainty and Risk. The United Nations World Water Development Report 4. 2012, Volume 1. Available online: http://www.unesco.org/new/fileadmin/ MULTIMEDIA/HQ/SC/pdf/WWDR4%20Volume%201-Managing%20Water%20under%20Uncertainty% 20and%20Risk.pdf (accessed on 25 November 2018).

130. Pirouz, B.; Maiolo, M. The Role of Power Consumption and Type of Air Conditioner in Direct and Indirect Water Consumption. *J. Sustain. Dev. Energy Water Environ. Syst.* **2018**, *6*, 665–673. [CrossRef]

131. Spurr, M. Cooling, power, water and waste. Why Middle East governments should integrate utility planning. *First Quart. 2007 Dist. Energy Mag. Permis. IDEA* **2007**, *93*, 56–57.

132. United Nations. Food and Agriculture of the United Nations. 6 Clean Water and Sanitation. Progress on Level of Water Stress. Global Baseline for SDG Indicator 6.4.2. 2018. Available online: http://www.unwater.org/ app/uploads/2018/10/SDG6_Indicator_Report_642-progress-on-level-of-water-stress-2018.pdf (accessed on 25 November 2018).

133. United Nations-Water. United Nations Development Programme. Water Governance in the Arab Region—UNDP. Arab Water Government Report. Water in the Arab Region: Availability, Status and Threats. Chapter 1. Available online: http://www.bh.undp.org/content/dam/rbas/doc/Energy% 20and%20Environment/Arab_Water_Gov_Report/Arab_Water_Report_AWR_Chapter%201.pdf (accessed on 25 November 2018).

134. TABREED. Company Overview. Available online: http://www.tabreed.ae/uploads/documents/Tabreed% 20Media%20Kit%20English%20and%20final%20%20last%20update%2022%2011%202012.pdf (accessed on 25 November 2018).

135. Younan, F. How will district cooling evolve in the GCC. In Proceedings of the District Cooling 2016 a Climate Solution, Dubai, UAE, 13–15 November 2016.
136. National Central Cooling Company PJSC (DFM: TABREED). Investor Presentation. May 2018. Available online: https://www.tabreed.ae/wp-content/uploads/2018/05/Tabreed-Corporate-Presentation-May-2018.pdf (accessed on 7 January 2019).
137. Brander, M.; Sood, A.; Wylie, C.; Haughton, A.; Lovell, J. *Technical Paper. Electricity-Specific Emission Factors for Grid Electricity*; Ecometrica: Edinburgh, UK, 2011; pp. 1–22.
138. Hassan, J. Next Generation District Cooling Network—A High-Level Overview. Clean Cooling. In Proceedings of the New "Frontier Market" for UAE & GCC Region, Dubai, UAE, 9–10 April 2018. Available online: https://www.eugcc-cleanergy.net/sites/default/files/events/20180410_dubai/clean_cooling_workshop_dubai_09.04.2018_session_3_hassan_javed_aateco.pdf (accessed on 25 November 2018).
139. Herold, K.E.; Radermacher, R.; Klein, S.A. *Absorption Chillers and Heat Pumps*, 2nd ed.; CRC Press, Inc.: Boca Raton, FL, USA, 2016; Volume 1, ISBN 9788578110796.
140. TABREED. 2016 H1 Results Presentation National Central Cooling Company PJSC. 26 July 2016. Available online: https://www.tabreed.ae/uploads/documents/H1%202016%20Analyst%20presentation.pdf (accessed on 25 November 2018).
141. Bhutto, A.W.; Bazmi, A.A.; Zahedi, G.; Klemes, J.J. A review of progress in renewable energy implementation in the Gulf Cooperation Council countries. *J. Clean. Prod.* **2014**, *71*, 168–180. [CrossRef]
142. Kilkis, B.; Kilkis, S. Hydrogen economy model for nearly net-zero cities with exergy rationale and energy-water nexus. *Energies* **2018**, *11*, 1226. [CrossRef]
143. Kilkis, S. A nearly net-zero exergy district as a model for smarter energy systems in the context of urban metabolism. *J. Sustain. Dev. Energy Water Environ. Syst.* **2017**, *5*, 101–126. [CrossRef]
144. Connolly, D.; Lund, H.; Mathiesen, B.V. Smart energy Europe: The technical and economic impact of one potential 100% renewable energy scenario for the European Union. *Renew. Sustain. Energy Rev.* **2016**, *60*, 1634–1653. [CrossRef]
145. AGO. Available online: http://www.ago.ag (accessed on 25 November 2018).
146. BROAD Group. Available online: http://en.broad.com/ (accessed on 30 September 2018).
147. Carrier Corporation. Available online: https://www.carrier.com/carrier/en/us/products-and-services/commercial-refrigeration/ (accessed on 25 November 2018).
148. Hitachi. Large-Tonnage Chiller. Available online: https://www.jci-hitachi.com/products/chillers/ (accessed on 25 November 2018).
149. Kawasaki Thermal Engineering. Available online: http://www.khi.co.jp/corp/kte/EN/index.html (accessed on 25 November 2018).
150. LG Catalogue Absorption Chillers. Available online: http://www.lg.com/global/business/download/resources/sac/Catalogue_Absorption%20Chillers_ENG_F.pdf (accessed on 25 November 2018).
151. SAKURA. Available online: http://www.sakura-aircon.com/Products/commercial/SAKURA_Absorption_Chiller_2015.html (accessed on 25 November 2018).
152. SHUANGLIANG ECO-ENERGY. Available online: http://sl-ecoenergy.com/products/0/163310/ (accessed on 25 November 2018).
153. THERMAX. Available online: http://www.thermaxglobal.com/thermax-absorption-cooling-systems/ (accessed on 25 November 2018).
154. TRANE. Available online: http://www.trane.com/commercial/north-america/us/en/products-systems/equipment/chillers/absorption-liquid-chillers.html (accessed on 25 November 2018).
155. Johnson Controls. Available online: http://www.johnsoncontrols.com/buildings/hvac-equipment/chillers (accessed on 25 November 2018).
156. Fischer Eco Solutions. Available online: www.fischer-group.com (accessed on 25 November 2018).
157. IPCC Fourth Assessment Report: Climate Change 2007. Climate Change 2007: Working Group I. The Physical Science Basis. Direct Global Warming Potentials. Available online: https://www.ipcc.ch/publications_and_data/ar4/wg1/en/ch2s2-10-2.html (accessed on 25 November 2018).
158. United Nations Environment Programme (UNEP). *Ozone Depletion Potential*; UNEP Ozone Secretariat: Nairobi, Kenya, 2006.
159. TABREED. Landmark Projects. Available online: https://www.tabreed.ae/landmark-projects/ (accessed on 25 November 2018).

160. Emicool Emirates District Cooling LLC Pre-Qualification Documents. Available online: https://www.emicool.net/en/Downloads/Pre_Qualification_Emicool.pdf (accessed on 25 November 2018).
161. Marafeq Qatar. Lusail Development. Available online: http://www.marafeq.com.qa/main.php?content=Projects&link=48&type=content (accessed on November 2018).
162. Qatar Cool West Bay and the Pearl Qatar. Available online: http://www.qatarcool.com/our-districts/ (accessed on 25 November 2018).

Article

Analysis of Different Strategies for Lowering the Operation Temperature in Existing District Heating Networks

Francesco Neirotti [1], Michel Noussan [1,2,*], Stefano Riverso [3] and Giorgio Manganini [3]

[1] Department of Energy, Politecnico di Torino, c.so Duca degli Abruzzi 24, 10129 Torino, Italy; francesco.neirotti@polito.it
[2] Fondazione Eni Enrico Mattei, c.so Magenta 63, 20123 Milano, Italy
[3] United Technologies Research Center, Penrose Wharf Business Centre, 4th floor, T23 XN53 Cork City, Ireland; RiversS@utrc.utc.com (S.R.); ManganG@utrc.utc.com (G.M.)
* Correspondence: michel.noussan@polito.it; Tel.: +39-011-090-4529

Received: 12 December 2018; Accepted: 18 January 2019; Published: 21 January 2019

Abstract: District heating systems have an important role in increasing the efficiency of the heating and cooling sector, especially when coupled to combined heat and power plants. However, in the transition towards decarbonization, current systems show some challenges for the integration of Renewable Energy Sources and Waste Heat. In particular, a crucial aspect is represented by the operating temperatures of the network. This paper analyzes two different approaches for the decrease of operation temperatures of existing networks, which are often supplying old buildings with a low degree of insulation. A simulation model was applied to some case studies to evaluate how a low-temperature operation of an existing district heating system performs compared to the standard operation, by considering two different approaches: (1) a different control strategy involving nighttime operation to avoid the morning peak demand; and (2) the partial insulation of the buildings to decrease operation temperatures without the need of modifying the heating system of the users. Different temperatures were considered to evaluate a threshold based on the characteristics of the buildings supplied by the network. The results highlight an interesting potential for optimization of existing systems by tuning the control strategies and performing some energy efficiency operation. The network temperature can be decreased with a continuous operation of the system, or with energy efficiency intervention in buildings, and distributed heat pumps used as integration could provide significant advantages. Each solution has its own limitations and critical parameters, which are discussed in detail.

Keywords: district heating; energy efficiency; optimization; heat pumps; low temperature networks

1. Introduction

District heating (DH) systems are a mature technology that shows multiple advantages for the heating and cooling sector in cities. The heating network can be connected to multiple generation units, including combined heat and power (CHP) plants and waste heat recovery (WHR) from industries, increasing the efficiency of the whole system. Moreover, DH has allowed in multiple cases the integration of renewable energy sources (RES) such as wood biomass and municipal solid waste (whose organic fraction is often considered renewable) to support the decarbonization of heating and cooling in large cities. Indeed, the possibility of operating large plants allows a better control of the local pollutants with respect to distributed generation.

Nowadays, the big challenge faced by fourth-generation DH [1] is the need of lowering the operational temperature of DH systems to foster the integration of additional technologies including solar energy [2], heat pumps and low-temperature waste heat [3]. This is a crucial aspect in the transition towards more sustainable heating systems, which should decrease the dependency on fossil fuels, which are currently the main source for DH systems in many countries. An additional driver is the major transition in electricity generation at country level: the large increase of the RES share is affecting the economic profitability of large CHP plants, which are often coupled to DH systems in Northern Europe [4].

The potential for low-temperature DH has already been evaluated in the literature [5], and a number of networks already operate at low or very low temperatures, especially in Northern Europe [6,7]. Existing low-temperature DH systems generally have a small size, the connected buildings have significant insulation, and often the heating system of each building has no hydraulic separation from the primary network. Traditional DH systems in large cities have no such characteristics, and therefore alternative approaches are needed to support an effective transition of existing systems to low-temperature solutions.

Some research works have focused on the benefits that can be reached by decreasing the network supply and return temperatures [8], indicating the heat demand density in the area as a key parameter for the evaluation of the competitiveness of the DH. An additional threat to the economic sustainability of future DH systems is an excessive insulation of buildings, which can lead to excessive payback times for the required infrastructure [9]. In some practical experiences, the importance of a proper monitoring and control of the actual operating parameters in the users' heating systems has been found to be a crucial aspect for the optimization of the DH network [10]. The use of simulation models can be an important support in evaluating the benefits that can be reached in existing systems [11], and different approaches have been used in modeling DH systems, including simplified RC models [12] and more complex models based on machine learning techniques [13]. More detailed models are generally applied to single buildings, as the simulation of complex algorithms over multiple buildings with different parameters generally leads to unacceptable time for model set-up and computation. In the sector of DH systems, few works have focused on the actual comparison of the operation of a given system at different temperature levels and with different operational logics.

This paper compares, by means of a dedicated simulation model, two different approaches for a low-temperature operation of an existing DH network. The first approach considers a 24-h operation instead of the traditional night set-back control (i.e., the heating system is shut down at night), while the second is based on a partial insulation of the buildings in order to operate the existing users' heating systems with a lower nominal temperature.

2. Methodology

A simulation model of a DH system was defined to evaluate the impact of decreasing, under different conditions, the water temperature of its distribution network. The model is composed by the final user and the distribution network of a DH central thermal plant, while the three separate use cases compare: (i) the current condition of the third-generation of DH system; (ii) a low-temperature network with continuous operations during the night; and (iii) a low-temperature network with energy efficient buildings.

2.1. Simulation Model

The simulation model was developed in Modelica using the Dymola programming environment [14]. The model was based on the library developed by IEA Annex 60 [15], an open source Modelica library that serves as the core of other Modelica libraries, and on the library "Buildings", developed by Lawrence Berkeley National Laboratory, for dynamic simulation of the energetic behavior of single rooms, buildings and whole districts [16].

The two main parts of the simulation model are the final users and the network model. The simulation model assumptions and components are explained in detail in the following sub-sections.

2.1.1. Final Users

The final users were simulated by a model of a building with its heating system connected to the DH heat exchanger as shown in Figure 1 by the Modelica–Dymola scheme.

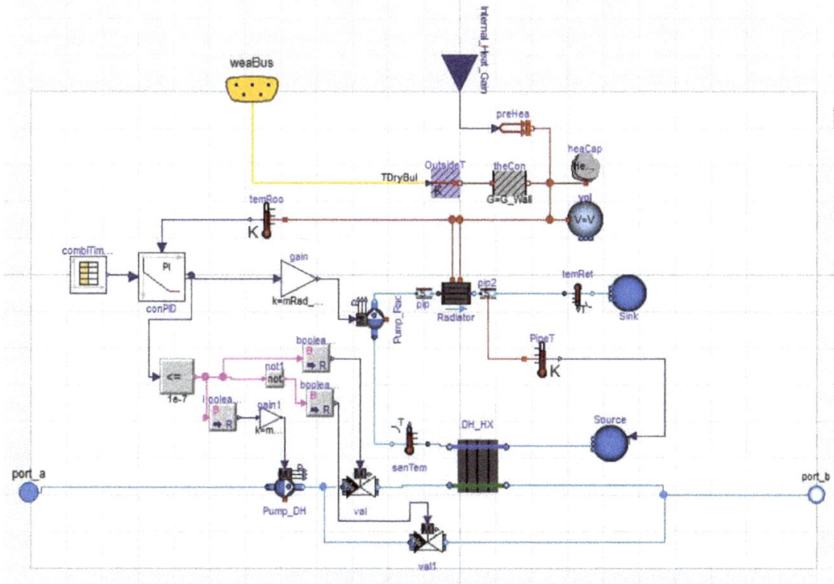

Figure 1. Simulation model for the final user, software: Modelica–Dymola [14].

The building is modeled by a volume of air with a specific heat capacity and it is connected to a thermal conductor representing the walls of the building, with a radiator used as heat source. The building volume is represented by the block *vol*, an instantaneously mixed volume, that has as main parameters the total volume and the nominal air mass flow rate. It can exchange heat through its *heatport*. The heat flows in the model are represented by *preHea* and the heat flow from the radiator by *Radiator*. The block *heaCap* is used to model the heat capacity of the building, assumed independent of the temperature and any specific geometry. This component changes the heat capacity value of the air to include the effect of energy storage in walls and furnitures. The overall dispersion of the building is modeled by using a single thermal conductor represented by *theCon* which takes as parameter the thermal conductance, assumed to be constant during the simulation. The block is connected, on one side, to the ambient temperature of the volume *vol* and, on the other side, to the outside temperature through the component *OutsideT*. This reproduces a variable temperature boundary condition derived from the weather data provided by *WeaBus* component (discussed in Section 2.1.2). Moreover, free gains are taken into account by means of the component *preHea*, which allows injecting or subtracting a specified amount of heat into the system. These free gains represent the passive heating by indirect heat sources (people, lights, appliances, solar, etc.) that are generally considered in buildings demand simulations. Summarizing, the volume of the building *vol* is connected to: (1) the thermal conductance to simulate the heat losses with the environment; (2) the heat capacity that approximate energy storage in furniture and building; (3) the radiator that is the main heat source; and (4) the free gain block.

The building heating system is modeled as a circuit between the *Radiator* and the DH heat exchanger *DH_HX*. The *Radiator* parameters consist of the data typically available from manufacturers compliant with the European Norm EN 442-2, including the nominal mass flow rate, the nominal inlet temperature and temperature difference, and the heating power. Furthermore, compared to the other components, the *Radiator* has two heat port connections with the building representing the convective heat flow rate and the radiative heat flow rate. For simplicity, it was assumed that the air and radiative temperature of the building are equal. Thus, there is no difference, in the calculation of the heat exchanged, between the radiative and the convective temperature. The transferred heat is modeled as follows. For each element $i = 1, \ldots, N$, where N denotes the number of elements used to discretize the radiator model, the convective and radiative heat transfer, Q_c and Q_r, are calculated as:

$$\dot{Q}_c = \frac{sign(T_i - T_a)(1 - f_r)UA}{N\,|T_i - T_a|^n} \tag{1}$$

$$\dot{Q}_r = \frac{sign(T_i - T_r)(f_r)UA}{N\,|T_i - T_r|^n} \tag{2}$$

where T_i is the water temperature of the element, T_a is the temperature of the room air, T_r is the radiative temperature, f_r is the fraction of radiant to total heat transfer, UA is the UA-value of the radiator and n is an exponent for the heat transfer. The sign function is used to give a direction to the heat flow: positive for heating. The model computes the UA-value solving the above equations with nominal values. Because the building is modeled as a single volume with homogeneous properties, a single radiator is used to minimize the computational time. The *DH_HX* is a model of a discretized coil made of two flow paths which are in opposite direction to model a counter-flow heat exchanger. The main parameters are the nominal mass flow rate of the two fluids (kg/s) and the thermal conductance at nominal flow (W/K). The water mass flow rate is controlled by a variable speed pump *Pump_rad* whose detailed description can be found in [17]. The nominal value of the mass flow rate is used to compute a default pressure curve (if no experimental pressure curve has been specified) that gives the electrical power consumption and the pump efficiency as function of the flow rate and the speed. The model computes the motor power consumption P_{ele}, the hydraulic power input W_{hyd}, the flow work W_{flo} and the heat dissipated Q. Based on the first law, the flow work is

$$W_{flo} = |\dot{V}\Delta P| \tag{3}$$

where $\dot{V}$ is the volume flow rate and ΔP is the pressure rise. The efficiencies are computed as

$$\eta = W_{flo}/P_{ele} = \eta_{hyd}\eta_{mot} \tag{4}$$

$$\eta_{hyd} = W_{flo}/W_{hyd} \tag{5}$$

$$\eta_{mot} = W_{hyd}/P_{ele} \tag{6}$$

where η_{hyd} and η_{mot} are, respectively, the hydraulic and motor efficiency. A PI controller *conPID* is used to track the temperature set point inside the building volume by controlling the mass flow rate, that can vary within 0–100% to simulate the thermostatic valves of the radiator. The water mass flow rate of the DH distribution network is considered to be constant. When the building heating system is switched off, the DH flow bypasses the heat exchanger thanks to the two valves *Val_01* and *Val_02*.

2.1.2. District Heating Network

The Dymola model of the DH network is represented in Figure 2. A detailed topological simulation of the flow distributions in the piping layout was beyond the scope of this work. Such simulations are usually performed in large networks where large distances may lead to complex flow distributions, especially in networks with looped circuits. In this model, the central plant is modeled as a simple block *IdealPlant_IP* that heats up the fluid to the set point temperature and calculates the transferred power: its parameters are the set point of water temperature supply, the nominal mass flow rate and the nominal pressure drop. The DH load is composed by nine users (*user_DH*) described in Section 2.1.1: each user can represent either a single building or an aggregate load by simply changing the design parameters. Each load is connected to the DH network in parallel by means of junctions *Junc* that takes as parameter the nominal mass flow rate per each branch and the relative pressure drop calculated as:

$$\dot{m} = K\sqrt{\Delta P} \tag{7}$$

where $\dot{m}$ is the mass flow rate, ΔP is the pressure drop and K is a constant that is calculated at the nominal values of the previous variables, namely $\dot{m}_{nominal}$ and $\Delta P_{nominal}$. No thermal losses are taken into account in this component. Right after the plant there are two pipe components for the supply and return water flows, namely *Pipe_Main_Flow* and *Pipe_Main_Return* that take as parameters the length, the insulation thickness and its thermal conductivity, the nominal mass flow rate and the velocity of the fluid. The diameter of the pipe can be arbitrary specified or automatically determined by the component according to the following equation:

$$d = \sqrt{\frac{4\dot{m}}{\rho \pi v}} \tag{8}$$

Unlike junctions, the pipe component takes into account thermal losses, based on the available geometric data and the outside temperature, while pressure drops are calculated as described in Equation (8).

The weather data component *WeaDat* reads Typical Meteorological Year (TMY) data, a statistical collection of weather data for a given location, listing hourly values of meteorological elements for a one-year period. TMY represents annual averages and it is used in buildings simulation to evaluate costs and expected energy consumption. However, TMY is not suited to design worst-case conditions, as an average reference year is considered for the analysis. In the current model, TMY data were obtained from the EnergyPlus web site [18].

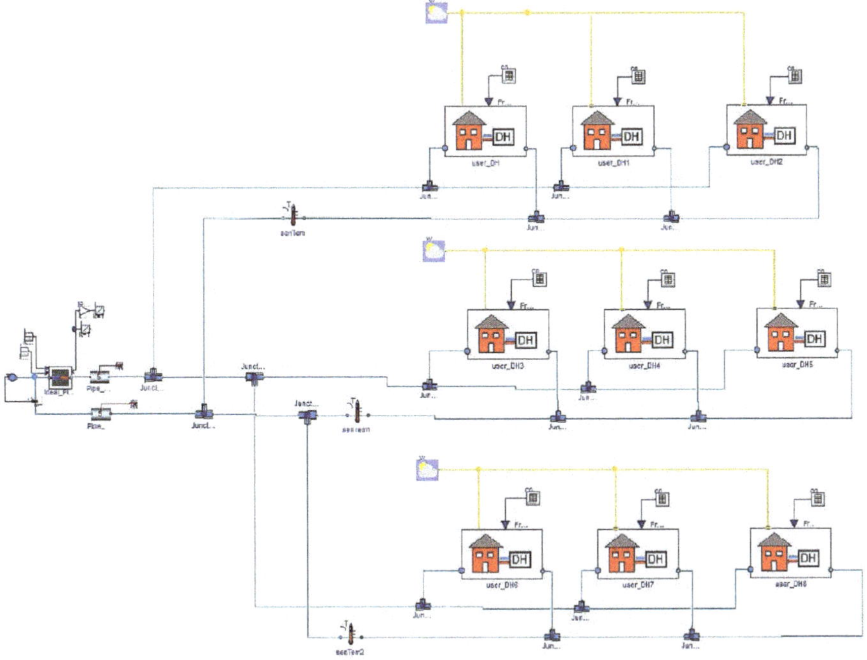

Figure 2. The simulation model for the district heating system, developed in Modelica–Dymola [14].

2.2. Case Studies

To evaluate the performance of the DH system in different conditions and the impact of lowering the operation temperature of the district, three case studies were studied and compared in the analysis. No case studies took into account domestic hot water production but only space heating.

2.2.1. Case 0: Current Conditions

This case represents a third-generation DH system and it was chosen as reference case. The network is operated with a nominal supply temperature of 90 °C, as typical for traditional DH systems. The users represent old residential buildings characterized by poor insulation, high energy demand and with the heating system sized for the worst case scenario (e.g., for the city of Turin, Italy, an outside temperature of −8 °C is usually considered). Users' heating systems are turned on during the day and shut down at night (night set-back control). In Italy, buildings are usually not heated during night in line with the Italian national regulation that limits the heating hours per day [19]. This operation produces a significant morning peak in the load profile, necessitating oversizing the capacity of the DH heat generators or installing a proper heat storage system.

The simulator was tuned using design parameters from the city of Turin. Users are sized to represent blocks of flats with the volumes 3×1800 m^3 , 3×2400 m^3, and 3×3000 m^3 for a total heated volume of 21,600 m^3 allocated to nine users. The radiators were sized by considering the usual design conditions for Turin, where old buildings' heating systems have an average installed nominal heat output of 30–40 W/m^3. The nominal power of radiators, which is usually defined by considering reference values of inlet temperature of 75 °C and outlet of 65 °C, was therefore calculated for each building as

$$\dot{Q}_{rad} = 40 \times V \tag{9}$$

being V the volume of the building. Using these design data, the nominal mass flow rate $\dot{m}$ could be calculated, for each radiator, by using the equation

$$\dot{m} = \dot{Q}_{rad} / (cp_W \times \Delta T_{rad}) \tag{10}$$

where Q_{rad} is the nominal power, cp_w (J/(kg K)) is the specific heat of water and ΔT_{rad} is the temperature difference between the inlet and the outlet of the radiator. Moving to the building parameters, free gains were set to zero as conservative hypothesis in order to investigate only the impact of the heating system parameters variation. The thermal conductance of the building, which is required by the simulation model, was calculated from the following equation:

$$U_{theCon} = f_{corr} \times \dot{Q}_{rad} / \Delta T_{worst} \tag{11}$$

where $\dot{Q}_{rad}$ is the nominal power of the radiator and ΔT_{worst} is the design delta temperature between the inside and the outside equal to 28 °C in Turin (given an outside temperature of −8 °C). f_{corr} is an empirical correction factor that is needed to correlate the nominal power of the heating systems with the real performance of the units when they are required to heat up the buildings in the morning in the worst conditions. In this study, this factor has been adjusted to a value of 0.5 in accordance with an acceptable transitory duration during the morning peak. The heat capacity of the building was chosen in order to have a temperature during the night around 16 °C. This empirical approach was chosen to be consistent and general enough with respect to the real operation of the heating systems, disregarding the unknown multitude of aspects that would affect the definition of the average building heat capacity, including the geometry, the materials of the walls and the furniture. Finally, as described in Section 2.1.1, the building temperature is controlled by a PI controller varying the radiator mass flow rate.

2.2.2. Case 1: Continuous Operation Without Night Set-Back Control

Lowering the network temperature is among the most important actions that has to be done in order to unlock the potentialities of the fourth-generation DH, e.g., RES integration and increasing the overall system efficiency. Nevertheless, in existing systems, a simple decrease of operational temperatures would not guarantee the comfort conditions for the users. In this case, a different schedule of the operations has to be considered to avoid the need of modifying the building or its heating system.

The aim of 1 was studying and understanding possible issues in lowering the temperature of the network, and which components of the building and its heating system might be affected. In particular, a continuous operation during day and night was analyzed as a viable way to decrease the network supply temperature by trying to avoid the morning peak request from the radiators. Moreover, daily load variations in DH systems, such as the morning and late afternoon peaks, lead to higher operational costs and higher losses, decreasing the overall efficiency of the system, as explained in [20,21]. In these two papers, the authors suggested different strategies to exploit heat storage systems to make the load profile flatter for the generation units, but they did not mention the possibility to modify the schedule of the operation of the users, as done in this study. The supply water temperature is then decreased from 90 °C to 50 °C and the building heating system is operated 24 h per day. To fairly compare the two different cases, the same heating system is used, with the same design parameters of Case 0. Thus, the temperature values of the heating systems of the buildings were calculated by the simulation model by operating the radiators in off-design conditions based on the nominal values of their design parameters.

The building temperature set point was constantly equal to 20 °C. Moreover the controller was changed: instead of a PI, a simple on/off controller is used, working within 20 °C $\pm$ 0.5 °C. This is due to difficulties on calibrating the PI controller for this specific case.

2.2.3. Case 2: Energy Efficiency Intervention in Buildings

An alternative approach for lowering the network temperature can be the partial insulation of buildings without the need for modifying the heating system operations. In this case, the aim was to understand if a partial insulation of the building allows meeting the comfort condition while keeping the same heating system day/night operation of Case 0. Indeed, due to the lower heat losses of the building, the radiators can be operated at a lower temperature, leading to a lower heat demand and consequently to the possibility of operating the radiators at a lower temperature.

The values of the design parameter for this case are the same as Case 1, except for the thermal conductance of the users that was reduced by 40% while maintaining the same heat capacity. The new conductance values were 0.78, 1.02, and 1.26 kW/K, respectively, for 1800, 2400, and 3000 m^3 users.

The reduction rate of the thermal conductance was set by considering a reasonable target obtained through energy efficiency measures (including the substitution of the windows and the insulation of the walls) performed on existing residential buildings in the context of Turin, based on the experience of the authors from real refurbishment interventions in the city. Due to the focus of this research work, a more detailed simulation of the building was not performed, leading to some approximations. For the same reason, the heat capacity was not varied accordingly, due to missing information of the specific characteristics of the buildings and the consequent need of an arbitrary new value. This approximation appeared to be acceptable in the context of the present work, and could be further analyzed in a future application with a more detailed definition of the features of each building.

2.3. Additional Analyses

Based on the results of the previous case studies, two additional analyses were performed: (1) a sensitivity analysis on different DH network temperatures; and (2) a further scenario with distributed Heat Pumps (HPs) to integrate the low temperature network and guarantee a better users' comfort.

2.3.1. Sensitivity Analysis for Supply Temperature

The previous cases were simulated with boundary temperatures in the network: 90 °C for Case 0 and 50 °C for Case 1 and Case 2. To better understand the impact of lowering the network temperatures, a sensitivity analysis was performed by considering three additional temperature levels: 45 °C, 55 °C and 60 °C. Moreover, to run a proper comparison, a "Comfort Factor" quantity to calculate the amount of hours in which the indoor temperature is in the comfort range from 20 °C to 21 °C was defined as

$$CF = \frac{h_{OSP}}{h_{heatingDay}} \tag{12}$$

where h_{OSP} is the number of hours that the indoor temperature is in the comfort range and $h_{heatingDay}$ is the number of hours where the heating system is on, namely between 7:00 and 22:00. It is important to notice that, even in Case 1, only the day hours were considered for the calculation, since the night operation had to be seen as an extra benefit.

2.3.2. Distributed Heat Pumps

The idea of this additional analysis was to substitute the heat exchanger between the network and the user with a heat pump to boost the temperature for space heating only where it is needed. The rationale here is that, even if in the near future most users live in high energy performance buildings, still some cities areas, e.g., protected historical buildings, would not be able to improve their energy performances by using low temperature heat sources. HPs use district heating water at the evaporator side to heat the user's heating system water (see Figure 3). Thanks to this configuration, it is possible to keep the network temperature at lower values, increasing the potential RES and WHR shares, and increase the temperature locally only for the users for which the temperature decrease is not a viable option.

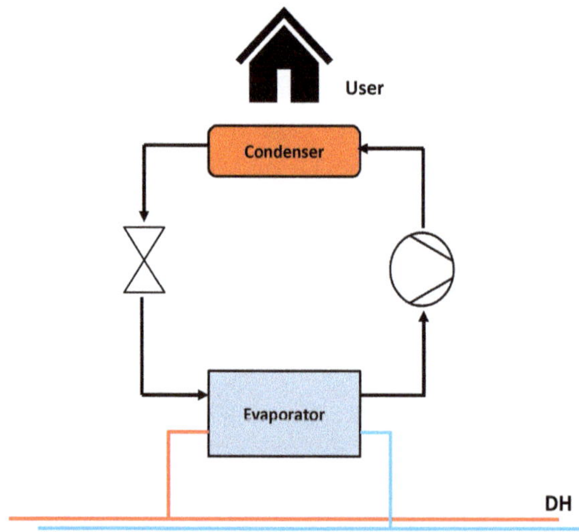

Figure 3. Distributed Heat Pump scheme.

The heat pump component, which is included into the Buildings library [16], represents a vapor compression HP that takes as main parameters the nominal mass flow rate, the nominal heat flow rate, the nominal temperature difference and the pressure drop at both evaporator and condenser sides. The condenser leaving water temperature represents a control inputs for the block and its COP is calculated as given by the following equations:

$$COP = \eta_{canot,0} COP_{carnot} \eta_{PL} \tag{13}$$

$$COP_{canot} = \frac{T_{condenser}}{T_{condenser} - T_{evaporator}} \tag{14}$$

$$\eta_{pl} = a_1 + a_2 y_{PL} + a_3 y_{PL}^2 + \ldots \tag{15}$$

where $\eta_{canot,0}$ is the Carnot Effectiveness, COP_{carnot} is the Carnot efficiency, η_{PL} is a polynomial expression to take into account partial load operation and y_{PL} is the partial load ratio. The Carnot Effectiveness coefficient can be set manually or it can be calculated as

$$\eta_{canot,0} = \frac{COP_0}{COP_{carnot,0}} \tag{16}$$

where COP_0 is the efficiency value in nominal conditions and $COP_{carnot,0}$ is the Carnot efficiency in nominal conditions. Since no accurate data were available, $\eta_{carnot,0}$ was set equal to 0.4 and η_{PL} equal to 1. Consequently, the COP was calculated as

$$COP = \eta_{canot,0} COP_{carnot} \tag{17}$$

In this scenario, the district heating heat exchanger was replaced by a heat pump with a condenser outlet temperature at 60 °C. The control logic of user is the same as in the previous cases.

Since the use of distributed heat pumps involves a significant electricity consumption, a proper comparison with the previous cases should be performed by considering the primary energy consumption, since heat and electricity cannot be simply summed up. In this comparison, the crucial aspect becomes the RES share in the power grid, which can vary from country to country but also shows a significant variability over time [22]. In this work, reference Primary Energy Factors for Italy was considered: 1.05 for natural gas and 2.42 for electricity. The DH heat was considered as produced by a natural gas boiler with a 90% efficiency (conservative approach).

3. Results

3.1. Simulation Results

Figure 4 shows the heat and temperature loads resulting from the model simulation for Case 0. The DH load profile is consistent with multiple examples of daily loads in many DH systems [23–25]. This aspect ensures a first qualitative validation of the proposed model, in the absence of reliable operation data of the case study under analysis. In particular, the reader can observe a morning peak (1 MW) that is 2–3 times higher than the stationary load in the afternoon (around 400 kW). This is a very well know problem that forces the heat providers to oversize thermal plants or to provide the network with large volumes of heat storage. The network temperature in the morning, where there is the highest energy request, has a temperature drop that is more or less equal to the design value of 15 °C (see bottom plot). During the rest of the day, the temperature drop is smaller due to the fact that the heating system's pump decreased the mass flow rate and so also the energy exchange rate with the grid drops.

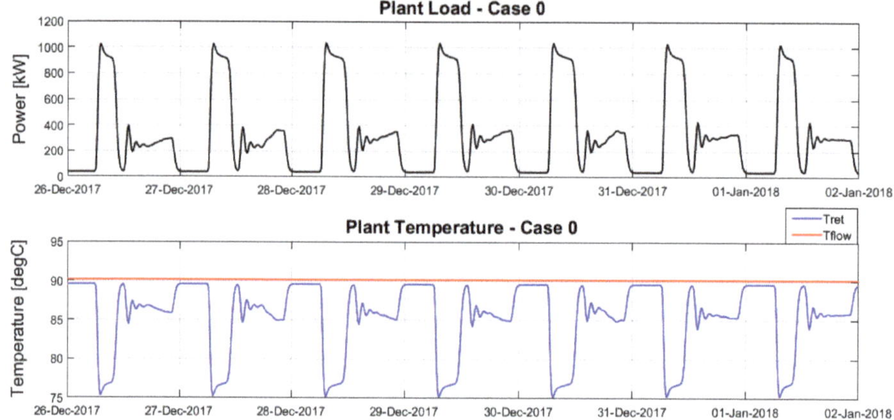

Figure 4. Baseline heat profiles of the heat supplied to the network.

Figures 5 and 6 compare the heat demand and temperature profiles of the three cases for a given day (1 January). The major difference is related with the morning peak, while some minor variances can be noticed throughout the day. Both Cases 1 and 2 allow significantly reducing the morning peak, approximately from 1 MW to 350–400 kW. This goal can therefore be reached with both approaches, although, without building insulation, the building needs to be heated also during the night. In particular, the controller lowers the room temperature in the afternoon in the alternative scenarios leading to a sharp decrease in heating demand. The deep drop is due to the fact that all the simulated users have been set with the same behavior. This is not compliant with the real world where multiple users have generally different behaviors. It is reasonable to think that, in such a situation, load and temperature profiles would be more stable, with small fluctuations around the rated power. This is a limit for the current simulator, but in future developments this model can be corrected by considering real data for different buildings.

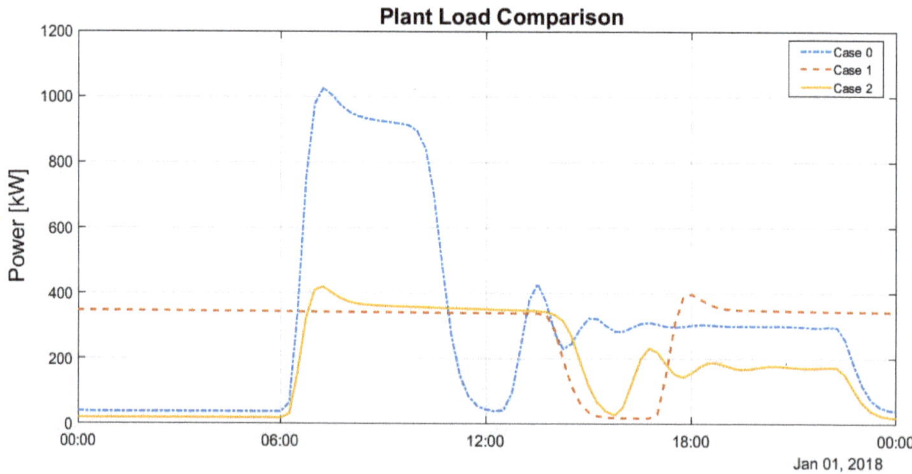

Figure 5. Comparison of heat profiles (1 January).

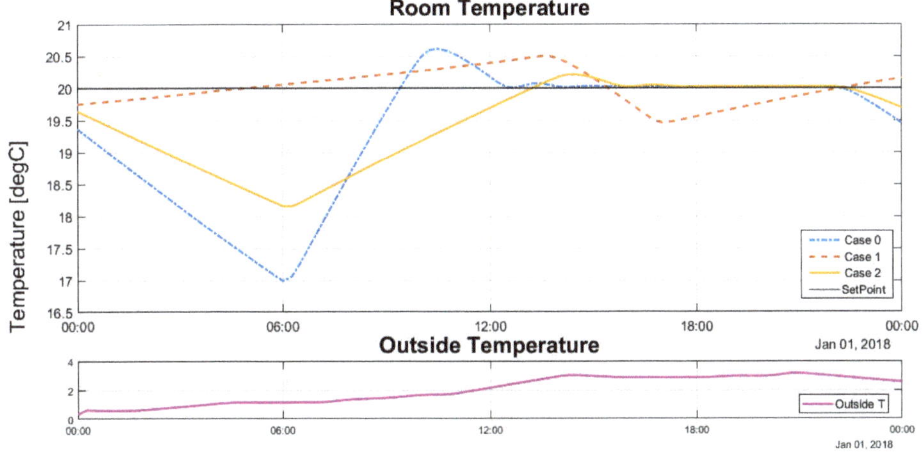

Figure 6. Comparison of temperature profiles (1 January).

While, to ease the discussion, the previous figures were limited to a single day simulation, Table 1 summarizes the performance of the system for the entire year. With respect to the reference case, Case 1 (continuous operation) leads to a slightly higher energy demand by the user, which is however compensated by lower network losses (due to the lower operation temperatures), thus resulting in a lower heat generation from the entire DH system. On the other hand, in Case 2, the user shows a substantial decrease of energy consumption due to the insulation, and the network losses are comparable to Case 1 (but obviously with a higher relative share).

Table 1. Annual energy performance of the different cases.

Case	User Demand (MWh)	DH Supply (MWh)	Network Losses (-)
Case 0	910	1075	18.0%
Case 1	941	1019	8.3%
Case 2	548	628	14.5%

The analysis thus far focused on comparing the three different cases from the DH operation point of view. However, to draw a fair comparison and gain more insight about the different configurations, the users's comfort has to be taken into account and discussed with a sensitivity analysis as done in the following sections.

3.2. Effect of the Supply Temperature

The sensitivity analysis on the effect of having different water supply temperatures is performed by considering as performance indicator (h_{OSP}) the share of hours in which the room temperature is equal to or higher than the set point (in the time frame 7:00–22:00, which is usually the time range in which heating systems are activated in Italy). The results are summarized in Figure 7, where Cases 1 and 2 were simulated for different supply temperatures and compared against Case 0.

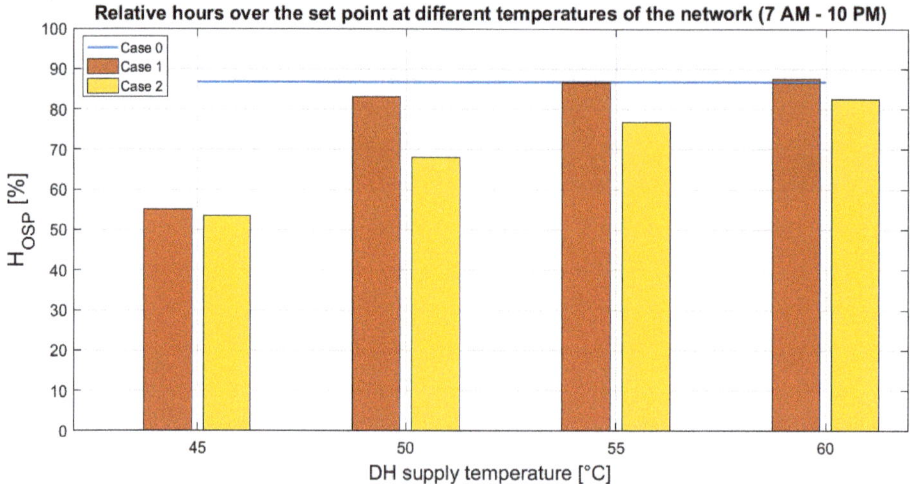

Figure 7. Comparison of annual performance for different supply temperatures.

The sensitivity analysis showed some interesting aspects. Firstly, it is important to notice that, in comparison with the other scenarios, not even Case 0 can guarantee a comfort factor of 100%, mainly due to the temperature transient in the morning towards the set point of 20 °C. Keeping in mind this aspect, Case 1 ensures an acceptable performance for a DH supply temperature as low as 50 °C, while a further decrease to 45 °C involves a drop of the Comfort Factor. The numerical analysis therefore suggest that such a low temperature requires some further actions in order to meet a reasonable comfort level, such as the installation of radiators with higher surface or the insulation of the building.

On the other hand, Case 2 appears to have a lower performance across all the analyzed temperatures, and for this reason cannot be considered a proper alternative to Case 0, even at medium temperatures of 55 °C or 60 °C. In this case, possible solutions can be found in a continuous operation similar to Case 1, or in the increase of radiators surface. Alternatively, to compensate for the network temperature decrease, distributed "booster" Heat Pumps can be exploited, as proposed in Section 2.3.2 and analyzed hereafter.

3.3. Alternative Layout: Distributed Heat Pumps

In this last simulation, the solution of distributed HPs was evaluated as an integrative solution for Case 2 and, more generally, as a viable option in networks that could be operated at low temperature but with specific users who require a high-temperature heat supply.

Figure 8 clearly illustrates the the HPs integration enables the network to operate at 45 °C and achieve an acceptable comfort for the users (Case 1 even outperforms Case 0). However, the HPs operation requires a significant power consumption, which cannot be ignored in a systemic analysis.

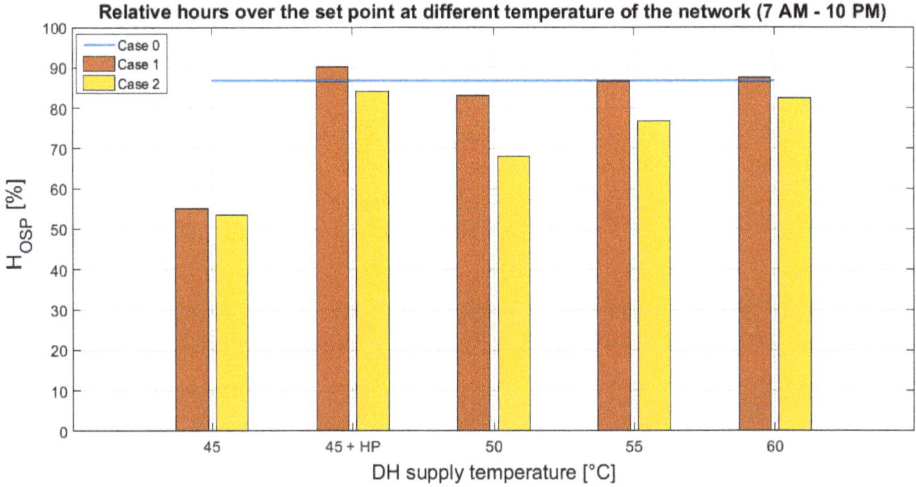

Figure 8. Comparison of annual performance of HP and different supply temperatures

The comparison of primary energy consumption in different cases is reported in Figure 9. With the assumptions made in this work, the HPs have a larger impact than other solutions in terms of total primary energy consumption, although further considerations may be done on the distinction between fossil primary energy and renewable primary energy. Since the primary energy consumption variation is quite low, it is expected that different assumptions might lead to opposite results.

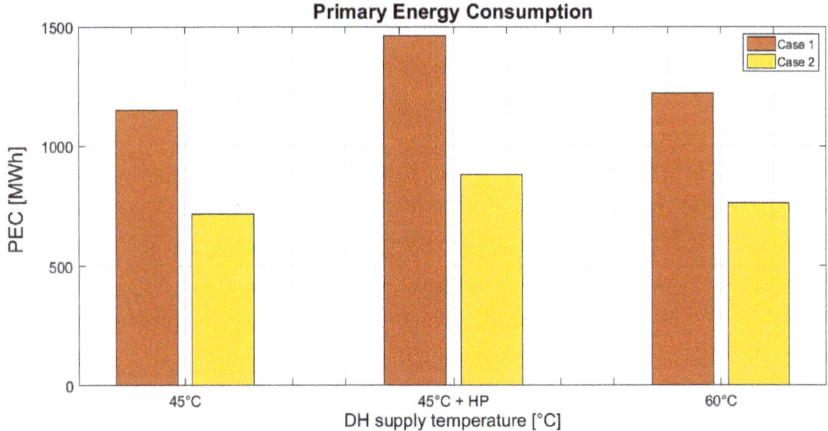

Figure 9. Comparison of annual primary energy consumption

For the sake of clarity, it is important to highlight that the focus of this work was on the energy analysis, and costs were not taken into account. From the economic point of view, a heat exchanger is much cheaper than a heat pump, both for investment costs and for maintenance. Moreover, in some countries, high electricity tariffs may hinder the use of HPs for space heating. On the contrary, this solution could be suitable for small areas in the city where energy efficiency actions are not possible for various reasons (e.g., historical buildings).

4. Discussion

The results of this study highlight the potential of optimizing the current heat generation logics in high-temperature DH systems. In particular, a common issue lays in the morning peak demand needed to heat up the buildings that have been cooled during the night, due to the shut down of the heat generation system. A continuous operation of the existing heating system could support a decrease of the network supply temperature, as the buildings would require a much lower heat rate that could be matched with existing radiators operating at a lower temperature.

The slight increase of energy demand of the users would be highly compensated by a reduction of the network losses, thanks to a lower average network temperature. This aspect should however be carefully evaluated when defining heat tariffs, as without proper actions this new operation strategy would increase the bills of the final users and lower the costs for the DH network operator. Proper regulation rules are required to support this transition by sharing the potential benefits among the final users and the DH system operators.

The refurbishment of buildings can lead to significant energy savings, but it not necessarily guarantee a shift towards a low-temperature DH operation. A critical parameter would be the amount of energy savings that are obtained, as without changing the existing radiators their heat supply is strictly related to their operation temperature. In this perspective, a buildings insulation strategy should be tailored on the features of the existing heating systems, unless a total substitution is included in the refurbishment intervention.

The analysis of the HPs behavior leads to a more complex model, as multiple parameters have an impact on their operation and different indicators can be chosen to evaluate their contribution. In particular, the shift towards low-temperature DH networks coupled to booster HPs is a promising solution where the integration of RES is of primary interest. On the other hand, if the DH generation plant remains fossil-based (considering a simple natural gas boiler), with the current Italian electricity mix, HPs do not lead to a decrease of the total primary energy consumption of the system. As already discussed, total primary energy consumption is just one of the possible indicators for evaluating the energy and environmental performance of a given technology: further research activities can underline the main parameters affecting these results.

Policy Indications

While this paper has been mostly focused on technical and operational aspects, energy policies are at the basis of the development of low-temperature DH systems. In particular, a well-designed policy support could foster a temperature decrease in existing systems, often without the need of economic-intensive actions.

DH systems are an effective and reliable solution for the heat supply in urban contexts, with a significant potential of integrating RES in heating and cooling. However, in many countries, there is a lack of specific regulations and targets to support this important transition, which involves the decrease of temperatures in existing networks. In some cases, DH systems would be compatible with low-temperature operation [10], but there is no interest for DH operators to reach lower temperatures and integrate RES with generally high investment costs.

However, some business models are already demonstrating that solar energy and centralized heat pumps for waste heat recovery are competitive solutions for RES integration in existing systems, although with generally higher payback times than usual industry applications. Specific targets set by regulators, with dedicated incentives, could support a wider diffusion of these technologies in countries that are showing a high unexploited potential.

5. Conclusions and Future Work

This work has presented a comparison of alternative strategies to decrease district heating network temperatures with the aim of improving the efficiency of DH systems and increase the potential RES integration. The main conclusions of this research work are the following:

- A continuous operation of the buildings' heating systems (Case 1) allows decreasing the network temperature without compromising the comfort until 50–60 °C, even without any action on building insulation or heating system configuration. Such a management logic is able to provide comfort levels that are comparable to traditional high-temperature operation of the network. However, temperatures below 50 °C do not guarantee an acceptable comfort level. On the other hand, the users energy consumption increases due to the continuous operation of the heating system by 3.5% with respect to the reference case, although the network losses are decreased.

- The results from Case 2 (buildings' insulation) confirm the great benefits of energy efficiency interventions in buildings. The energy consumption is much lower with respect to the other cases, the fluctuations of the indoor temperature are smaller, and day and night operation could be used instead of continuous operation. However, unlike the previous case, the network temperature must be at least 60 °C to guarantee an acceptable comfort level under the hypotheses of this study. The cause is the lower temperature difference between the heating system's water and the indoor temperature, which makes the morning transient longer with respect to the reference case.

- The combination of heat pumps and district heating systems seems to have a noteworthy potential, although some parameters are critical for their success. In fact, booster HPs can lower the energy demand from the DH network (and the supply temperature as well) but at the expense of a non-negligible local electricity consumption. Their environmental benefit is therefore strictly dependent on the source of these electricity, with reference to its renewable share and CO_2 emission factor. HPs are very efficient devices, if well designed, but they are more expensive and fragile than simple heat exchangers. Nevertheless, the simulation highlights the potential advantages of this combination that have to be evaluated through a detailed business and management plan.

These findings highlight that, from the technical point of view, alternative strategies exist for the evolution of current DH systems to low-temperature DH systems. Moreover, in some cases, a simple modification of the heat supply schedule could allow reducing the network temperatures without the need for further actions. Distributed heat pumps used as temperature boosters can be a technical solution for specific users who do not accept a decrease of their temperature supply.

Future Developments

This work can be the basis for the future development of a more detailed model of both user and DH systems. Design data were used to characterize the different components but, most of them, accept real data as input. This features could be interesting, for example, for a comparison between large HPs on the return line of the DH plant or small distributed HPs near the users (as considered in this work). Real COP data, pump consumption curves and thermal load data can be used to further develop this model by considering different real case studies. Moreover, the present study used average building features obtained from real cases, but without a detailed modeling of each building. As a result, some approximations were made due to some missing information. A specific building simulation model would be needed to increase the reliability of the results, which in turn leads to a more complicated model with higher computational resources. In addition, future work will include detailed piping of the network and increasing size of the DH system to demonstrate DH network balance for both temperature and flow.

Moreover, the possibilities to simulate multi physics systems can be useful to address pro and cons for high penetration of electrical and thermal grids. Thanks to the "encapsulation of knowledge" property, it is possible to develop and validate single-physics systems (PV plants, wind plants, buildings, etc.) and to combine the different blocks together to study the interactions between components and to find the optimum scheme to increase energy efficiency and RES utilization.

Author Contributions: conceptualization, M.N. and F.N.; methodology, F.N., G.M. and M.N.; simulation, F.N. and G.M.; writing—original draft preparation, F.N. and M.N.; writing—review and editing, G.M., M.N. and S.R.; supervision, M.N. and S.R.; and project administration, S.R.

Funding: The authors gracefully acknowledge financial support by the European Commission project E2DISTRICT under grant number 696009.

Conflicts of Interest: The authors declare no conflict of interest.

Abbreviations

The following abbreviations are used in this manuscript:

CHP	Combined Heat and Power
COP	Coefficient Of Performance
DH	District Heating
HP	Heat Pump
PEC	Primary Energy Consumption
PI	Proportional Integral
RES	Renewable Energy Sources
TMY	Typical Meteorological Year
WHR	Waste Heat Recovery
A	area (m^2)
CF	comfort factor (-)
COP	Coefficient Of Performance (-)
d	diameter (m)
f_{corr}	empirical correction factor (-)
f_r	fraction of radiant heat (-)
h_{OSP}	comfort factor (-)
$\dot{m}$	mass flow rate (kg/s)
ΔP	pressure difference (Pa)
$\dot{Q}$	thermal power (W)
T	temperature (K)
U	transmittance (W/(m^{2}K))
V	volume (m^3)
$\dot{V}$	volume flow rate (m^3/s)
v	velocity (m/s)
W	power (W)
ρ	density (kg/m^3)
η	efficiency (-)

References

1. Lund, H.; Werner, S.; Wiltshire, R.; Svendsen, S.; Thorsen, J.E.; Hvelplund, F.; Mathiesen, B.V. 4th Generation District Heating (4GDH): Integrating smart thermal grids into future sustainable energy systems. *Energy* **2014**, *68*, 1–11. [CrossRef]
2. Reiter, P.; Poier, H.; Holter, C. BIG Solar Graz: Solar District Heating in Graz—500,000 m^2 for 20% Solar Fraction. *Energy Procedia* **2016**, *91*, 578–584. [CrossRef]
3. Wahlroos, M.; Pärssinen, M.; Manner, J.; Syri, S. Utilizing data center waste heat in district heating—Impacts on energy efficiency and prospects for low-temperature district heating networks. *Energy* **2017**, *140*, 1228–1238. [CrossRef]
4. Helin, K.; Zakeri, B.; Syri, S. Is District Heating Combined Heat and Power at Risk in the Nordic Area?—An Electricity Market Perspective. *Energies* **2018**, *11*, 1–19. [CrossRef]
5. Schmidt, D.; Kallert, A.; Blesl, M.; Svendsen, S.; Li, H.; Nord, N.; Sipilä, K. Low Temperature District Heating for Future Energy Systems. *Energy Procedia* **2017**, *116*, 26–38. [CrossRef]

6. Yang, X.; Svendsen, S. Ultra-low temperature district heating system with central heat pump and local boosters for low-heat-density area: Analyses on a real case in Denmark. *Energy* **2018**, *159*, 243–251. [CrossRef]

7. Vetterli, N.; Sulzer, M.; Menti, U.P. Energy monitoring of a low temperature heating and cooling district network. *Energy Procedia* **2017**, *122*, 62–67. [CrossRef]

8. Nord, N.; Løve Nielsen, E.K.; Kauko, H.; Tereshchenko, T. Challenges and potentials for low-temperature district heating implementation in Norway. *Energy* **2018**, *151*, 889–902. [CrossRef]

9. Nielsen, S.; Grundahl, L. District Heating Expansion Potential with Low-Temperature and End-Use Heat Savings. *Energies* **2018**, *11*, 277. [CrossRef]

10. Østergaard, D.S.; Svendsen, S. Experience from a practical test of low-temperature district heating for space heating in five Danish single-family houses from the 1930s. *Energy* **2018**, *159*, 569–578. [CrossRef]

11. Schweiger, G.; Larsson, P.O.; Magnusson, F.; Lauenburg, P.; Velut, S. District heating and cooling systems—Framework for Modelica-based simulation and dynamic optimization. *Energy* **2017**, *137*, 566–578. [CrossRef]

12. Ogunsola, O.T.; Song, L.; Wang, G. Development and validation of a time-series model for real-time thermal load estimation. *Energy Build.* **2014**, *76*, 440–449. [CrossRef]

13. Idowu, S.; Saguna, S.; Åhlund, C.; Schelén, O. Applied machine learning: Forecasting heat load in district heating system. *Energy Build.* **2016**, *133*, 478–488. [CrossRef]

14. DassaultSystem. Dymola. Available online: https://www.3ds.com/products-services/catia/products/dymola/ (accessed on 1 September 2018).

15. IEA. *New Generation Computational Tools for Building and Community Energy Systems*; International Energy Agency: Paris, France, 2017; p. 60.

16. Lawrence Berkeley National Laboratory. Buildings Library. Available online: http://simulationresearch.lbl.gov/modelica/ (accessed on 1 September 2018).

17. Wetter, M. Fan and pump model that has a unique solution for any pressure boundary condition and control signal. In Proceedings of the 13th Conference of International Building Performance Simulation Association, Chambery, France, 26–28 August 2013; pp. 3505–3512.

18. EnergyPlus. Weather Data. Available online: http://energyplus.net/weather (accessed on 15 September 2018).

19. Legislatura. Decreto Del Presidente Della Repubblica 26 Agosto 1993, n. 412. Available online: http://www.gazzettaufficiale.it/eli/id/1993/10/14/093G0451/sg (accessed on 30 September 2018).

20. Verda, V.; Colella, F. Primary energy savings through thermal storage in district heating networks. *Energy* **2011**, *36*, 4278–4286. [CrossRef]

21. Gadd, H.; Werner, S. Daily heat load variations in Swedish district heating systems. *Appl. Energy* **2013**, *106*, 47–55. [CrossRef]

22. Noussan, M.; Roberto, R.; Nastasi, B. Performance indicators of electricity generation at country level— The case of Italy. *Energies* **2018**, *11*, 650. [CrossRef]

23. Weissmann, C.; Hong, T.; Graubner, C.A. Analysis of heating load diversity in German residential districts and implications for the application in district heating systems. *Energy Build.* **2017**, *139*, 302–313. [CrossRef]

24. Dahl, M.; Brun, A.; Kirsebom, O.S.; Andresen, G.B. Improving Short-Term Heat Load Forecasts with Calendar and Holiday Data. *Energies* **2018**, *11*, 1678. [CrossRef]

25. Noussan, M.; Jarre, M.; Poggio, A. Real operation data analysis on district heating load patterns. *Energy* **2017**, *129*, 70–78. [CrossRef]

Article

Reuse of Data Center Waste Heat in Nearby Neighborhoods: A Neural Networks-Based Prediction Model

Marcel Antal [1], Tudor Cioara [1,*], Ionut Anghel [1], Radoslaw Gorzenski [2], Radoslaw Januszewski [3], Ariel Oleksiak [3], Wojciech Piatek [3], Claudia Pop [1], Ioan Salomie [1] and Wojciech Szeliga [3]

[1] Computer Science Department, Technical University of Cluj-Napoca, Memorandumului 28,
    400114 Cluj-Napoca, Romania; marcel.antal@cs.utcluj.ro (M.A.); Ionut.anghel@cs.utcluj.ro (I.A.);
    claudia.pop@cs.utcluj.ro (C.P.); ioan.salomie@cs.utcluj.ro (I.S.)
[2] Faculty of Civil and Environmental Engineering, Poznan University of Technology, 60-965 Pozanan, Poland;
    radoslaw.gorzenski@put.poznan.pl
[3] Poznan Supercomputing and Networking Center, 60-965 Poznan, Poland; radekj@man.poznan.pl (R.J.);
    ariel@man.poznan.pl (A.O.); piatek@man.poznan.pl (W.P.); wojteks@man.poznan.pl (W.S.)
*   Correspondence: tudor.cioara@cs.utcluj.ro; Tel.: +40-264-202-352

Received: 22 January 2019; Accepted: 25 February 2019; Published: 1 March 2019

Abstract: This paper addresses the problem of data centers' cost efficiency considering the potential of reusing the generated heat in district heating networks. We started by analyzing the requirements and heat reuse potential of a high performance computing data center and then we had defined a heat reuse model which simulates the thermodynamic processes from the server room. This allows estimating by means of Computational Fluid Dynamics simulations the temperature of the hot air recovered by the heat pumps from the server room allowing them to operate more efficiently. To address the time and space complexity at run-time we have defined a Multi-Layer Perceptron neural network infrastructure to predict the hot air temperature distribution in the server room from the training data generated by means of simulations. For testing purposes, we have modeled a virtual server room having a volume of 48 m^3 and two typical 42U racks. The results show that using our model the heat distribution in the server room can be predicted with an error less than 1 °C allowing data centers to accurately estimate in advance the amount of waste heat to be reused and the efficiency of heat pump operation.

Keywords: data center; heat reuse; Computational Fluid Dynamics; prediction algorithm; neural networks

---

1. Introduction

Nowadays data centers (DCs) are subjected to significant pressure to perform more efficiently from an environmental perspective towards generating carbon-neutral benefits. The DC industry is investing in finding effective ways to improve energy efficiency. The challenge is how to turn the environmental focus into a long-term business opportunity by finding new revenue streams. The electricity consumed by the IT infrastructure to implement the DCs' core mission, which is to reliably execute their clients' workload, is almost completely converted into heat. Additionally, even in well designed DCs, the cooling system consumes almost 37% of the total energy demand to maintain the temperature set points for the servers' safety [1]. The H2020 CATALYST project [2] vision is to achieve cost and environmental efficiency by integrating DCs with the districting heating infrastructure transforming them into active players in the thermal energy value chain.

Naturally, in this context solutions for re-using the otherwise wasted heat of a DC in nearby neighborhoods have been proposed, however, even though there is a big potential for DC heat reuse there are still technological challenges that need to be addressed. The first challenge is the low quality of waste heat extracted from DC, while the second one is the efficiency in the operation of the heat pumps [3]. This is, in particular, true for air-cooled DCs using electrical cooling systems where the heat extracted from the server rooms has a temperature below 40 °C [4]. To make heat usable and marketable the DCs have to install heat pumps that are able to raise heat temperature to around 80 °C which allows heat transportation over longer distances to nearby buildings [5]. In the latter case, the heat pump consumes energy in the process of increasing heat quality [6]. The more efficient a heat pump is, the less energy it will consume and it will be more cost-effective for a DC to operate it. One of the factors influencing this is air temperature in the server room. The higher the temperature the less energy the heat pump will consume. Thus, methods for increasing the server room temperature set points have emerged allowing the temperature of the air extracted in the server room to increase. However, this leads to the third issue related to the generation of server room hotspots, that can lead to emergency stops or server failure. To avoid such hazardous situations, the server room and cooling system settings and configurations need to be properly evaluated by simulations allowing the DC operators to properly assess their impact in terms of heat distribution, prior of making them effective. We address the above-presented issues by bringing the following novel contributions:

- Definition of a heat reuse model for DCs allowing them to estimate in advanced the amount of generated waste heat and the impact on the efficiency of the heat pump operation;
- Definition of Computational Fluid Dynamics (CFD) models to simulate the thermodynamic processes inside the server room and estimate the temperature of the hot air generated;
- Development of neural networks algorithm to predict the heat distribution in the server room from training data generated using the CFD simulations, making our model feasible for near real-time decision making;

By using the proposed approach, the DC operators will be able to accurately forecast the temperature of the hot air recovered from the server room and the amount of waste heat that might be reused. They will also be able to compare and contrast additional investment costs with incremental revenues that can be achieved from valorizing forecasted waste heat.

The rest of the paper is organized as follows: Section 2 describes the relevant related work in the area of DC heat grid integration focusing on the approaches for modeling the thermal processes and predicting the heat distribution inside the server room. Section 3 presents an analysis of the heat reuse potential of the Poznan Supercomputing and Networking Center (PSNC) Section 4 details our proposed DC waste heat reuse model which combines the Computational Fluid Dynamics with neural network based prediction infrastructure and Section 5 presents evaluation results for a virtual server room with two racks. Section 6 concludes the paper and presents suggestions for relevant future work.

2. Related Work

Several solutions proposed in the literature tackle the reutilization of DCs' waste heat for heating-up closely located houses, apartments and offices [6–8]. District heating (DH) networks, already identified as an important need the intelligent heat distribution process that must take advantage of the third party-generated heat (DCs are candidates for this process) [9]. Other approaches are aimed at providing heat-oriented ancillary services which involve forecasting methods to identify heat demand patterns [10,11]. Modeling and simulation techniques are defined in [12,13] with the objective of reusing and transporting thermal energy within DH networks. These can help to evaluate limitations, benefits, and costs and serve as a preliminary feasibility study before actual solution implementation. Another aspect to be considered is the reduction of emissions generated for peak load production of thermal energy usually produced with fossil fuels [2].

In the direction of DCs integration with heat networks, modeling thermal processes related to the computing infrastructure and discovering their impact on the surrounding environment has recently gained in popularity. A constant increase in spatial and thermal density of computing systems make their evaluation more important but also more challenging. For now, Computational Fluid Dynamics (CFD) simulations are seen as the most suitable solution. In contrast to experimental examinations, CFD allows obtaining the volumetric field of many physical variables. It is also much more convenient and cheaper to simulate varying scenarios than evaluate them in the real world. Numerous examples of studies of airflow inside the server room with CFD analysis can be found in the literature. A detailed CFD analysis of various air distribution systems and their cooling efficiency was described in [14], while in [15] the impact of air conditioning failures and fluctuations of servers' power were considered. Other studies [16] were focused on the analysis of optimal airflow angle through supply tiles for server rooms with raised floor cooling system. Nevertheless, the CFD-based solutions are effort-intensive for model preparation and time consuming for gaining good results, which makes them inadequate for complex system simulations and analysis of numerous configurations. As an alternative, the Potential Flow Model [17] and orthogonal decomposition methodology [18] have been proposed to reduce the complexity of the initial CFD model. Another approach is leveraging on analytical models which are applied directly within the process of DC thermal evaluation or to the simulation toolkits. The power and thermal models proposed in literature correspond to different levels in the hierarchy of resources within DC. In [19] the authors discussed the power models of a server and its subcomponents, while the thermal behavior of a server is analyzed in [20]. The models define the temperature of the processing unit, as well as the changes in the temperature at the server's outlet. Extension to these models, including the power leakage phenomena is described in [21]. Moreover, in [20] also the power models for the whole data center are proposed, together with the corresponding cooling models [22]. A simplified version of the cooling model was presented [23], while [24] introduced the concept of a heat distribution matrix specifying the heat recirculation between the servers. It defines the impact of hot air leaving the server to its inlet. Authors showed also how a combination of CFD simulation, together with a heat flow model and analytical data can contribute to the overall thermal analysis of a DC. An example of the complex simulation environment for the assessment of DC energy and thermal efficiency is the SVD Toolkit being, the result of the CoolEmAll project [25]. The toolkit integrates analytical simulations and the CFD modeling to the impact of different DC's management policies, hardware configurations and intensity of workloads. In this way, it enables energy-efficiency and heat-efficiency optimization with respect to the common metrics.

However, to perform a complex evaluation of the DC, including analysis of the great number of possible states and with plenty of parameters, another approach, which benefits from the aforementioned solutions, is required. Thus, in this paper, we built upon the existing state of the art to present a combination of CFD simulations with a neural network-based prediction infrastructure to allow the forecast of temperature distributions in server rooms considering a high number of different cases. Our approach is rolling in the black-box prediction methodologies which allow for learning a prediction model at of training data without any information on the underlying physical processes. The prediction model, once learned, features low computational and time overhead, and could be integrated with a proactive DC management strategy to control workload allocation and cooling system settings for adapting the DC heat generation in order to meet different heating goals [1].

Few learning-based approaches are described in the research literature. In [26,27] such algorithms are trained to predict the behavior of the DC cooling system. In [27] data measured from a real DC is used for training a feedforward and a dynamic recurrent artificial neural network and results are compared against a CFD simulation. The proposed system has comparable accuracy with the CFD simulation but a much faster convergence time and can be incorporated in real-time control strategies. Similarly, in [28], a neural network solution is proposed for predicting DC temperatures. The neural network approach has the advantage that it can learn the environment continuously by adapting its parameters each time new environment data is taken from sensors. The neural network can be initially trained either with real environmental data or with data generated by a CFD simulation for the DC model. In [29] a novel thermal forecasting method is developed which can predict server temperatures using data streams acquired from sensors. Compared with CFD-based solutions the proposed one is simpler and uses a gray-box model of the underlying physical systems to model the thermodynamic processes and sensor data together with an algorithm that continuously adapts the models to new monitored data. Evaluation results show better prediction accuracy than standard approaches driven only by data and identifying risky situation 4 minutes before they appear. Similarly in [30] the authors use the trained model as input to a thermal-aware scheduling algorithm and evaluated its performance through simulations.

3. PSNC Data Centre Example

The DC IT infrastructure executes the clients' workload and as result, heat is generated and it accumulates in the server room. All consumed electricity of DC IT infrastructure is ultimately converted into heat that needs to be dissipated by an electrical cooling system in order to maintain the temperature inside the server room under pre-defined set points for safe operation of servers. As the IT servers design is continuously improved for operating at higher temperatures and the server room density continues to rise, the DCs will become important producers of waste heat.

One of such DC is the PSNC [31] which already uses part of the heat produced by its DC to provide heating for offices (for around 300 people) located within the same building. PSNC is located nearby a campus of the Poznan University of Technology (PUT) so further potential use of the remaining waste heat was identified. Analysis of this case provides motivation and requirements for the models and methods of heat reuse prediction and optimization proposed in this paper.

3.1. Heat Reuse Within the Building

PSNC's DC covers an area of 1600 m^2 with a maximum (possible) power capacity equal to 2 MW (and with the possibility to extend the supplied power up to 16 MW). However, the usual, average power drawn by DC is around 0.9 MW. IT infrastructure is both liquid and air-cooled (where the liquid cooling is used for the HPC part of the DC) resulting in a Power Usage Effectiveness (PUE) of 1.3. Currently, around 400 kW is used by the system using a Direct Liquid Cooling (DLC) approach. Shortly, the DC will be extended by another DLC system leading to an increase the mean power usage (>1 MW). The average resource usage oscillates around 70%, however, the utilization of nodes of the largest HPC system reaches often 90%. In the liquid cooling system, the inlet/outlet temperatures of the coolant are up to 35/45°C in the summer and 20/30 °C, respectively, in the rest (around 70%) of the year. Figure 1 shows the amount of accumulated heat, generated in DC within the period of two months between September and November.

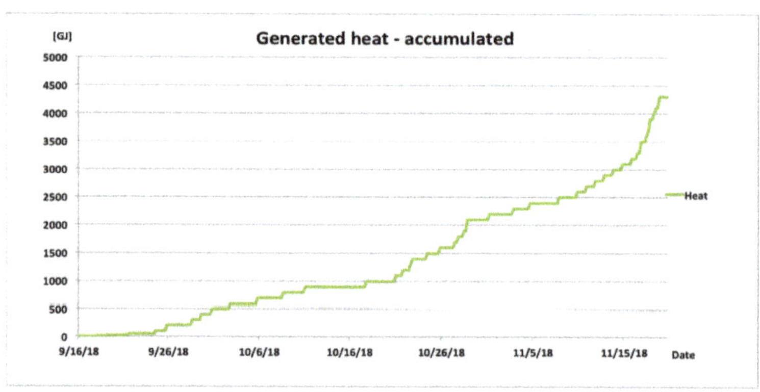

Figure 1. Heat characteristics data for the PSNC's DC.

One should note that the total increase in the amount of heat within the evaluated period was around 4200 GJ, which corresponds to the average power drawn by the DC. Currently, the heat from

DC is used for heating the whole PSNC building. PSNC's facilities are equipped with a 300 kW heat exchanger. The total heat reused is within 200–300 kW in the winter season. The heat is reused from both low and high-temperature loops. The following chart (Figure 2) shows the heat supplied (from the DC) to the PSNC building and the external temperature characteristics within the evaluated time slot.

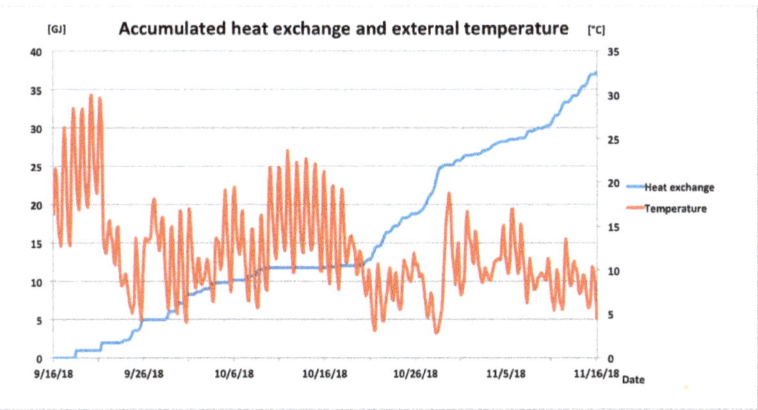

Figure 2. PSNC's heat exchanger data.

Also, the more rapid increases in the amount of heat exchanged correspond the drops in the external temperatures. The total amount of heat exchanged in the analyzed period was around 40 GJ, which is far below the capabilities of the current heat exchanger.

3.2. Heat Reuse in the Nearby Neighbourhood

In the neighborhood of the PSNC building, there is a PUT campus. The PSNC building is thus located near three main buildings: the Faculties of Architecture and Engineering Management Building (building C), Faculty of Chemical Technology Building (building B) and Library and Lecture Center Building (building A), all only at most 400 meters away. This can be seen in Figure 3. Buildings A and B are equipped with air cooled water chillers and coupled with heating substations connected to the Poznan district heating network, with a contracted power demand of 1027 kW and 500 kW respectively. Building B is also equipped with 360 kW heating power Ground Source Heat Pumps (GSHPs). In building C, a nearly zero-energy one, cooling and space heating is provided by a low temperature thermally activated building system and heat is generated only with the GSHPs.

Heating energy demand is covered with the use of GSHP by 100% for building C and by 80% for building B. Heat generated with GSHPs costs 10 €/GJ, while heat from the district heating network costs twice this price. 100% of the heating energy for building A (5430 GJ annually) and 20% for building B (843 GJ annually) is provided by heating substations, resulting in a total of 6273 GJ annually.

A water loop of 250 kW and 5 °C supply/return temperature difference between buildings B and C was designed, as both B heating station and water chillers are oversized. Two operating modes are considered:

- Heat peak source for building C—building C heating system is assisted with heat from building B during winter;
- Building C cooling—chilled water from building B chillers and heat pumps is supplied to building C cooling system.

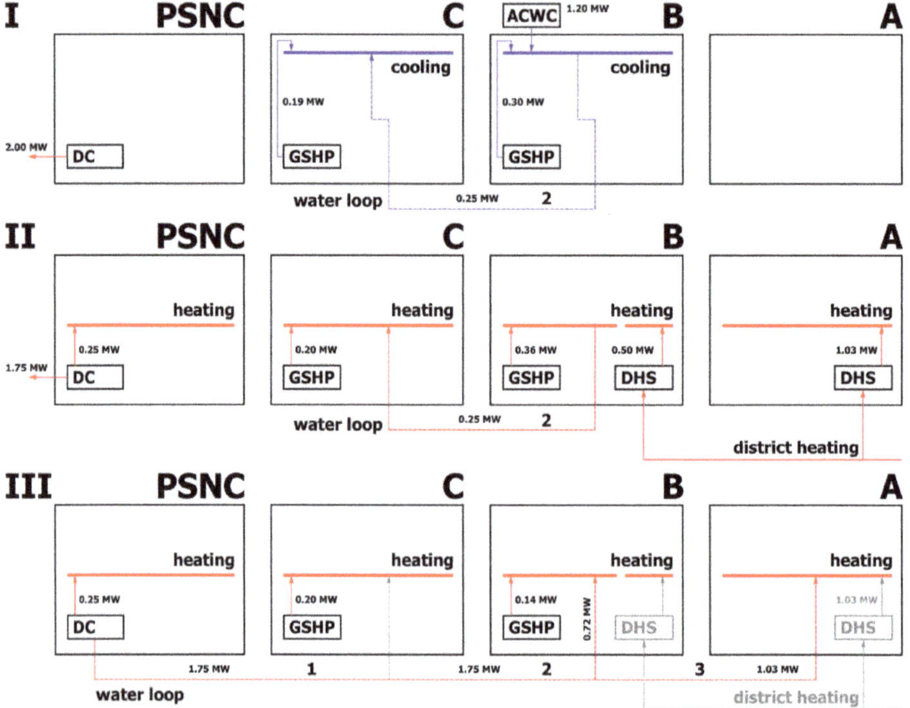

Figure 3. Main heat fluxes for PSNC and PUT A, B and C buildings, (**I**)—summer period, (**II**)—winter period with water loop #2, (**III**)—winter period with #1, #2 and #3 loops, DHS—district heating substation, ACWC—air cooled water chiller.

A detailed design for the water loop was prepared and it is to be built in the first half of 2019 when building C will be put into use.

As the heat production of PSNC exceeds its energy need subsequent considerations on capabilities of heat transfer between PSNC and PUT buildings were conducted. It was assumed that 700 or up to 1000 kW heat could be transferred to buildings B and C. Two subsequent water loops between PSNC and building C and between buildings A and B should be built together with local storage tanks and pumping stations (Figure 4).

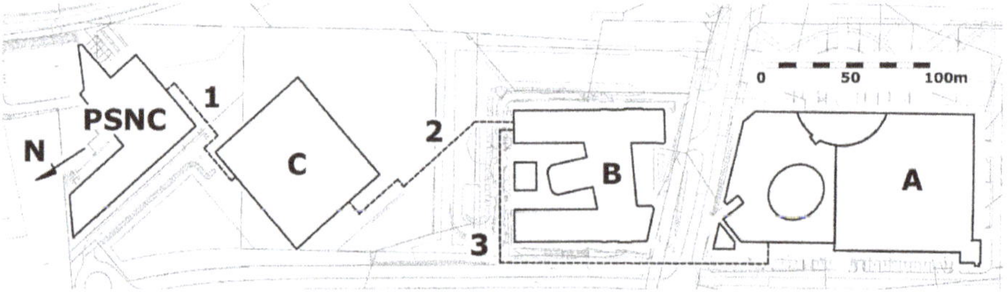

Figure 4. Loop connection schema for the considered buildings.

Supply/return water temperature for building B heating system is 60/45 °C and 80/60 °C for building A, built earlier with less thermal protection. Building C heating system with a thermally activated building system is a very low-temperature one with temperatures of 35/30 °C.

PSNC utilizes DC waste heat with a local multi-split and variable refrigerant volume systems. Water-to-water or air-to-water heat pump should be used in PSNC to increase the temperature of air or chilled water returning from the DC to slightly above the building supply temperatures to thus cover the heat losses in water loops and in the heat exchangers in each building. Exchangers are necessary to separate the water systems. To obtain higher supply temperatures additional boilers or second heat pump could be used in a particular building. The total length of the loop, made with pre-insulated pipes, will be less than 500 m.

Taking into account the current heat demands of the PSNC building and the upcoming enlargement of the DC's computational capabilities, there are promising prospect to fulfill the PUT heat demands at least partially by using the PSNC facilities. However, the heat demand of the neighborhood may vary significantly depending on external temperatures. Additionally, DC heat generation and outlet temperatures may differ depending on load, temperatures, and cooling system configuration. Therefore, the optimal configuration of such a system requires accurate prediction of heat production and the state of the system. Models and methods for such accurate prediction of the DC thermal flexibility are studied in the following sections.

4. Optimizing DC Heat Reuse

4.1. Heat Energy Harvesting Efficieny

The potential value of residual heat is partially determined by applied cooling technology from where the energy has been gathered. The heat produced by the servers has low temperature, due to the low temperature range of their safe operation ($T_{HOTAIR} \leq 30$ degrees Celsius) but at the same time it is the simplest and most efficient method to recover the waste heat and reuse it for domestic heating. However, district heating networks usually require temperatures above 60–70 degrees Celsius. The thermal energy must either be harvested at a higher temperature; this being possible only for liquid cooling systems (that can reach a temperature of 60 degrees Celsius) or passed through a heat pump to increase its temperature ($T_{DISTRICT}$) by using a refrigerant cycle that consumes electrical energy. Consequently, DCs that rely mostly on air cooling systems are equipped with heat pumps to raise the temperature of the reused heat.

The heat pump has two coefficients of performance indicators (COP): one defined for the cooling process of the air to be fed back to the DC server room through the perforated floor and one defined for the heating process to deliver in nearby neighborhoods:

$$COP_{Cooling} = \frac{\Delta Q_{Cool}}{E_{Compressor}} \leq \frac{T_{COLDAIR}}{T_{HOTAIR} - T_{COLDAIR}} \tag{1}$$

$$COP_{Heating} = \frac{DQ_{Hot}}{E_{Compressor}} \leq \frac{T_{DISTRICT}}{T_{DISTRICT} - T_{HOTAIR}} \tag{2}$$

where $E_{Compressor}$ is the heat pump compressor energy consumption. In other words $COP_{cooling}$ characterizes the process done by the pump to cool the server room, while $COP_{Heating}$ characterizes the process done by the heat pump to increase temperature of supplied heat and to transfer it.

When an air-cooled DC uses heat pumps to dynamically generate thermal energy for heating purposes, the server room is used as a thermal energy buffer that allows increasing or decreasing the heat produced (see Figure 5). This is achieved by modifying the temperature set-points in the server room and deploying and executing more workload on the IT servers to allow thermal energy to accumulate. This will also improve the efficiency of the heat pump operation which operates better at higher temperatures. As it can be seen in relations (1) and (2) minimizing the compressor's dissipated work $E_{Compressor}$ leads to maximizing both $COP_{Heating}$ and $COP_{Cooling}$ and this happens when the temperature of the air extracted from the server room T_{HOTAIR} is high.

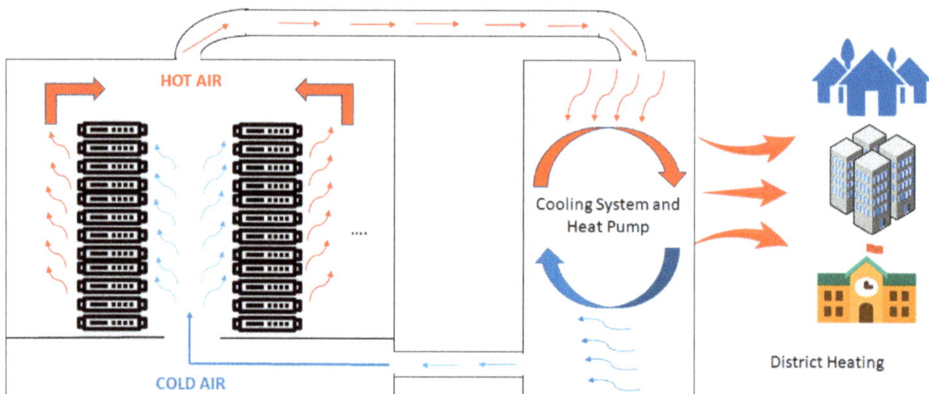

Figure 5. DC electrical air cooling infrastructure and heat reuse.

However, increasing the server room temperature set points can lead to potentially disastrous situations, such as equipment overheating and malfunctioning due to hotspot formation in certain areas of the server room. To avoid these risks, before making any decisions a simulation should be performed to validate the new temperature set-points and workload distribution of the IT servers, eliminating the risks of hotspot formation and equipment malfunction. Thermal processes are highly complex and need dedicated simulation tools to achieve accurate and realistic results. The most used methods are based on Computational Fluid Dynamics (CFD) techniques that use numerical simulations to compute the flow of fluids (liquids or gasses) in an area defined by boundary surfaces. Applied in DC thermal distribution simulation, CFD tools report an error of about 1 degree Celsius compared to the real environment, thus being suited for decision analysis regarding server room temperature set-point.

At the same time, CFD-based simulation methods have their own limitations when applied to thermal flexibility studies. Calculations related to a single scenario require a parallel environment and a significant amount of time. Performing CFD for multiple scenarios for every server room in a large-scale DC is costly and time-consuming, being an approach that cannot be considered as affordable and effective in near real time. Thus, to overcome such limitations, in our study we propose to combine CFD simulation with neural networks based methods to predict the temperature distribution in the server room aiming to increase the amount of heat recovered. In this case, the simulations result of numerous configurations of the single simplified setup will constitute the training data set for the neural network based prediction process, which will be used at run time for decision making.

4.2. Server Room Thermal Model

CFD simulations of a virtual server room are used to model the input cold air in terms of pumped airflow and temperature as well as the generated hot air and heat distribution in the server room model for multiple setups. Both the model and the subsequent simulations are prepared with an in-house tool dedicated to the server room CFD analysis, based on the open-source OpenFOAM v4.1 software [32]. To obtain the dynamics of the server room airflow, unsteady Reynolds-averaged Navier-Stokes (URANS) equations together with the two-equational k-ε turbulence model are resolved.

The utilized solver, buoyantBoussinesqPimpleFoam, allows capturing buoyancy effects, taking into account pressure gradient along the vertical dimension, describing static pressure p as:

$$p = p_{rgh} + \rho gh \tag{3}$$

where: p_{rgh} is the pseudo hydrostatic pressure, q is the air density, g the gravitational acceleration, h the height in the opposite direction to gravity. The solver introduces also the Boussinesq approximation which assumes linear dependency of density and temperature variations. It is applicable to the server room airflow due to relatively slight density fluctuations expected. The approximation reduces nonlinearity and simplifies solved problem. The effective density q_k, present in the gravity term of momentum equation, is expressed with the following equation:

$$\rho_k = 1 - \beta\left(T - T_{ref}\right) \tag{4}$$

where β is the thermal expansion coefficient, T the temperature, T_{ref} the reference temperature. Such an approach makes the solver incompressible. The solver uses the PIMPLE pressure-velocity coupling algorithm to obtain a transient solution, which is a combination of Pressure Implicit with Splitting of Operator (PISO) and Semi-Implicit Method for Pressure-Linked Equations (SIMPLE) schemes.

The server room is modelled considering its size (width × depth × height) as well as the racks deployed in it. In general, the utilized CFD tool allows modelling a rack both as a set of separate servers of any occupation pattern or as a simplified single device. Due to performance reasons, we chose the second approach. The racks are considered as fully and homogeneously occupied with servers, their geometry is simplified in a way that each of the racks is treated as single airflow and heat source. They are provided with one inlet and one outlet each, allowing the use of less computationally expensive mesh. Cold air is supplied into the room through the floor surface whereas the return air stream leaves through the ceiling. The intention of such Computer Room Air Conditioning (CRAC) modelling was to create a possibly generic case, independent from location and number of floor supply tiles and return outlets. The virtual server room model used in this study is presented in Figure 6.

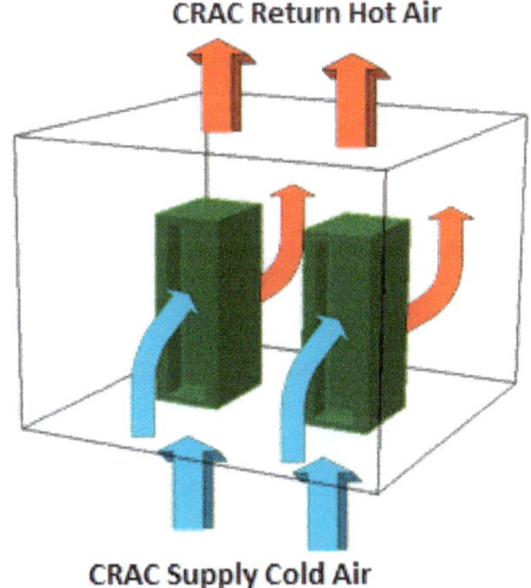

Figure 6. Arrangement of virtual server room in case of 2 racks.

To properly model each rack as heat sources and to keep the power of racks at a constant level during every simulation execution, swak4Foam [33] library (Swiss Army Knife for Foam) was utilized. It allows binding boundary conditions with functional dependency of one from another with groovyBC utility. While at racks inlets the temperature was forming freely, at racks' outlets, on the other hand, the temperature boundary condition was characterized as a function of inlet temperature, to fulfil below heat balance equation:

$$\rho_k = 1 - \beta\left(T - T_{ref}\right) \tag{5}$$

where P_r is the heat output of rack, ρ_{ref} is the reference density (1.1884 kg/m^3 for reference $p_{rgh} = 100$ kPa and $T_{ref} = 20\ °C$), Q_r is the rack airflow volume (2.0 m^3/s, ≈0.048 m^3/s per U), C_p is the specific heat capacity (1003.98 J/kgK), T_{HOTAIR} is the rack outlet temperature and finally $T_{COLDAIR}$ is the rack inlet temperature. For this study, it is assumed, that 100% of racks power consumption is responsible for heat generation. The final form of the outlet temperature equation is stated below:

$$T_{HOTAIR} = T_{COLDAIR} + \frac{P_r}{\rho_{ref} Q_r C_p} \tag{6}$$

4.3. Predicting the Heat Distribution

The CFD experiments described above compute the temperature of the hot air generated in the server room (T_{HOTAIR}) considering M virtual probes, the simulation outputs $T_{HOTAIR}^{Probe[m]}[t]$ where $m \in \{1 \ldots M\}$ and t is the time instance when the data was taken (see Figure 7). The probes are deployed at the inlet and outlet of the K racks as well as at the outlet of the server room. The simulation input is defined by the initial room temperature ($T_{ROOM-INITIAL}$), the temperature of the air pumped by the cooling system in the server room ($T_{COLDAIR}$) and its flow (air_{FLOW}) and the heat generation of the K racks modelled ($Heat_{Rack}[k]$, $k \in \{1 \ldots K\}$).

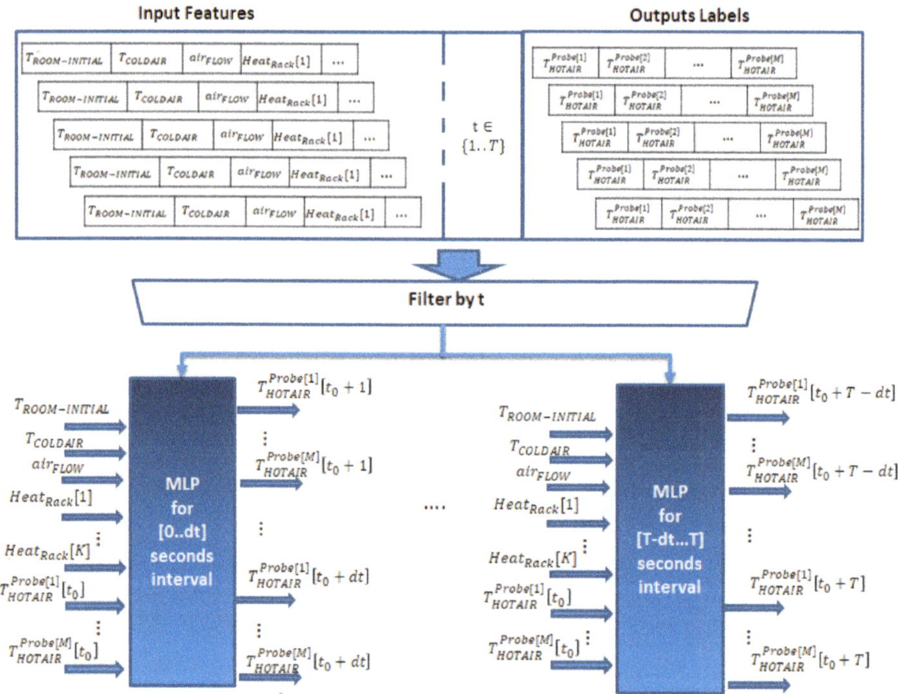

Figure 7. Structure of the neural network infrastructure and data set used in the forecasting process.

The CFD input parameters are varied to generate N different scenarios, each run over an interval of T seconds to compute the temperatures for the M probes. The generated data are fed to a neural network based infrastructure with the goal of predicting the heat distribution in the server room and the temperature of the hot air generated. The forecasting infrastructure is based on neural networks of type Multi-Layer Perceptron (MLP) featuring 4 fully-connected layers of neurons of type ReLU (Rectified Linear Units). The input layer has $K + 3 + M$ neurons, the other two hidden layers have $\alpha(K + 3 + M) + \beta$ neurons, while the output layer has $dt \times M$ neurons, where dt is the number of seconds for which each MLP will predict the probe temperatures.

Because we need predictions for every second in the $[0 \ldots T]$ forecasting interval, the developed infrastructure has $M_T = \frac{T}{dt}$ MLP modules trained with data filtered for a specific second (t in Figure 7) to compute temperature predictions of the M probes for that time instance. Each of the T MLPs has as input the $K + 3$ parameters ($T_{ROOM-INITIAL}$, $T_{COLDAIR}$, air_{FLOW}, $\{Heat_{Rack}[i], i \in \{1 \ldots K\}\}$) describing the initial situation of the simulation and additional M parameters describing the temperature probe values at the beginning of the prediction window $T_{HOTAIR}^{Probe[m]}[t_0]$. Each MLP module aims at predicting the temperature of the M key points representing the probes deployed in the server room for every timestamp t in each time interval of the form $MLP_{domain}(i) = [i \times dt, (i+1) \times dt]$, where $i \in \{0 \ldots M_T - 1\}$. Thus, for computing the all the hot air probe temperatures at second t in the future, the index i of the interval $MLP_{domain}(i)$ has to be computed as the integer part of the ratio $i = [\frac{t}{dt}]$, and the temperatures will be predicted by the ith MLP as its $(t - i \times dt)$ outputs:

$$T_{HOTAIR}^{Probe[m]}(t_0 + T) = MLP_i\left(T_{ROOM-INITIAL}, T_{COLDAIR}, air_{FLOW}, Heat_{Rack}[K], T_{HOTAIR}^{Probe[M]}[t_0]\right)[t - i \times dt] \quad (7)$$

For looking forward into the future and predicting the server temperature at deployed probes over periods larger than the CFD simulations of T seconds, we proposed the iterative algorithm shown below (Algorithm 1). The algorithm inputs are the initial parameters of a simulation and the time for which the prediction has to be computed, while the outputs are the M simulation output parameters after the simulation time $Time_{prediction}$.

Algorithm 1. Heat Prediction Algorithm for Long Time Periods

Input: $T_{ROOM-INITIAL}$, $T_{COLDAIR}$, air_{FLOW}, $Heat_{Rack}[k]$, $Time_{prediction}$
Output: $T_{HOTAIR}^{Probe[m]}\left[Time_{prediction}\right]$, $m \in M$
Begin
$Intervals_T = \frac{Time_{prediction}}{T}$ // iterate over the interval $\left[0 \ldots Time_{prediction}\right]$ with time step of length T
$Index_T = Time_{prediction} \% T$ // compute time interval remaining that cannot be covered by time steps of length T
$I_{MLP}^{iterations} = \frac{T}{dt}$ // compute MLP index within T interval
$Output_{MLP}^{iterations} = T \% dt$ // compute output index within the MLP
$I_{MLP}^{final} = \frac{Index_T}{dt}$ //compute MLP index within last $Index_T$ interval
$Output_{MLP}^{final} = Index_T \% dt$ //compute output index within last MLP
$$t_0 = 0$$
For ($time = 1$ to $Intervals_T$) Do
For ($m = 1$ to M) Do
$\quad T_{HOTAIR}^{Probe[m]}[t_0 + T] = MLP_{I_{MLP}^{iterations}}\left(T_{ROOM-INITIAL}, T_{COLDAIR}, air_{FLOW}, Heat_{Rack}[K], T_{HOTAIR}^{Probe[m]}[t_0]\right)[Output_{MLP}^{iterations}]$
$t_0 = t_0 + T; m = m + 1$
End For
$\quad T_{HOTAIR}^{Probe[1\ldots M]}\left[Time_{prediction}\right] = MLP_{I_{MLP}^{final}}\left(\begin{array}{c} T_{ROOM-INITIAL}, T_{COLDAIR}, air_{FLOW}, Heat_{Rack}[K], \\ T_{HOTAIR}^{Probe[M]}[Intervals_T \times T] \end{array}\right)[Output_{MLP}^{final}]$
Return $T_{HOTAIR}^{Probe[1\ldots M]}\left[Time_{prediction}\right]$
End

The algorithm starts by computing the number of T second intervals included in the prediction period $Time_{prediction}$. Furthermore, it computes the index of the MLP used to predict T seconds in the future, and the index of the output of the MLP corresponding to this prediction. It also computes the length of the last interval that cannot be covered by intervals of length T, the index of the MLP and the index of the MLPs output to compute the last prediction. Then, it performs iterations over these intervals and for each interval it computes the average room temperature in the M probes, while updating the time for the next prediction. Finally, it calls the MLP for the last time interval of length I_{MLP}^{final} to compute the final values of the probes.

5. Evaluation Results

We had used the proposed DC heat reuse model to investigate the thermal energy flexibility potential of a virtual server room from PSNC DC modeled with CFD aiming to predict the temperature of the air inside which can be recovered using a heat pump and delivered to nearby neighborhoods. The virtual server room modeled with CFD has the size of 4 m × 4 m × 3 m (width × depth × height) and is equipped with two typical 42U racks (0.8 m × 1.07 m × 2.05 m respectively). They are provided with one inlet and one outlet (width: 0.45 m, height: 42U ≈ 1.8669 m), allowing the use of less computationally expensive mesh. Figure 8 reflects the virtual server room in OpenFoam environment. It was assumed, that between all the scenarios, five key parameters are changed, according to Table 1.

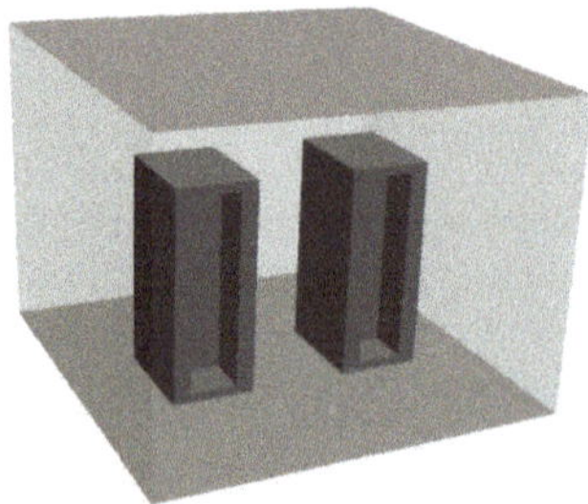

Figure 8. The CFD model of the server room implemented in OpenFoam.

Table 1. List of parameters variation for simulation scenarios.

Parameter	Min	Max	Step
Initial room temperature: $T_{INITIAL-ROOM}$ [°C]	18	26	2
Air conditioning volume flow rate: air_{FLOW} [m³/s]	0.6	1.8	0.3
Air conditioning outlet temperature: $T_{COLDAIR}$ [°C]	10	18	2
Power consumption of rack no 1: P_1 [kW] $\cong HEAT_{Rack-1}$	2	10	2
Power consumption of rack no 2: P_2 [kW] $\cong HEAT_{Rack-2}$	2	10	2

From all received results, the temperature values at specific locations were extracted for every 1-s time interval of total 600-s (10 min) time range of every run. Virtual temperature probes were located at I/O surfaces in the domain, except CRAC supply outlet, where the temperature was fixed during the execution. List of probes together with their count and location is placed in Table 2.

Table 2. List of temperature probes in virtual server room.

Probed I/O Surface	Probe Count	Probe Location
CRAC return duct inlet	4	(1,1,3), (1,3,3), (3,1,3), (3,3,3)
Racks outlets	1 (× 2)	1.025 m above floor surface (half of the rack height, at outlet vertical symmetry axis)
Racks inlets	4 (× 2)	0.465 m, 0.838 m, 1.212 m, 1.585 m above floor surface (evenly spaced along vertical inlet dimension, at inlet vertical symmetry axis)

All possible CFD-based simulations (considering the Table 1 parameters) were executed and this generated over 200 GB of output data, from which probe data were extracted. Taking into account the number of probes $N_p = 14$ and the number of collected time steps per execution $N_t = 600$, the total number of temperature records per simulation is equal $N_{ts} = N_p \times N_t = 8400$ and the sum for whole described study $N_{tt} = N \times N_p \times N_t \approx 26\ M$. The generated data will constitute the training data of our prediction algorithm. Exemplary results for different air conditioning volume flow rate can be seen in Figure 9.

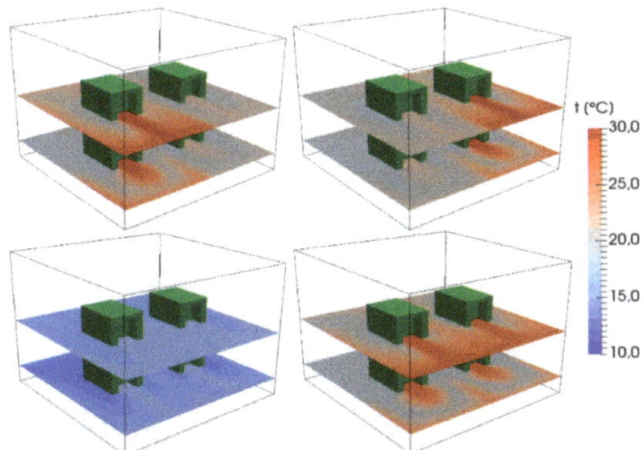

Figure 9. Various scenarios of air volumetric flow value with other parameters preserved constant; top-left: 1 m^3/s, top-right: 2 m^3/s, bottom-left: 3 m^3/s, bottom-right: 4 m^3/s.

We had varied the five input parameters of the CFD simulation defined in Table 1 to generate 3125 different scenarios, each scenario being run over a period of 600 s to compute the temperatures for the 14 probes defined in Table 3 at each second thus generating 8400 output values. Thus we had chosen to split the main dataset into 60 sub-datasets, each used to train the neural networks based algorithm to predict the 14 outputs of the simulation for every second of a 10 s time interval within the 600 s prediction interval. In total, we create 60 datasets for each scenario, that will be used to train 60 MLPs. Thus, for each second within each 10 s simulation interval we created 3125 training pairs, associating the parameters of each simulation with the output to be predicted at a given timestamp in the future. We train the MLP-based prediction infrastructure defined by splitting the data set in 90% training data and 10% test data.

Table 3. MSE of simulation outputs for the tested scenarios.

Output	Average MSE
$Rack^1_{inlet-1}$	0.028
$Rack^1_{inlet-2}$	0.029
$Rack^1_{inlet-3}$	0.034
$Rack^1_{inlet-4}$	0.046
$Rack^1_{outlet}$	0.019
$Rack^2_{inlet-1}$	0.025
$Rack^2_{inlet-2}$	0.036
$Rack^2_{inlet-3}$	0.040
$Rack^2_{inlet-4}$	0.041
$Rack^2_{outlet}$	0.035
$Room^1_{outlet}$	0.091
$Room^2_{outlet}$	0.087
$Room^3_{outlet}$	0.032
$Room^4_{outlet}$	0.038

The first evaluation aims at determining the temperature prediction performance over the test scenarios. For each scenario, the Mean Square Error (MSE) was computed for each predicted output defined in Table 3 over the 600 s simulation interval and it was compared against the reference CFD scenario:

$$MSE(output) = \frac{1}{312} \times \sum_{i=1}^{312} \frac{1}{600} (\sum_{t=1}^{600} (output(t) - MLP_{\frac{i}{60}}[i\%10])) \tag{8}$$

As it can be seen from the table above, the predictions exhibit an error of less than 0.1 that corresponds to less than 1 degree Celsius, being suited for real time temperature evaluation to avoid hot spots.

The temperature evolution prediction compared with the one generated by the CFD can be seen in Figure 10 for one of the scenarios run. The red lines represent the predicted values while the blue lines represent the real temperature values taken from the simulation. It can be seen for the sample probes even for the worst MSE the predicted temperature profile follows closely the real temperature profile, exhibiting errors of roughly 1 degree Celsius.

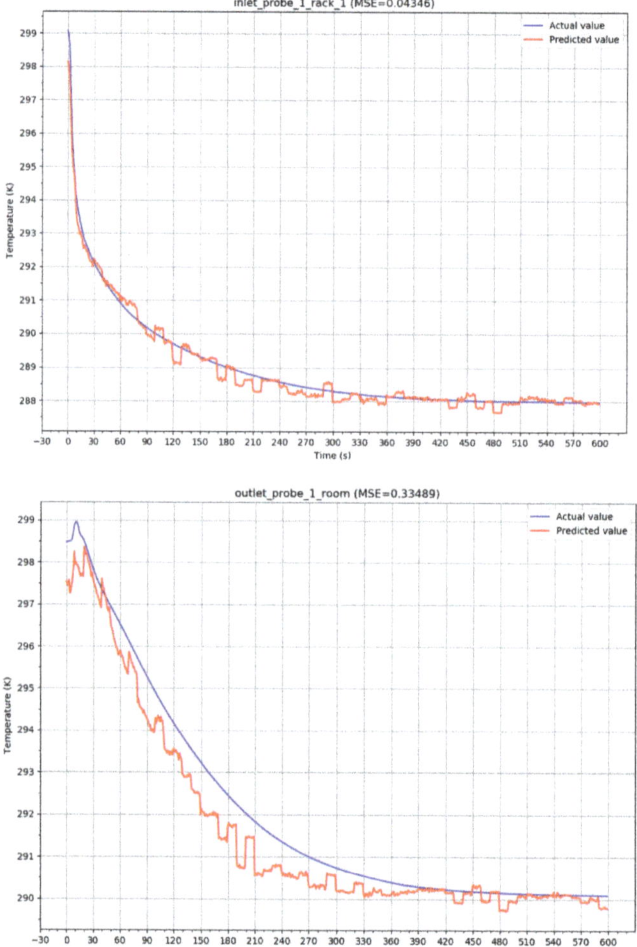

Figure 10. Predicted temperature (with RED) compared with the temperature determined with CFD simulation (with BLUE): TOP—T_{HOTAIR} at rack 1, inlet 1, probe 1; BOTTOM—T_{HOTAIR} at room outlet probe 1.

Figure 11 show the impact of prediction errors onto amount of waste heat estimated to be delivered to the heat grid in comparison with the actual ones. We have considered a heat pump featuring a $COP_{Cooling} = 3.8$ and a $COP_{Heating} = 2.3$ and the hot air temperatures in the server room reported in Figure 10 (BOTTOM). As it can be seen the impact is almost negligible showing that our solution could be successfully used by DC managers to accurately estimate the amount of available heat to be injected at future time frames. At the same time, the correlation between the amount of heat actually recovered from the server room the one delivered and indirectly with the temperature of the hot T_{HOTAIR} at server room outlet can be clearly seen. The higher the temperature of the hot air extracted is the higher the temperature of the heat delivered to the heat grid (assuming a constant energy consumption of the heat pump compressor) is too.

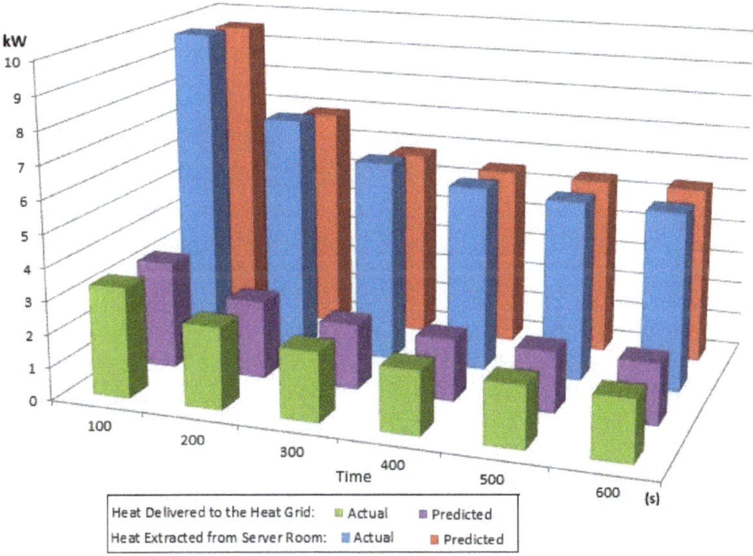

Figure 11. Amount of waste heat delivered to the heat grid and the heat extracted by means of heat pumps

In terms of the economics of heat reuse with respect to the requirements and example presented in Section 3.1, we have estimated the potential cost and savings related to the installation of the heat transfer system. The calculation was made on the following assumptions: (i) DC generated heat for winter period (based on data presented in Figure 1, linear estimation of winter period heat production is 7400 GJ annually) is much higher than total heat demand of PSNC buildings A and B (annual heat demand, based on the metered average value is less than 7000 GJ) and (ii) the DC waste heat is transferred to buildings A and B. Waste heat could be potentially be reused in building C, but detailed calculations of heat cost produced by heat pumps in the building will be conducted within a year, based on real exploitation, and then the final decision will be made. As the current design of buildings B-C water loop assumes only 250 kW power the pipes diameters should be increased to obtain higher power, about 1.75 MW.

District heating heat cost in PSNC, Poland is 20 €/GJ which gives 125,460 € annually since the yearly heating demand is 6273 GJ. There are three options of the heat consumption from district heating providers. In the first option the heat is still delivered by a district heating operator. For the second option, even if no energy is consumed from a district heating, maintaining the current levels of contracted power demand values, gives the annual heating station costs of 52,750 €. In the third option the contracted power demand is left on the same level as in present but only for a short period of time to make sure the system works properly. As the certainty of heat supply is important for PUT, this is the best option. Renouncing at the heating stations in the third option will reset the annual heating cost from the district heating to 0 €. The price of the heat from PSNC was assumed at 2 €/GJ and the cost of energy appliances (pumps, heat pumps) as well as maintenance costs at the level of 11,628 € annually. These are examples of values based on the assumption that the investment in the required infrastructure is done by PUT (hence even low PSNC heat price is profitable). In the next steps, other options will be considered, including separate ROI estimations for both PSNC DC and the PUT campus. A cost summary of these three options is presented in Table 4.

Table 4. Operational costs analysis for the three heat reuse options considered.

Annual Cost Components	Option 1 Contracted Power Demand 1.527 MW, DH Heat Demand 6273 GJ/year	Option 2 Contracted Power Demand 1.527 MW, DH Heat Demand 0 GJ/year	Option 3 Contracted Power Demand 0 MW, DH Heat Demand 0 GJ/year
A+B buildings district heating cost	125,460 €	52,750 €	0 €
PSNC DC heat revenues (6273 GJ annually, 2€/GJ)	N/A	12,546 €	12,546 €
Water loop running cost (pumps—electricity, maintenance)	N/A	11,628 €	11,628 €
Savings	N/A	48,536 €	101,286 €

In the consideration of capital expenditures, we assume that the investment in the required infrastructure is done by PUT. Water loop cost with heat pumps and pumping stations estimation is 1,150,000 €. Operational cost estimation was based on current district heating and electricity cost, whereas capital cost was based on real expenditures of the ongoing building C investment process. For Option 2 presented above (and in Table 4) with annual savings of 101,286 €, Simple Pay Back Time (SPBT) is 11.4 years, while Net Present Value (NPV) with 3% discount rate is 59,000 € in 15 years' period. With 50% EU investment subsidy SPBT would be 5.7 years while NPV 56,000 € in 7 years' period. Without subsidies and with district heating contracted power demand of 1.527 MW assurance (option 2 above) the installation of the heat transfer system is unreasonable due to SPBT of 23.7 years. Summary of these values is presented in Table 5 below. The potential use of existing pipes network of Poznan district heating could significantly reduce project costs and will be further investigated during future works.

Table 5. Capital expenditures analysis results for the three heat reuse options considered.

Option	SPBT	NPV (3% Rate)
No subsidies, DH contracted power demand 1.527 MW—Option 2 -	23.7 years	N/A
No subsidies, No DH connection (contracted power demand 0 MW)—Option 3 -	11.4 years	59,000 € (15 years)
50% subsidies, No DH connection (contracted power demand 0 MW)—Option 3 -	5.7 years	56,000 € (7 years)

6. Conclusions

In this paper, we have defined models and techniques for predicting the temperature of the air inside a server room which can be recovered and transferred by means of heat pumps to nearby neighborhoods. We proposed CFD simulations to evaluate a large number of scenarios which impact the temperature of the air. We defined a neural network-based prediction method which takes as input a large amount of data generated by means of CFD simulations to forecast the server room temperature overcoming the simulations' time and computational overhead at run time. The prediction process showed good performance having an error rate of less than 1% in predicting the server room temperature which represents less the 1 degree Celsius.

For the next steps, we plan to extend our work to the entire Poznan Supercomputing and Networking Center. This is in line with the current objectives of the collaboration between PSNC and Poznan University of Technology to perform a feasibility study of the heat transfer system between the PSNC DC to the PUT facilities. For now, we analyzed the current heat demands of PUT facilities, described existing infrastructure and showed corresponding economic analysis. The analysis has shown high potential. The return on investment time was initially estimated as 5.7–11.4 years depending on concrete heat demand/supply and price of heat defined by PSNC. However, current

PSNC heat production is not sufficient to cover the worst-case demand (around 1.5 MW when the temperature falls to $-18\,^{\circ}$C) even if it fits the typical demands (0.5–0.75 MW for temperatures within 0–8 $^{\circ}$C). For this reason, an approach with a connection to the heating network will be also considered, and accurate prediction and control of thermal flexibility presented in this paper is needed to achieve high efficiency and reliability.

Author Contributions: Conceptualization, T.C., M.A. and A.O.; Methodology, I.A., I.S. and R.G.; Software, M.A., C.P. and R.J.; Validation, M.A., W.P. and W.S.; Formal Analysis, T.C. and I.S.; Investigation, I.A. and C.P.; Resources A.O. and I.A.; Data Curation, R.G. and R.J.; Writing-Original Draft Preparation, T.C. and A.O.; Writing-Review & Editing, I.A. and W.P.; Visualization, C.P.; Supervision, I.S. and T.C.; Project Administration, T.C. and A.O.; Funding Acquisition, I.A. and T.C.

Funding: This research was funded by European Union's Horizon 2020 research and innovation programme grant number 768739 and from Polish National Science Center grant number 2013/08/A/ST6/00296. The APC was funded by the Technical University of Cluj-Napoca, Romania.

Acknowledgments: The results presented in this paper are partially funded from European Union's Horizon 2020 research and innovation programme under grant agreement No 768739 (CATALYST) and from Polish National Science Center under grant number 2013/08/A/ST6/00296.

Conflicts of Interest: The authors declare no conflict of interest. The funders had no role in the design of the study; in the collection, analyses, or interpretation of data; in the writing of the manuscript, and in the decision to publish the results.

References

1. Antal, M.; Cioara, T.; Anghel, I.; Pop, C.; Salomie, I. Transforming Data Centers in Active Thermal Energy Players in Nearby Neighborhoods. *Sustainability* **2018**, *10*, 939. [CrossRef]
2. H2020 Catalyst Project. Available online: http://project-catalyst.eu/ (accessed on 20 January 2019).
3. Lind, J.; Rundgren, E. Industrial Symbiosis in Heat Recovery Collaborations between Data Centers and District Heating and Cooling Companies. Available online: http://www.diva-portal.org/smash/record.jsf?pid=diva2%3A1130513&dswid=3389 (accessed on 10 January 2019).
4. Wahlroos, M.; Pärssinen, M.; Manner, J.; Syri, S. Utilizing data center waste heat in district heating—Impacts on energy efficiency and prospects for low-temperature district heating networks. *Energy* **2017**, *140*, 1228–1238. [CrossRef]
5. Antal, M.; Cioara, T.; Anghel, I.; Pop, C.; Salomie, I.; Bertoncini, M.; Arnone, D. DC Thermal Energy Flexibility Model for Waste Heat Reuse in Nearby Neighborhoods. In Proceedings of the 8th International Conference on Future Energy Systems (e-Energy '17), Hong Kong, China, 16–19 May 2017; pp. 278–283.
6. Davies, G.F.; Maidment, G.G.; Tozer, R.M. Using data centres for combined heating and cooling: An investigation for London. *Appl. Ther. Eng.* **2016**, *94*, 296–304. [CrossRef]
7. Brenner, P.; Go, D.B.; Buccellato, A.P.C. Data Center Heat Recovery Models and Validation: Insights from Environmentally Opportunistic Computing. In Proceedings of the ASHRAE Winter Conference Technical Program, Dallas, TX, USA, 26–30 January 2013.
8. Sarkar, J.; Bhattacharyya, S.; Ramgopal, M. Performance of a transcritical CO_2 heat pump for simultaneous water cooling and heating. *Int. J. Appl. Sci. Eng. Technol.* **2010**, *6*, 57–64.
9. Gelažanskas, L.; Gamage, K.A.A. Forecasting hot water consumption in residential houses. *Energies* **2015**, *8*, 12702–12717. [CrossRef]
10. Measurement of Domestic Hot Water Consumption in Dwellings. Available online: https://www.gov.uk/government/publications/measurement-of-domestic-hot-water-consumption-in-dwellings (accessed on 11 January 2019).
11. Parker, D.S.; Fairey, P.W. Estimating daily domestic hot-water use in North American homes. *ASHRAE Trans.* **2015**, *121*, 258.
12. Arce, I.H.; López, S.H.; Perez, S.L.; Rämä, M.; Klobut, K.; Febres, J.A. Models for fast modelling of district heating and cooling networks. *Renew. Sustain. Energy Rev.* **2018**, *82*, 1863–1873. [CrossRef]
13. Goumba, A.; Chiche, S.; Guo, X.; Colombert, M.; Bonneau, P. Recov'Heat: An estimation tool of urban waste heat recovery potential in sustainable cities. *AIP Conf. Proc.* **2017**, *1814*, 020038.
14. Cho, J.; Yang, J.; Park, W. Evaluation of air distribution system's airflow performance for cooling energy savings in high-density data centers. *Energy Build.* **2014**, *68*, 270–279. [CrossRef]

15. Alkharabsheh, S.; Sammakia, B.; Shrivastava, S.; Schmidt, R. Dynamic models for server rack and CRAH in a room level CFD model of a data center. In Proceedings of the 2014 IEEE Intersociety Conference on Thermal and Thermomechanical Phenomena in Electronic Systems (ITherm), Orlando, FL, USA, 27–30 May 2014.
16. Ni, J.; Jin, B.; Zhang, B.; Wang, X. Simulation of Thermal Distribution and Airflow for Efficient Energy Consumption in a Small Data Centers. *Sustainability* **2017**, *9*, 664. [CrossRef]
17. VanGilder, J. Real-Time Data Center Cooling Analysis. Available online: http://www.electronics-cooling. com/2011/09/real-time-data-center-cooling-analysis/ (accessed on 12 December 2018).
18. Ghosh, R.; Joshi, Y. Rapid Temperature Predictions in Data Centers using Multi-Parameter Proper Orthogonal Decomposition. *Numer. Heat Transf. Part A Appl.* **2014**, *66*, 41–63. [CrossRef]
19. Oleksiak, A.; Da Costa, G.; Piatek, W. Energy and Thermal Models for Simulation of Workload and Resource Management in Computing Systems. *Simul. Model. Pract. Theory* **2015**, *58*, 40–54.
20. Basmadjian, R.; Ali, N.; Niedermeier, F.; De Meer, H.; Giuliani, G. A methodology to predict the power consumption of servers in data centres. In Proceedings of the ACM SIGCOMM 2nd International Conference on Energy-Efficient Computing and Networking (e-Energy 2011), New York, NY, USA, 31 May–1 June 2011.
21. Piatek, W.; Oleksiak, A.; vor dem Berge, M. Modeling Impact of Power- and thermal-Aware Fans Management on Data Center Energy Consumption. In Proceedings of the 2015 ACM Sixth International Conference on Future Energy Systems, Bangalore, India, 14–17 July 2015.
22. Oleksiak, A.; Piatek, W.; Salom, J.; Siso, L.; Costa, G.D. Minimization of costs and energy consumption in a data center by a workload-based capacity management. In Proceedings of the 3rd International Workshop on Energy-Efficient Data Centres Co-Located with the ACM e-Energy, Cambridge, UK, 10 June 2014.
23. Chase, J.; Ranganathan, P.; Sharma, R.; Moore, J. Making scheduling "cool": Temperature—Aware workload placement in data centers. In Proceedings of the 2005 USENIX Annual Technical Conference, Anaheim, CA, USA, 10–15 April 2005.
24. Tang, Q.; Mukherjee, T.; Gupta, S.K.S.; Cayton, P. Sensor-based fast thermal evaluation model for energy efficient high-performance datacenters. In Proceedings of the Fourth International Conference on Intelligent Sensing and Information Processing, Bangalore, India, 15–18 December 2006; pp. 203–208.
25. Cupertino, L.; Da Costa, G.; Oleksiak, A.; Piatek, W.; Pierson, J.-M.; Salom, J.; Sisó, L.; Stolf, P.; Sun, H.; Zilio, T. Energy-efficient, thermal-aware modeling and simulation of data centers: The CoolEmAll approach and evaluation results. *Ad Hoc Netw.* **2015**, *25*, 535–553. [CrossRef]
26. Blarke, M.B.; Yazawa, K.; Shakouri, A.; Carmo, C. Thermal battery with CO_2 compression heat pump: Techno-economic optimization of a high-efficiency Smart Grid option for buildings. *Energy Build.* **2012**, *50*, 128–138. [CrossRef]
27. Kumar, V.A. Real Time Temperature Prediction in a Data Center Environment Using an Adaptive Algorithm. Master's Thesis, University of Texas Arlington, Arlington, TX, USA, December 2013. Available online: https://rc.library.uta.edu/uta-ir/handle/10106/24083 (accessed on 12 January 2019).
28. Kansara, N.; Katti, R.; Nemati, K.; Bowling, A.P.; Sammakia, B. Neural Network Modeling in Model-Based Control of a Data Center. In Proceedings of the International Electronic Packaging Technical Conference and Exhibition, San Francisco, CA, USA, 6–9 July 2015.
29. Li, L.; Liang, C.J.M.; Liu, J.; Nath, S.; Terzis, A.; Faloutsos, C. ThermoCast: A cyber-physical forecasting model for datacenters. In Proceedings of the 17th ACM SIGKDD International Conference on Knowledge Discovery and Data Mining, San Diego, CA, USA, 21–24 August 2011; pp. 1370–1378.
30. Wang, L.; Laszewsk, G. Task scheduling with ANN-based temperature prediction in a data center: A simulation-based study. *Eng. Comput.* **2011**, *27*, 381–391. [CrossRef]
31. Poznan Supercomputing and Networking Center. Available online: http://www.man.poznan.pl/online/en/ (accessed on 12 December 2018).
32. OpenFOAM. Available online: https://www.openfoam.org (accessed on 12 December 2018).
33. Gschaider, B. swak4Foam. Available online: https://openfoamwiki.net/index.php/Contrib/swak4Foam (accessed on 12 December 2018).

 *energies*

Article

An Influencing Parameters Analysis of District Heating Network Time Delays Based on the CFD Method

Jing Zhao * and Yu Shan

School of Environmental Science and Engineering, Tianjin University, Tianjin 300072, China; shanyu@tju.edu.cn
* Correspondence: zhaojing@tju.edu.cn; Tel.: +86-022-87402072

Received: 11 March 2019; Accepted: 2 April 2019; Published: 4 April 2019

Abstract: With the expansion of cities, district heating (DH) networks are playing an increasingly important role. The energy consumption due to the time delay caused by the transport of the medium in the DH network is enormous, especially in large networks. The study of time delay is necessary for the operation and optimization of DH networks. Compared with previous studies of constant flow rates and ideal pipeline (without regard to branches, elbows, variable pipe diameters, etc), this paper simulates a DH network in Tianjin University, China, by establishing the actual engineering model in a Computational Fluid Dynamics (CFD) method to analyze the time delay. The CFD model has great advantages in terms of computational cost and application range compared to theoretical calculations. The peak-valley method was used to verify the correctness of the time delay simulation model. Results show that the time delay calculated by the CFD model is consistent with the actual time delay obtained from the measured data. Based on this model, the parameters that affect the time delay are furtherly analyzed. Four key parameters, including flow rate, pipe length, pipe diameter, and water supply temperature are summarized. The results show that the flow rate, pipe length and pipe diameter have a great influence on the time delay of the DH network, while the temperature has little effect on the time delay. The time delay of the DH network system has a significant impact and can provide services for optimal control of the system.

Keywords: district heating (DH) network; CFD model; optimal control; time delay; parameter analysis

---

1. Introduction

DH (District heating) networks are considered a very efficient option for providing heating and domestic hot water to buildings, particularly for densely populated areas [1]. DH systems are community energy systems which can provide long-term achievements in terms of greenhouse gas emission abatement, energy security, local economic development, energy, and exergy efficiencies increase and exploitation of renewable energy [2]. Improving the heating performance of DH networks, one of the most important components in district heating systems, has a significant influence on the promotion of energy efficiency [3].

The behavior and characteristics of the DH network system are influenced by its various parameters [4]. The dynamic features of DH consumers and networks usually influence the operational effect of the DH system heavily because of the time delays in the DH system [5]. It may take a huge amount of time for the further located ends to response to heat injection. The time delay of heat transportation can reach minutes or hours [6]. The existence of time delays usually results in a mismatch between the heat source output and the users' heating load [7], and considerable heat loss [8]. The optimization of system control [9,10] to reduce the energy losses caused by time delays has attracted more and more scholars' attention.

In the published papers, many researchers focus on the dynamic characteristics of heating networks. A method for calculating the time delay of the network was presented by Jie, et al [11] who using physical and mathematical models, proved that in the case of a single straight pipe with constant pipe diameter, the time delay was approximately equal to the flow time of the heat medium. Zheng [12] proposed a new physical method (function method) and corresponding dynamic temperature simulation computer code to analyze the thermal dynamic characteristics of the district heating network, focusing on the dynamic temperature distribution. The results showed that the function method could be recommended to simulate the dynamic temperature of district heating network accurately and quickly for efficient system performance. Wang and Meng [13] used the third-order numerical solution method to investigate the influence of thermal inertia of the pipe wall on the thermal transient prediction model of a long pipe. The results show that for pipes with a diameter greater than 200, it is not necessary to consider the influence of thermal inertia in the modeling process.

Some scholars also focused on research on control strategies in terms of time delay. Li [14] proposed a special control strategy called Smith predictor which took the transport time delay out of the closed-loop feedback control system to deal with the effects of transport delay problem. Sartor [15] proposed a dynamic model of the heat wave in the network to determine the temperature and mass flow in some key locations to solve the problem of excessive delay time in the DH network. Additionally, some authors focus on the dynamic simulations of DH systems. In [16,17], the validation of the simplified approaches based on practical comprehensive DH systems was formulated, but the case was validated for a single pipe or several branches, the validations performing on a full system are limited. Gabrielaitiene [18] modeled a DH system in Denmark by a node method and TERMIS software and indicated that the local discrepancies between the predicted and measured supply temperatures were pronounced for consumers located at distant pipelines containing numerous bends and other fittings. Mattias [19] used a simulation tool developed in MATLAB/Simulink to analyze the flow distribution in the district heating network behavior, the results showed that the tool was a valuable instrument to simulate and analyze the behavior of meshed district heating network.

It can be seen that the function method is mainly based on the mathematical model and the construction of the physical model to calculate the time delay and temperature distribution of the DH network system. However, the proposed function method is only suitable for the single-pipe solution, which is limited by the computational cost and the complexity of the pipe network for large DH networks. Therefore, the calculation of the time delay of the DH network system caused by temperature change is inaccurate. The method of simulation such as MATLAB/Simulink, which also can prove that the mathematical method used to determine the temperature distribution of the elbow and other fittings of the long pipe is unreasonable, has a potential for simulating the temperature distribution in the DH network system and calculating the time delay caused by the flow. However, the establishment of the Simulink module of each node requires the basis of the mathematical model and the flow state parameters. Some simulation research on the optimal control of the DH network system to reduce the impact of time delay on system control, ignoring the main factors causing the time delay of the DH network system. This paper focused on the calculation method of the time delay of the DH network system, which was determined by establishing a CFD model. The combination of steady-state calculation and transient calculation solves the established DH network model. Compared with the calculation method of time delay theory in the previous articles, the relative degree of CFD method reduces the calculation cost and breaks through the constraints of the pipeline model, which is not only limited to a single pipe, but the time delay of the fluid through the varying diameters, bends, and branches can also be calculated. Compared with some other simulation methods, only simple boundary conditions need to be obtained without the operation parameters inside the pipe of the DH network system. The DH network of Tianjin University was taken as the research object. The time delay of the DH network obtained by using the CFD model was compared with the measured data to verify the correctness of the model. The results show that the proposed CFD model is superior in calculating the time delay of the DH network. Based on the CFD model, the parameters affecting the time delay were analyzed. The time delay of the DH network system obtained by the CFD simulation method can be used as a reference for the optimization of a DH network system to ensure that the operating parameters of the DH network and the heat load requirements of the heat consumers can be matched in time.

2. DH Network System Description

The studied system is located at Tianjin University. The heat source is composed of a ground-source heat pump and boiler, which serves more than 265600 square meters of building area. The DH network system supplies heat to 15 consumers, which includes a library, teaching building, laboratories and so on. Figure 1 shows the distribution network of the Tianjin University subsystem. In this case, there is no substation and the consumer is a terminal building. The heat source directly transfers the heat through the DH network to the terminal building. The network is constructed of pre-insulated pipes with a total length of approximately 1.8 km. The pipe diameter ranges from 125 mm to 600 mm. The distance between the heat consumer and the heat source is shown in Table 1. The closest and farthest to the heat source are the heat consumer 1 and the heat consumer 5, which are 145 m and 633 m, respectively.

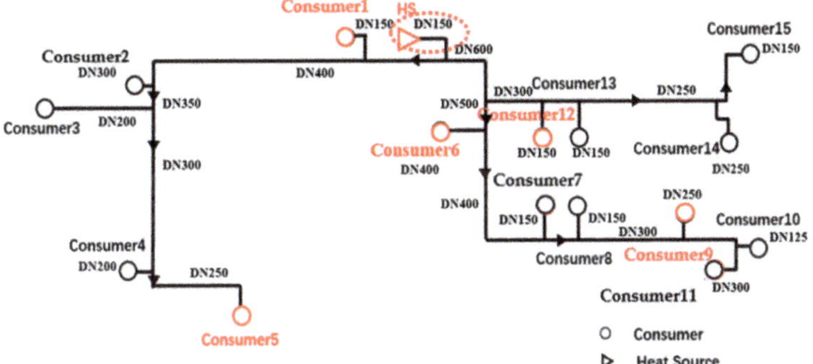

Figure 1. The district heating network in Tianjin University.

Table 1. Building function and distance from the heat source about each consumer.

Consumer	Building Function	Distance from the Heat Source (m)
Consumer1	Information and network center	145.00
Consumer5	teaching building 44	633.00
Consumer6	library	156.00
Consumer9	laboratory building	451.10
Consumer12	teaching building 46	179.00

Considering the similarities of the building functions and the symmetry of the actual distribution of the pipe network, five typical consumers were selected for test verification. The functions of each consumer and the distance from the heat source are shown in Table 1.

3. Methodology

3.1. Computational Fluid Dynamics Modelling

3.1.1. CFD Model and Preliminary Settings

Since numerical simulation has unique advantages over experimental studies, such as low cost and short cycle time, complete data can be obtained and various measured data states can be simulated during actual operation [20]. Fluent is a commercial software package that uses a finite volume method to couple mass equations, momentum equations, and energy equations. The flow in the pipe is an incompressible fluid that satisfies the general governing equation describing the state of the flow field as a flow pattern. The basic governing equations are shown in Equations (1)–(3). A coupled solution based on three equations of a powerful workstation can be solved. In this study, Fluent 17.0 was used for our numerical simulation calculations. The workstation for the simulation calculation is an Asus Z10PA-D8 2.20 GHz, 64GB of RAM and 120 GB of hard drive memory. The CPU of this workstation is an Inter(R) Xeon(R) CPU E5-2630 v4 @ 2.20 GHz. A parallel computing technique was implemented in ANSYS Fluent solver:

$$\frac{\partial \rho}{\partial t} + \nabla(\rho \vec{v}) = 0 \tag{1}$$

$$\frac{\partial}{\partial t}\left(\rho \vec{v}\right) + \nabla\left(\rho \vec{v}\,\vec{v}\right) = -\nabla p + \nabla \vec{\tau} + \rho \vec{g} \tag{2}$$

$$\frac{\partial(\rho C_v T)}{\partial t} + \nabla\left(\rho C_p \vec{v} T\right) = \nabla(K_T \nabla T) \tag{3}$$

where ρ is the density, t the time, v the velocity, p the pressure, τ the shear stress, g the gravity acceleration, c_v and c_p the specific heats, T the temperature and k_T the thermal conductivity.

The workflow of the model is mainly composed of the following aspects:

- The STEP AP214 format in SOLIDWORKS was used to generate simplified 2D CAD
- A structured mesh based on ICEM was generated. The maximum mesh size was defined as 30. The minimum orthogonal quality of the mesh is 0.932 and the maximum orthogonal skewness is 0.068

3.1.2. Setting the Fluent Solver

ANSYS 17.0 was used to generate the simplified 2D CFD model. There is no doubt that in simulation calculations, the accuracy of 2D simulation results is worse than that of 3D simulations. However, considering the hugeness of the DH network system, there is a high demand for simulation cost both in terms of calculation time and computational intensity. Therefore, the system was subjected to simulation. In order to obtain a high-quality mesh, a structural mesh was generated in the 2D CAD model, which can easily realize the boundary fitting of the region and is suitable for calculation of fluid and surface stress concentration. Structured meshes generate slower than unstructured meshes, but the number of structured meshes is much smaller than that of unstructured meshes, which can reduce computational intensiveness. The maximum mesh size was defined as 30 mm. In order to ensure the accuracy of the model, it was intended to increase the density of the mesh at the tee junction of the pipe. A partial view of the mesh from the heat source to the first elbow (see the red dotted ellipse area in Figure 1) is shown in Figure 3. The total number of meshes in the calculation domain is 252,7918.

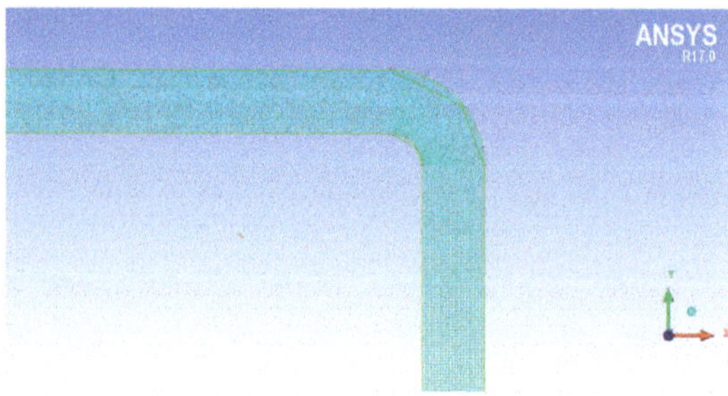

Figure 2. The local mesh of the CFD model.

In order to prove the reasonableness and accuracy of the number of grids, the model needs to be verified for grid independent. The temperature and velocity contours comparisons from the heat source to the first elbow in five conditions of 1, 2, 2.5, 4 and 6 million are shown in Figure 3.

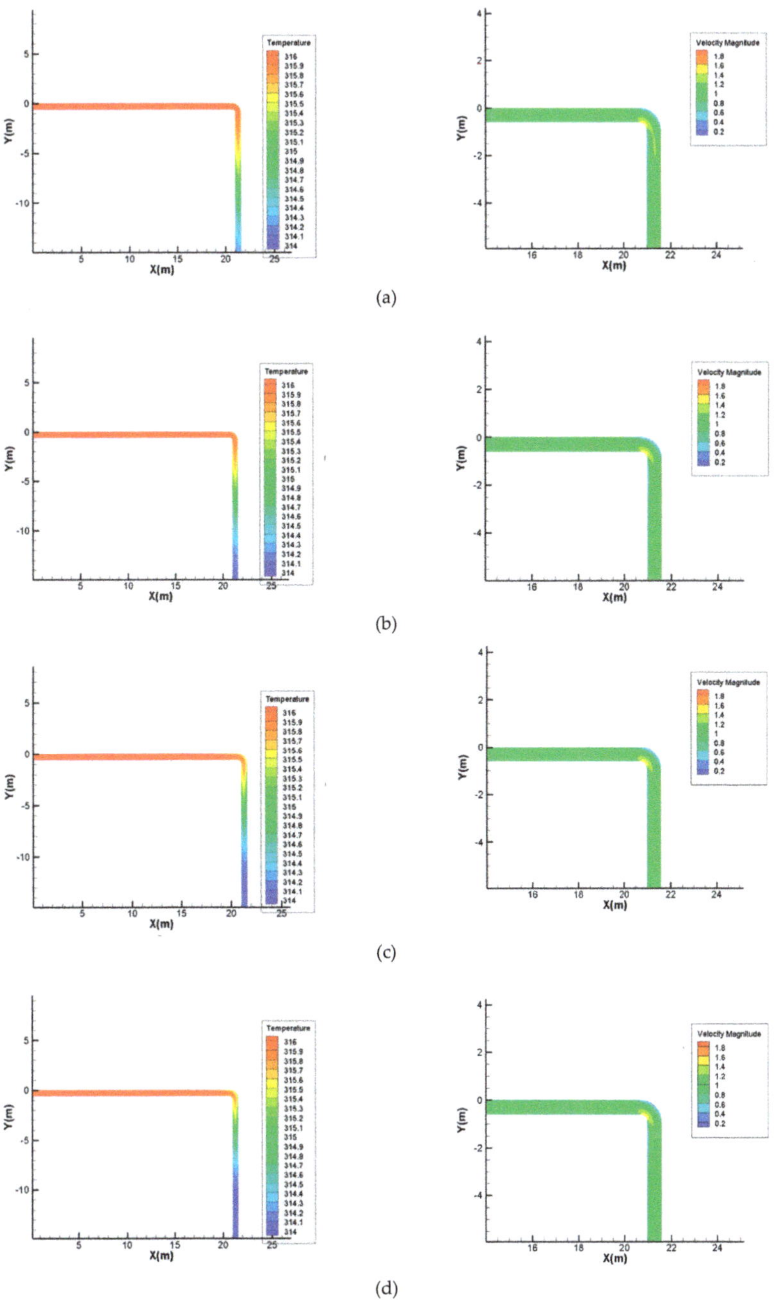

Figure 3. *Cont.*

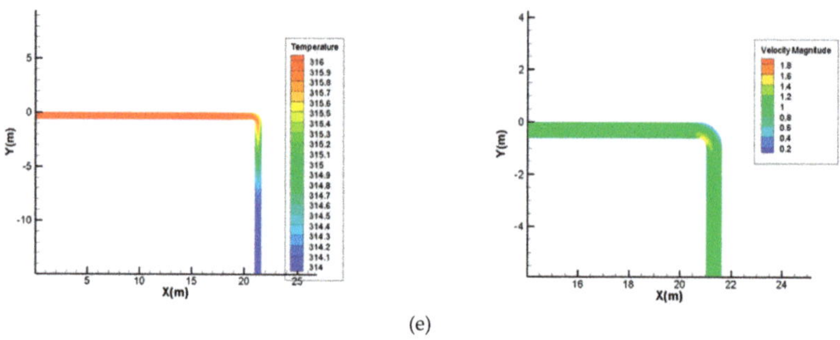

(e)

Figure 3. (**a**) Temperature contours and velocity contours under 1 million grids; (**b**) Temperature contours and velocity contours under 2 million grids; (**c**) Temperature contours and velocity contours under 2.5 million grids; (**d**) Temperature contours and velocity contours under 4 million grids; (**e**) Temperature contours and velocity contours under 6 million grids.

The test results of grid independence are shown in Figure 4. As can be seen from the grid independence results, when the number of grids reaches 2.5 million, the temperature and velocity changes will not be obvious. The computational cost and intensity are affected by the number of grids, and the computational cost increases as the number of grids increases. It is reasonable to have a grid number of 2.5 million.

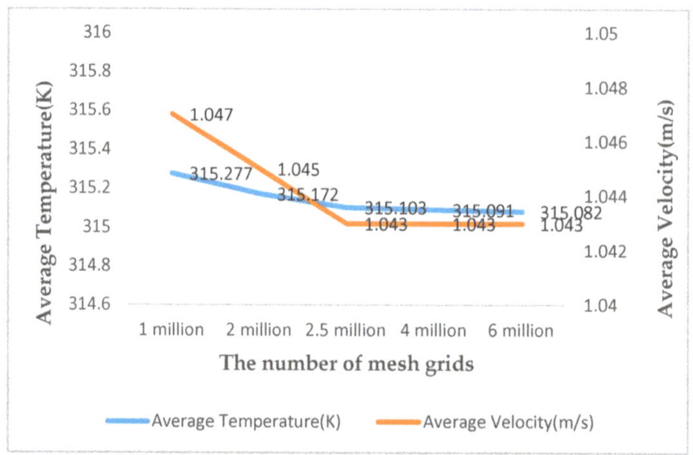

Figure 4. The test results of grid independence.

In a large DH networks system, the temperature change of the supply water of the heat source takes a period of time to affect the heat consumers. This is the time delay due to changes in the water supply temperature. The Fluent solver was set up as a combination of steady-state calculations and transient calculations. The steady-state calculation makes the temperature in the DH network reach a certain value, and then the transient calculation was performed based on the steady state calculation to prove the influence of the water supply temperature change on the time delay of the DH network. The energy model was activated and the standard k-ε model equation was used as the turbulence model. Compared with the RNG model suitable for rotating flow calculation, the standard k-ε model is more suitable for solving the problem in this case. The turbulence governing equation is shown below:

$$\frac{\partial(\rho k)}{\partial t} + \frac{\partial(\rho k u_i)}{\partial x_i} = \frac{\partial}{\partial x_j}\left[\left(\mu + \frac{\mu_t}{\sigma_k}\right)\frac{\partial k}{\partial x_j}\right] + G_k + G_b + \rho\varepsilon - Y_M \tag{4}$$

$$\frac{\partial(\rho\varepsilon)}{\partial t} + \frac{\partial(\rho\varepsilon u_i)}{\partial x_i} = \frac{\partial}{\partial x_j}\left[\left(\mu + \frac{\mu_t}{\sigma_\varepsilon}\right)\frac{\partial k}{\partial x_j}\right] + G_{1\varepsilon}\frac{\varepsilon}{k}(G_k + G_{3\varepsilon}G_b) - G_{2\varepsilon}\rho\frac{\varepsilon^2}{k} \tag{5}$$

where G_k is the production term of turbulent energy k caused by average velocity gradient, G_b is the production term of turbulent energy k caused by buoyancy, Y_M represents the contribution of pulsation expansion in compressible turbulence, For incompressible fluids, $G_b = Y_M = 0$. $G_{1\varepsilon}$, $G_{2\varepsilon}$, $G_{3\varepsilon}$, σ_k and σ_ε are empirical constants, $G_{1\varepsilon} = 1.44$, $G_{2\varepsilon=}1.92$, $G_{3\varepsilon} = 0.09$, $\sigma_k = 1.0$ and $\sigma_\varepsilon = 1.3$, ρ is fluid density, k is turbulent kinetic energy, ε is turbulent kinetic energy dissipation rate, u is fluid relative velocity, μ is fluid dynamic viscosity.

The SIMPLE algorithm was used as the pressure-velocity coupling method. The pressure discrete difference format uses the standard discrete difference format. For solving the flow problem, the other variables use the 1st upwind style. The 1st upwind is less accurate than the 2nd upwind, but it can make the steady-state calculation easier to converge, especially for this case. In order to improve the calculation accuracy, the 1st upwind convergence under steady-state conditions was used as the initial condition for the 2nd upwind transient calculation. The results of the comparison of the velocity and temperature profiles from the heat source to the first elbow are shown in Figure 5 when the momentum, turbulent model, and energy were discretized in 2nd order upward and 1st, respectively. The results show that it is reasonable to use the 1st upwind for steady-state calculations. The under-relaxation factors of pressure, density, momentum, turbulent kinetic energy, and energy were 0.3, 1, 0.7, 0.8 and 0.8, respectively. All of the given time steps are iteratively solved in a separate manner until the convergence condition was met. Two control monitors of the iterative process were defined to check convergence: a monitor for the residuals of the iterative process for the equations solved and the surface monitor of the outlet velocity. When the residual decreases at a value of 10^{-5}, the simulation process was considered to be convergent. However, the oscillation occurs when the residual is calculated to be below 10^{-3}. At this point, the surface monitor of the outlet velocity shows that the outlet velocity at this time remains constant, so the calculation is considered to have reached convergence. The Fluent 17.0 solver was calculated at steady state and the number of iterations is 4000. Another major step in CFD modeling was the establishment of boundary conditions. In the CFD model, the heat source and the heat consumers were defined as a velocity-inlet and pressure-outlet, respectively. The parameters of the boundary conditions were obtained from the measurements. Moreover, the wall of the pipe in the model was defined as an adiabatic wall to ignore the heat loss in the radial direction of the pipe.

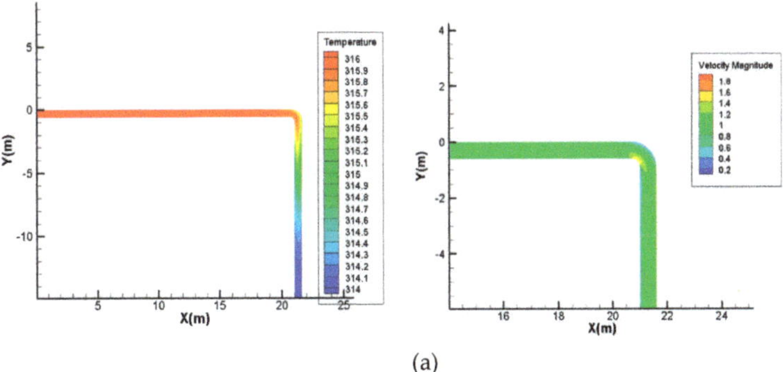

(a)

Figure 5. *Cont.*

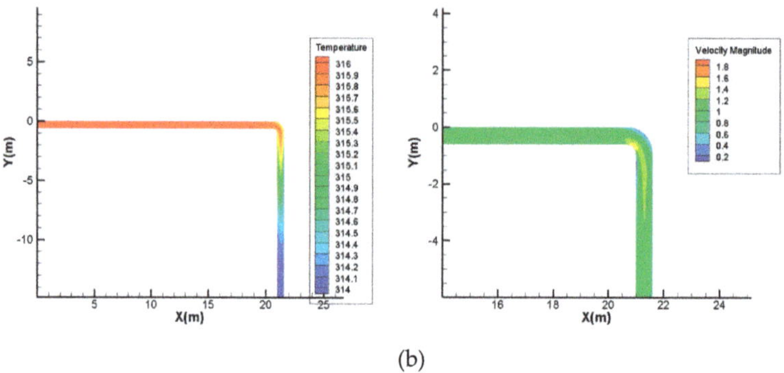

(b)

Figure 5. (a) Temperature contours and velocity contours in 1st order upward; (b) Temperature contours and velocity contours in 2nd order upward.

After the steady state calculation converges, the temperature of the flow field in the model at this time reached the input temperature of the pipe velocity-inlet. Meanwhile, the Fluent 17.0 solver was subjected to transient calculations based on the steady-state simulation, with the temperature input of the velocity-inlet being changed. The second-order upwind space discretization algorithm was used for transient algorithms. Compared to first-order algorithms, second-order algorithms can achieve the best results because they can significantly reduce interpolation errors and false-value diffusion [21,22]. The time step size of the transient solver was set to 1s and the total time of the simulation was set to 4000 s. The maximum iterations were 20. The automatically save time for Fluent transient calculation was set to be automatically saved every 10 s. The temperature at each outlet of the DH network will vary with the input temperature of the velocity-inlet. The automatically saved Fluent data was read and analyzed the change of the temperature of the whole DH network system, which the time of the automatically saved Fluent data corresponding to each outlet temperature reaching the inlet temperature setting value to determine the time delay from the heat source to each consumer due to the temperature variation. By this method, the time delay of each heat consumer caused by the change in the water supply temperature of the heat source can be determined.

3.1.3. Boundary Conditions of the CFD Model

Boundary conditions need to be defined as part of the CFD model. In this case, the inlet and outlet of the model were set as the velocity-inlet and pressure-outlet, respectively. Parameters such as temperature and flow rate were obtained through experimental measurements and used as input parameters to define the boundary conditions.

Temperature and flow measurements were performed at the supply pipe for the heat source and each heat consumer. The average of the supply water temperature and flow rate at the heat source were set as the inlet temperature and the inlet flow rate. The probe of the temperature recorder was installed on the water supply pipe at the heat source, and the water supply temperature was measured every 10 minutes. Similarly, the ultrasonic flowmeter was used to measure the flow rate of the water supply pipeline at the heat source every 10 minutes. The measured water supply temperature and flow rate at the heat source were shown in Figure 6. The average flow rate and water supply temperature at the heat source is 1.04 m/s and 39.2°C, which were used as a boundary condition for calculating the steady-state velocity-inlet of the CFD model. In the process of transient calculation, the velocity-inlet temperature was defined as 40.9°C (The maximum temperature value on test day). In the CFD model, each consumer building heating entry was considered as an outlet, and the boundary condition of the outlet was defined as pressure-outlet. However, in an actual DH network system, the consumer building heating entry is not an outlet that is in communication with the atmosphere but rather has a pressure sufficient to provide hot water to the most unfavorable end of the building. This means that the pressure at the pressure-outlet cannot be set to 0 Pa. The outlet pressure can be obtained by reading the pressure gauge at the consumer building heating entry. A relative pressure of 0.4 MPa was defined as the input parameter of the pressure-outlet. The input parameters of the CFD model boundary conditions are shown in Table 2.

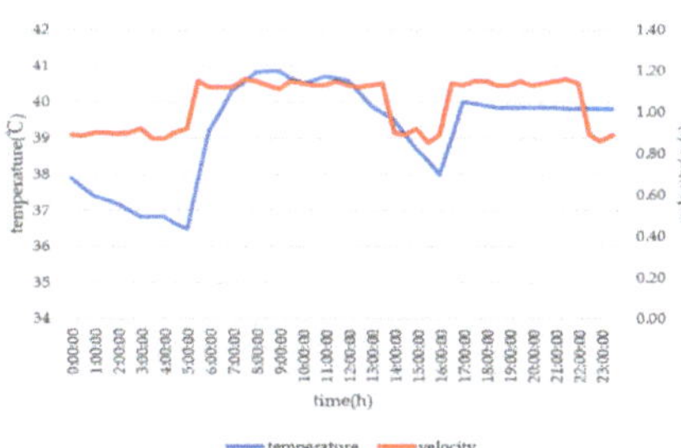

Figure 6. The measured temperature and velocity at HS.

Table 2. The input parameters of the CFD model boundary conditions.

		Steady State	**Transient**
Velocity-inlet	Flow rate (m/s)	1.04	1.04
	Temperature (K)	314	316
Pressure-outlet	Pressure (MPa)	0.4	0.4

3.2. Equivalent Pipe Diameter

In the DH network, there is a strong coupling between the influencing parameters. The time delay of the heat consumer was determined by the combination of pipe length and pipe diameter. The diameter of the pipe from the heat source to the heat consumer is constantly changing with distance and is not a fixed value. In order to analyze the influence of various parameters on the time delay, complex practical engineering problems were transformed into ideal pipeline models. Therefore, the concept of equivalent pipe diameter (d_R) was proposed.

The pipe length is the superposition of the length of the heat medium flowing from the heat source to each consumer. The flow rate in the equivalent pipe diameter of each consumer is the flow rate to each consumer. Since pipe diameters are constantly changing along with the flow of heat medium, simple overlays summation is not possible. The equivalent pipe diameters (d_R) are defined based on practical frictional and local resistance. Under the same pipe length, the summation of the frictional resistance and local resistance of the heat medium from the heat source to the consumer through all the pipe sections is equivalent to a fixed pipe diameter of the single straight pipe, as Equation (6) shows. The fixed pipe diameter, namely the equivalent pipe diameter, can be calculated as Equation (7), which is derived from Equation (6):

$$\sum_{i=1}^{n}\left(h_{l,i} + h_{f,i}\right) = \left(\sum_{i=1}^{n} L_i\right) \cdot \frac{\lambda_R}{d_R} \frac{\rho v^2}{2} \tag{6}$$

$$d_R = \frac{\left(\sum_{i=1}^{n} L_i\right) \cdot \lambda_R \cdot \rho v^2}{2 \times \sum_{i=1}^{n}\left(h_{l,i} + h_{f,i}\right)} = \frac{\left(\sum_{i=1}^{n} L_i\right) \cdot \lambda_R \cdot \rho v^2}{2 \times \sum_{i=1}^{n}\left(L_i \cdot \frac{\lambda_i}{d_i} \cdot \frac{\rho v_i^2}{2} + h_{f,i}\right)} \tag{7}$$

where, h_l is the frictional resistance (pa), h_f is the local resistance (pa), L is the pipe length (m), v is the flow rate in each pipeline (m/s), ρ is the density of heat medium (kg/m^3), λ is the friction coefficient, i is the pipe number and n is the total number of pipe sections.

The above Equation can calculate the equivalent diameter of each consumer in this case (see Section 4.3.3), and prepare for the subsequent parameter sensitivity analysis.

4. Results

4.1. A Solution of the CFD Model

The steady state calculation results were used as initial conditions for transient calculations. The transient temperature change in the pipeline of the DH network system can be obtained by reading the temperature contours. As can be seen from Figure 7, taking Consumer 1 as an example, the temperature distribution in the pipeline ranges from 314.2 K to 315.9 K. 314.2 K is the temperature of the velocity-inlet under steady-state calculation, while the 315.9 K is the velocity-inlet temperature under transient simulation. Observe the change of the temperature band in the simulation result. When the temperature at the outlet reaches 315.9 K, the corresponding time is the time delay caused by the temperature change.

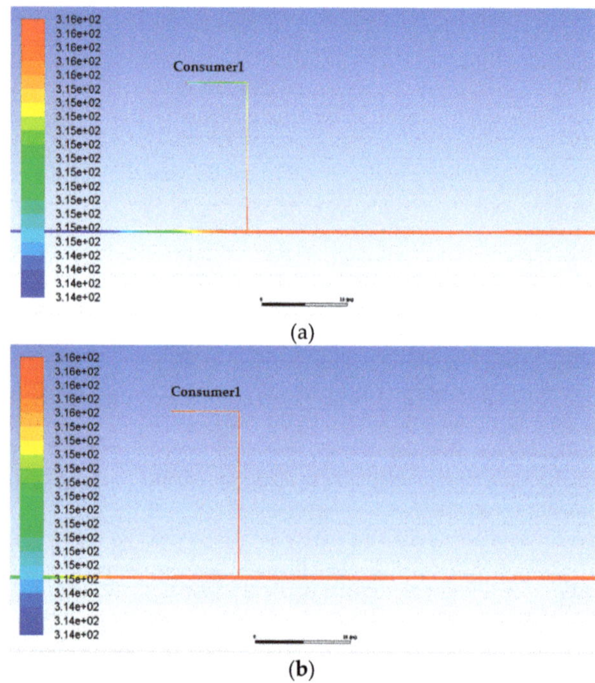

Figure 7. (**a**) The temperature of Consumer 1 at 220 s; (**b**) the temperature of Consumer 1 at 240 s.

In Figure 7, the results represent the temperature at the build heating entry of Consumer 1 at 220 s and 240 s. As can be seen from the change of temperature of Figure 7a, the inlet temperature of the Consumer 1 reached 315 K at 220 s. However, from Figure 7b, we can see that the temperature of Consumer 1 has fully reached 315.9 K (the inlet temperature setting) at 240 s. Since the model is adiabatic and does not consider heat loss, it is considered that the time of setting temperature of the consumer inlet temperature is due to the time delay caused by the flow. The time delay of consumer1 is 240 s. Similarly, the time delay of other heat consumers can be obtained by reading the temperature contours of the simulation results. The time delay of other heat consumers is shown in Table 3.

Table 3. The delay time of the heat consumer.

Consumer	Time Delay (s)	Distance from Heat Source (m)
1	240	145.00
2	1040	339.00
3	1720	439.70
4	2680	514.30
5	3830	633.00
6	420	156.00
7	1050	316.00
8	1150	339.60
9	1720	451.10
10	2240	534.10
11	2630	563.10
12	610	179.00
13	720	204.20
14	2090	367.80
15	3010	420.60

4.2. Model Validation

4.2.1. Measurement Data

The correctness of the CFD model with variable temperature delay was verified based on the test data. The peak-valley method [23] was used to verify the correctness of the time delay. The temperature at the outlet of the heat consumer changes with the change of the inlet temperature of the heat source, and the temperature fluctuation curves of the two are consistent. The time corresponding to the peak and valley of the water supply temperature curve of the heat source and the inlet of each consumer was read separately. The difference in time between the peak (valley) of the heat source and the consumer is the time delay due to the change in the water supply temperature. Therefore, the water supply temperature of both the heat source and each of the heat consumers was measured.

A temperature recorder was installed on the water supply pipe of the heat source and the water supply pipe at the build heating entry of each heat consumer. The temperature of the water in the pipe was measured by the temperature probe of the temperature recorder and automatically recorded once every ten minutes.

The hourly variation curve of water supply temperature at the HS and the consumer in one day is shown in Figure 8. The basic shapes of the hourly temperature curves are similar. The water temperature at the consumers changes after the water temperature changes at the heat source. The change of water temperature in the consumer lags behind the temperature of the water supply in the heat source.

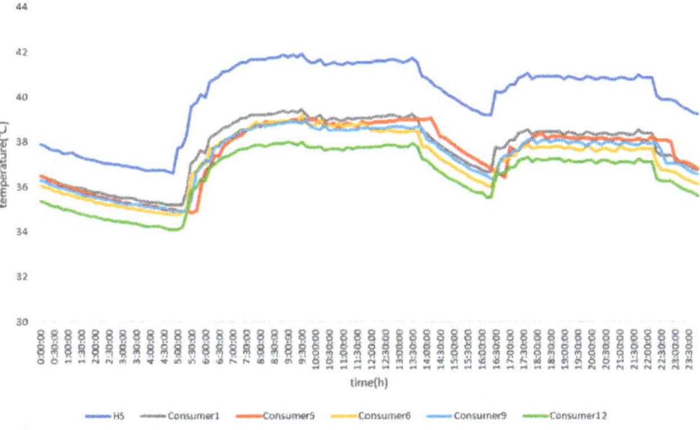

Figure 8. Hourly water supply temperature at HS and consumer.

Three days of test data were selected for model validation, with a total of six temperature peaks and six temperature valleys. The frequency statistics of the water supply temperature peak and valley of each consumer are shown in Table 4. The weighted average equation is as follows:

$$\overline{\xi} = \frac{\sum \xi * n}{\sum n} \tag{8}$$

where, $\overline{\xi}$ is the time delay of the weighted average (s), n is the frequency of the time delay, ξ is the time delay (s).

Table 4. Frequency statistics of the time delay of consumers.

Consumer	Time Delay (s)	Water Temperature Valley	Water Temperature Peak
	210	1	0
Consumer 1	240	4	5
	300	1	1
	3300	1	0
Consumer 5	3600	4	4
	3900	1	2
	420	4	4
Consumer 6	480	2	1
	600	0	1
	1500	1	0
Consumer 9	1680	3	2
	1800	2	4
	300	3	1
Consumer 12	600	2	2
	900	2	2

The average time delay of each consumer can be calculated. The calculation results show the average time delay of each consumer: $\overline{\xi}_{con1} = 245$ s $\overline{\xi}_{con5} = 3650$ s, $\overline{\xi}_{con6} = 450$ s, $\overline{\xi}_{con9} = 1725$ s, $\overline{\xi}_{con12} = 600$ s.

4.2.2. Model Verification and Error Analysis

The calculation time delay with the test data and the time delay with the CFD model are shown in Table 5. In order to verify the accuracy of the model, relative error was introduced to compare two methods. As seen in Table 5, the results obtained by CFD model agree well with the test data. The maximum relative error is 7.14%. The minimum relative error is 0.29%. The correctness of the CFD model was proved by the result of a relative error.

Table 5. Comparison of the CFD method with the peak-valley method.

Consumer	The Time Delay Calculated by the CFD Method (s)	The Time Delay Calculated by the Peak-Valley Method (s)	Relative Error
Consumer 1	240	245	2.08%
Consumer 5	3830	3650	4.70%
Consumer 6	420	450	7.14%
Consumer 9	1720	1725	0.29%
Consumer 12	610	600	1.64%

The reasons for the error between the CFD model and the measured data are as follows:

- The CFD model was built based on the actual size of the system. However, due to the lack of data on some of the construction drawings, some pipelines are determined by the equivalent replacement method. During the actual construction process, the layout of the pipe network might end up being different from that described in the engineering data.

- The boundary condition of the CFD model based on test results may have some small gaps with the actual pipe network operation.

4.3. Parameter Sensitivity Analysis

4.3.1. Influencing Parameters

The time delay was defined as the time from when the inlet water temperature changes to the time when the pipe outlet water temperature begins to change. For fixed pipe insulation characteristics, the time delay is related to the distance of water flow and water velocity, while the distance of water flow is related to pipe length, and water velocity is related to flow and nominal diameter. In addition, the change in water supply temperature is another important parameter affecting the time delay. Therefore, the time delay of water temperature caused by hot water transmission in the main pipes and branch pipes is mainly affected by flow rate, pipe length, pipe diameter, and supply water temperature. The influence of four parameters on the time delay was studied, respectively.

4.3.2. Flow Rate

The time delay of each heat consumer in the actual case was obtained under the condition that the velocity-inlet is 1.04 m/s. The input flow rate at the velocity-inlet was changed to simulate under five working conditions of 0.8, 0.9, 1.0, 1.1 and 1.2 m/s, respectively, and the time delay of each heat consumer under five different flow rates was obtained. The time delay of each heat consumer under five different flow rates is shown in Figure 9. As can be seen from the simulation results, the time delay decreases as the flow rate increases. However, the time delay decreases rapidly at the beginning and then gradually slows down. This means that the change in time delay will no longer be apparent when the flow increases to a certain value. From the equation of resistance (see Equation (6)), it can be seen that the resistance is proportional to the square of the flow rate, and the resistance increases rapidly with the increase of the flow rate, which is the reason why the increase of the flow rate causes the rate of decrease of the time delay to continuously decrease.

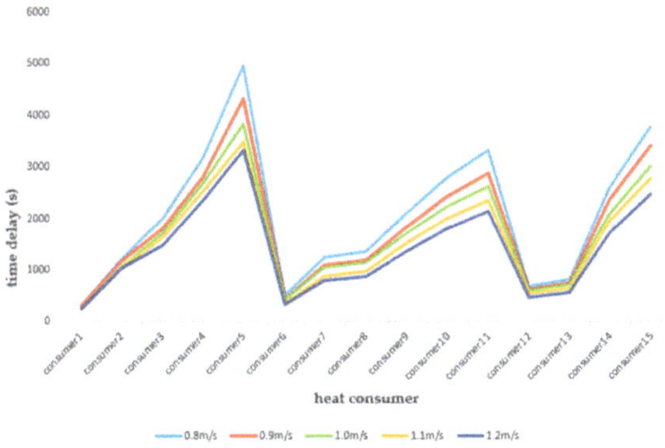

Figure 9. The time delay of each heat consumer under five different flow rates.

4.3.3. Pipe Length and Pipe Diameter

The relationship between the pipe length from the heat source to each heat consumer and the time delay is shown in Figure 10, where the abscissa represents the increase of the pipe length, from which it can be seen that the time delay of the heat consumer increases with the increases of pipe length. There are, however, two distinct breaks in the curve. The reason for this phenomenon is that the research on the relationship between the time delay of each heat consumer and the pipe length was not carried out under the condition of the same pipe diameter. The equivalent pipe diameter of each heat consumer was calculated by Equations (6) and (7). The calculation results of pipe length, equivalent pipe diameter, and flow rate are shown in Table 6.

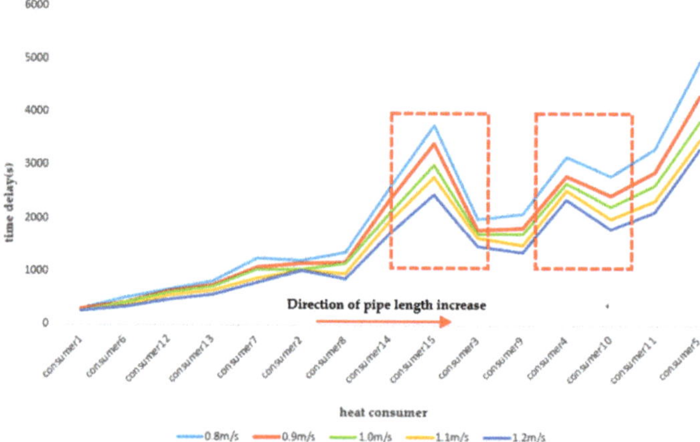

Figure 10. The relationship between the pipe length and the time delay of each heat consumer.

Table 6. Correlation pipeline parameters of each consumer.

Consumer	Time Delay (s)	Pipe Length (m)	Equivalent Pipe Diameter (mm)	Flow Rate (m^3/h)
1	240	145.00	159.2	38.07
2	1040	339.00	155.7	24.08
3	1720	439.70	192.4	41.80
4	2680	514.30	208.4	56.66
5	3830	633.00	231.8	69.80
6	420	156.00	289.6	126.89
7	1050	316.00	164	30.02
8	1150	339.60	174.5	33.51
9	1720	451.10	181.2	37.23
10	2240	534.10	135.1	19.82
11	2630	563.10	258.8	78.42
12	610	179	151.1	31.11
13	720	204.2	153	32.28
14	2090	367.8	232.9	82.29
15	3010	420.6	246.4	57.02

The pipe length of heat Consumer 3 is greater than that of heat Consumer 15, but the time delay of heat Consumer 15 is greater than that of Consumer 3 because the equivalent pipe diameter of heat Consumer 15 is greater than that of heat Consumer 3. Similarly, the heat Consumer 10 has a larger pipe length than the heat Consumer 4, but its diameter is smaller than that of the heat Consumer 4. It can be seen that the influencing factors of the time delay of the DH network system are the result of the combined effect of pipe diameter and pipe length. The sensitivity analysis results of pipe length and pipe diameter are shown in Table 7. It can be seen from Table 7 that the effect of pipe diameter on the time delay of each heat consumer is greater than that of pipe length.

4.3.4. Supply Water Temperature

In the actual case, the water supply temperature changed from 314.2 K to 315.9 K. In order to prove whether the time delay of each heat consumer is related to the value of water supply temperature change, the input temperature of velocity-inlet of CFD model needs to be changed. Therefore, the temperature of the velocity-inlet was changed from 315.9 K to 318 K for transient simulation. The two water supply temperatures were simulated separately under five different flow rate conditions, and the obtained time delay of each heat consumer was shown in Figure 11a–e.

Table 7. Sensitivity analysis of pipe length and pipe diameter.

Variable Parameters	Variation Rate of Parameters (%)	Time Delay to Each Heat Consumer (s)				
		Consumer 1	Consumer 5	Consumer 6	Consumer 9	Consumer 12
	−20	150	2530	270	1100	390
	−10	180	3080	310	1380	460
d	0	240	3830	420	1720	610
	10	280	4600	490	2060	720
	20	330	5500	580	2470	850
	−20	180	3010	320	1340	470
	−10	200	3420	350	1530	520
L	0	240	3830	420	1720	610
	10	260	4120	460	1870	670
	20	280	4590	490	2060	720

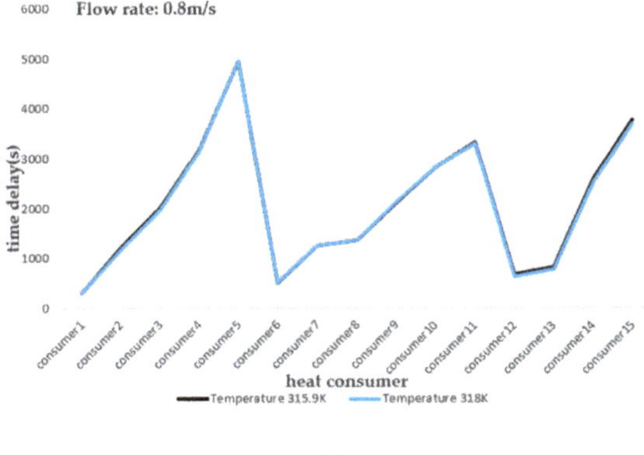

(a)

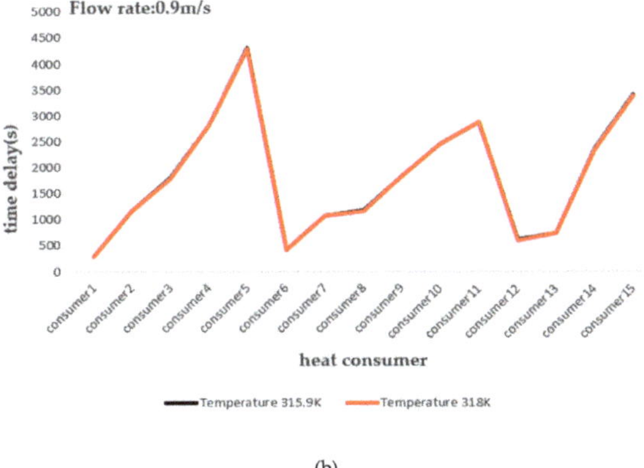

(b)

Figure 11. *Cont.*

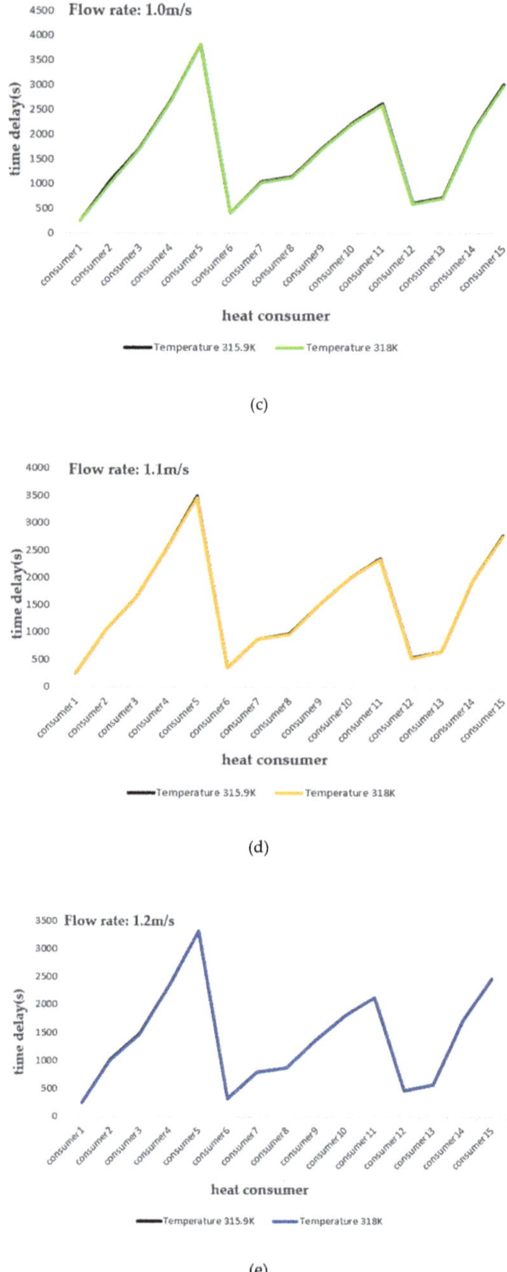

(c)

(d)

(e)

Figure 11. (**a**) Comparison of the time delay obtained by the two different temperature simulations under the condition of 0.8 m/s; (**b**) Comparison of the time delay obtained by the two different temperature simulations under the condition of 0.9 m/s; (**c**) Comparison of the time delay obtained by the two different temperature simulations under the condition of 1.0 m/s; (**d**) Comparison of the time delay obtained by the two different temperature simulations under the condition of 1.1 m/s; (**e**) Comparison of the time delay obtained by the two different temperature simulations under the condition of 1.2 m/s.

It can be seen from the simulation results that the change of the water supply temperature has little effect on the time delay. The time delay of each heat consumer is almost the same before and after changing the water supply temperature for both the DH network system with large flow rate (1.2 m/s) and the DH network system with small flow rate (0.8 m/s). However, the fitting degree of the two temperature curves of the DH network system with a large flow rate is higher than that of the DH network system with a small flow rate. It means that the change of water supply temperature has a greater impact on the DH network system with a small flow rate than the DH network system with a large flow rate. The reason for the analysis is that the inlet water temperature rises to make the water temperature in the pipeline higher, and the high water temperature is favorable for accelerating heat transfer. Therefore, the time delay is shortened when the inlet water temperature changes.

5. Discussion

According to the parameter sensitivity analysis, the time delay of each heat consumer is different for the DH network system, which proves that the time delay of each heat consumer was affected by the interaction of various influencing parameters. In order to regulate and optimize the DH network system, the relationship between the time delay of each heat consumer and the influence parameters in the DH network system must be considered. The CFD simulation method proposed in this paper can calculate the time delay of the established DH system and be applied to the control operation of the DH network system. Therefore, the analysis of the time delay of each heat consumer is very important and meaningful for the implementation of the control strategy of the DH network system.

The water in the DH network is regarded as an incompressible fluid, and the change of flow and pressure at the inlet can be regarded as a non-delayed response to the terminal. However, under the influence of the combined effects of flow and heat transfer, the temperature has a certain time delay response to the terminal, which is a dynamic response problem. This paper only analyzes the influence parameters of time delay under the static working condition through the change of boundary conditions. Based on the CFD model established in this paper, the dynamic response of the DH network after temperature change will continue to be deeply analyzed in the future.

The CFD method was used to obtain the time delay of each heat consumer of the DH network system and they can provide services for the optimization of the control strategy. However, the CFD method was limited to computation time and intensity. The current simulation is suitable for the small-scale DH network systems without any range limitation of temperature and flow rate. For large-scale DH network systems, the computational cost will also increase. In addition, this method only provides an idea to simulate the calculation of the system time delay at the time when the flow rate and temperature at the heat source are constant. The method is limited for the calculation of the time delay of the system where the temperature and flow rate of the heat source is constantly changing.

6. Conclusions

A model based on the CFD method to determine the time delay of the DH network was proposed. By changing the boundary conditions of the CFD model, the influence parameters of the DH network time delay were analyzed. The different time delay of each heat consumer is due to the combined effect of various influencing parameters. The results show that the flow rate, pipe length and pipe diameter have a significant impact on the time delay of the DH network system.

The increase in the operating flow rate of the DH network system can reduce the time delay of each heat consumer. However, when the flow rate increases to a certain value, the reduction of time delay will no longer be obvious. The time delay of the DH network system is proportional to the change in pipe length and pipe diameter. The pipe diameter has a greater impact on the time delay of the DH network system than the effect of the pipe length. The water supply temperature has little effect on the time delay of the DH network system. The time delay before and after the water supply temperature change in the DH network system is almost the same.

In the future, the proposed method can simulate the CFD model to calculate the time delay of the DH network system. Compared to other simulation methods and mathematical methods, the method can quickly calculate the time delay of the DH network system under different operating parameters by changing the boundary conditions. The relationships between the influencing parameters and the time delay are very important for the control strategy of the DH network system. The time delay obtained by the CFD method is of great significance for the implementation of the operation control strategy of the DH network system and the reduction of system energy consumption. The calculation method of time delay is potential research for the optimization control of the DH network system in the future.

Author Contributions: Methodology, J.Z.; Software, Y.S.; Validation, Y.S. Formal Analysis, J.Z.; Investigation, J.Z.; Resources, J.Z.; Writing-Original Draft Preparation, Y.S.; Writing-Review & Editing, Y.S.

Acknowledgments: This research was supported by Natural Science Foundation of China, Project No. 51678398 and the Natural Science Foundation of Tianjin, Project No.18JCQNJC08400.

Conflicts of Interest: The authors declare no conflict of interest.

References

1. Lund, H.; Möller, B.; Mathiesen, B.V.; Dyrelund, A. The role of district heating in future renewable energy systems. *Energy* **2010**, *35*, 1381–1390. [CrossRef]
2. Persson, U.; Werner, S. Heat distribution and the future competitiveness of district heating. *Appl. Energy* **2011**, *88*, 568–576. [CrossRef]
3. Xu, Y.C.; Chen, Q. An entransy dissipation-based method for global optimization of district heating networks. *Energy Build.* **2011**, *43*, 50–60. [CrossRef]
4. Neuman, P.; Pokorny, M.; Weiglhofer, W. Principles of Smart Grids on the generation electrical and thermal energy and control of heat consumption within the District Heating Networks. *IFAC Proc. Vol.* **2014**, *47*, 1–6. [CrossRef]
5. Benonysson, A.; Bøhm, B.; Ravn, H.F. Operational optimization in a district heating system. *Energy Convers. Manag.* **1995**, *36*, 297–314. [CrossRef]
6. Sartor, K. Experimental validation of heat transport modeling in district heating networks. In Proceedings of the 29th International Conference on Efficiency, Cost, Optimisation Simulation and Environmental Impact of Energy Systems (ECOS 2016), Portorož, Slovenia, 19–23 June 2016.
7. Wang, N.; You, S.J.; Zheng, W.D.; Zhang, H.; Zheng, X.J. A Simple Thermal Dynamics Model and Parameter Identification of District Heating Network. *Procedia Eng.* **2017**, *205*, 329–336. [CrossRef]
8. Pirouti, M.; Bagdanavicius, A.; Ekanayake, J.; Wu, J.; Jenkins, N. Energy consumption and economic analyses of a district heating network. *Energy* **2013**, *57*, 149–159. [CrossRef]
9. Hong, S.L.; Kim, G. Prediction model of Cooling Load considering time-lag for preemptive action in buildings. *Energy Build.* **2017**, *155*, 53–65.
10. Li, X.M.; Zhao, T.Y.; Zhang, J.L. Predication control for indoor temperature time-delay using Elman neural network in variable air volume system. *Energy Build.* **2017**, *154*, 545–552. [CrossRef]
11. Jie, P.F.; Tian, Z.; Yuan, S.S.; Zhu, N. Modeling the dynamic characteristics of a district heating network. *Energy* **2012**, *39*, 126–134. [CrossRef]
12. Zheng, J.F.; Zhou, Z.G.; Zhao, J.N.; Wang, J.D. Function method for dynamic temperature simulation of a district heating network. *Appl. Therml. Eng.* **2017**, *123*, 682–688. [CrossRef]
13. Wang, H.; Meng, H. Improved thermal transient modeling with new 3-order numerical solution for a district heating network with consideration of the pipe wall's thermal inertia. *Energy* **2018**, *160*, 171–183. [CrossRef]
14. Li, L.; Zaheeruddin, M. A control strategy for energy optimal operation of a direct district heating system. *Int. J. Energy Res.* **2004**, *28*, 597–612. [CrossRef]
15. Sartor, K.; Thomas, D.; Dewallef, P. A comparative study for simulation of heat transport in large district heating network. In Proceedings of the 28th International Conference on Efficiency, Cost, Optimization, Simulation and Environmental Impact of Energy Systems (Ecos2015), Pau, France, 29 June–3 July 2015; Volume 56, pp. 174–180.

16. Gabrielaitiene, I.; Sunden, B.; Kacianauskas, R.; Bohm, B. Dynamic modelling of the thermal performance of district heating pipelines. *Am. J. Physiol.* **2003**, *211*, 95–98.

17. Palsson, H.; Larsen, H.V.; Ravn, H.F.; Bohm, B.; Zhou, J. *Equivalent Models of District Heating Systems*; Technical University of Denmark and Risø National Laboratory: Roskilde, Denmark, 1999.

18. Gabrielaitiene, I.; Bøhm, B.; Sunden, B. Modelling temperature dynamics of a district heating system in Naestved, Denmark—A case study. *Energy Convers. Manag.* **2007**, *48*, 78–86. [CrossRef]

19. Vesterlund, M.; Toffolo, A.; Dahl, J. Simulation and analysis of a meshed district heating network. *Energy Convers. Manag.* **2016**, *122*, 63–73. [CrossRef]

20. Liu, X.; Ge, X.F. FLUENT software and its application in China. *Energy Res. Util.* **2003**, *02*, 36–38.

21. Ferziger, J.H.; Peric, M. Computational methods for fluid dynamics. *Phys. Today* **1997**, *50*, 80–84. [CrossRef]

22. Blazek, J. *Computational Fluid Dynamics: Principles and Applications*; Elsevier: Oxford, UK, 2001; Volume 55, pp. 1–496.

23. Östlund, K.; Samuelsson, C.; Mattsson, S.; Rääf, C.L. The influence of 134Cs on the 137Cs gamma-spectrometric peak-to-valley ratio and improvement of the peak-to-valley method by limiting the detector field of view. *Appl. Radiat. Isotopes* **2017**, *128*, 249–255. [CrossRef] [PubMed]

Article

Thermal-Hydraulic Performance Analysis of Twin-Pipes for Various Future District Heating Schemes

Milad Khosravi [1] and Ahmad Arabkoohsar [2,*

[1] Department of Civil Engineering, Razi University, Kermanshah, Iran; khosravi.mi@razi.ac.ir
[2] Department of Energy Technology, Aalborg University, 6700 Esbjerg, Denmark
* Correspondence: ahm@et.aau.dk

Received: 7 March 2019; Accepted: 1 April 2019; Published: 4 April 2019

Abstract: Future energy systems will come with a 100% share of renewable energy and high integration of energy systems. District heating and cooling systems will be undeniable parts of the future energy systems, as they pave the bed for high-efficiency, low cost, and clean production. District heating systems may come into a wide range of designs in the future. Currently, most of the world's district heating systems are based on the third generation design while everything in this framework is on the verge of a transition to the fourth generation. A large number of technologies for the future district heating systems has been proposed so far, among which low-, ultralow- and variable-temperature systems seem more of qualification. This study employs computational fluid dynamics to make a comprehensive examination of the compatibility of regular twin-pipes with various potential district heating schemes for future energy systems. The results show that both low- and ultralow-temperature systems could efficiently use regular twin-pipes commonly used in the third generation district heating systems, though the insulation of the pipe could be proportionally strengthened based on a techno-economic trade-off. In contrast, the results show that the thermal inertia of the pipe does not allow the variable-temperature district heating system to effectively operate when the transmission pipeline is longer than a limited length. Therefore, a regular heat distribution network may not be an appropriate host for a variable-temperature district heating scheme unless decentralized heat production units come into service.

Keywords: low-temperature district heating; ultralow-temperature district heating; variable-temperature district heating; twin-pipe; thermal-hydraulic performance; thermal inertia

---

1. Introduction

District heating and cooling systems will be the key elements of future smart energy systems [1]. According to [2], in a smart energy system, there must be strong synergies between the district energy systems and other energy sectors. As such, district energy systems (including heating) show high potential for the utilization of renewable sources, where a further certain feature of the smart energy systems is the high integration of renewable energy [3].

Today, most of the world's district heating systems are based on an advanced design and characteristics so-called the third generation district heating systems [4]. Two of the main features of the third generation district heating systems are the material/components employed and their lower operating temperatures compared to the previous generations. Here, the supply temperature is below 100 °C and pressurized water is the heat carrier. The circulation is carried out by central pumps, and medium-temperature radiators or under-floor heating are the main space heating methods. Domestic hot water is supplied either directly by plate heat exchangers or by employing storage tanks and plate heat exchangers [5]. Apart from the revision of the employed technologies in the heat production chain in order to increase the efficiency, reliability, and sustainability (bringing CHP units, biomass and waste incineration plants, heat pumps, etc.), a revolution of the distribution/transmission components took place for this generation of district heating systems as well. Here, the main components are the prefabricated compact substations, the compact stainless steel plate heat exchangers, the valves, and more importantly the pre-insulated twin-pipes buried directly into the ground while in the first two generations, regular pipes with separate insulations for supply and return lines were used [6].

District heating technology is, however, on the verge of transition to its fourth generation where the temperature of supply is to be as low as possible, there is a concrete integration between the district heating system and the other energy sectors, the share of renewable energy is high, etc. [3]. Reducing the supply temperature not only reduces the rate of losses from the system (and subsequently increases the efficiency of supply and distribution) but also enables for further integration of low-temperature renewable energy and waste heat sources [7]. Generally, the lower the supply temperature, yet being sufficient to cover the demanded temperature, the more it is desirable, though it might have some challenges as well. Properly meeting the comfort standard requirements, the risk of legionella generation and technical/societal foundation preparation for this transition are of the main challenges in this regards [8]. Therefore, for the fourth generation of district heating, as a necessity, a major revision of the system components is also required to adequately address the transition challenges.

There is a large number of studies in the literature introducing and investigating new designs of district heating systems for future energy systems. Low-temperature (LT), ultralow-temperature (ULT), and variable-temperature (VT) district heating systems are some of the most interesting schemes investigated over the last years. Some of the most recent works in this framework are introduced hereunder.

Generally, the concept of LT district heating (LTDH) system offers to decrease the supply/return temperatures to about 50–55/30–35 °C [9]. Nord et al. [8] comprehensively discussed the challenges of employing LTDH systems in Norway. They concluded that the heat loss could be reduced by lowering the supply temperature from 80 °C to 55 °C, though competitiveness of LTDH might be decreased for the heat densities lower than 1 MWh/m. Im and Liu [10] presented a feasibility study on the LTDH systems with bilateral heat trade model. They proved that their proposed solution can alleviate the inefficiency arising from mismatching of heat demand and supply in the consumer side significantly. Cai et al. [11] investigated the use of electric boilers in LTDH systems and showed that by such a measure, i.e., a lower supply temperature and intelligent components, can improve the heat distribution system efficiency and turn district heating into an integrated part of the future smart energy systems. Averfalk and Werner [12] proposed a highly novel LTDH scheme where along with the lower supply and return temperatures, the system consists of three principal changes: three-pipe distribution networks, apartment substations, and longer thermal lengths for heat exchangers. Li et al. [13] investigated the integration of LTDH systems with industrial parks for the utilization of low-grade waste heat available in such sites.

In a ULT district heating (ULTDH) system, the supply and return temperatures come even to lower ranges of about 40–45 °C and 25 °C, respectively [14]. Yang and Svendsen [15] proposed the integration of a central heat pump and local boosters with a ULTDH system and investigated this proposal for a low-heat-density area in Denmark. The results showed that the proposed design of the ULTDH system has a better performance compared to the existing methods in terms of energy, exergy, and economy due to substantial savings from the distribution heat loss. Vivian et al. [16] evaluated the cost of heat for the end-users in ULTDH networks. It was concluded that the system is already competitive with individual gas boilers if a local low-temperature heat source can be recovered with minor marginal costs. Zuhlsdorf et al. [17] proposed using zeotropic mixtures for improving the performance of booster heat pumps in ULTDH systems. Although they found the mixed working fluids resulting in a higher investment cost, the economic performance was comparable to the pure fluids while the mixtures showed similar performance as the pure fluids at off-design conditions. Ommen et al. [18] analyzed the performance of ULTDH systems with utility plant and booster heat pumps, observing a performance improvement of 12% for the reference calculations when the system was supplied by central heat pumps. In another work, they designed a new heat pump furnished substation configuration for multifamily buildings in ULTDH systems, finding it energetically and economically interesting [19]. Knudsen et al. [20] developed a model predictive control for demand response of domestic hot water preparation in ULTDH systems.

In a VTDH system, however, the supply temperature is not stable and fluctuates between low and high values periodically during a day to keep the heat losses as low as possible [21]. A team of researchers in the Netherlands has continued to work on this concept. In this project, the low supply temperature is 45 °C and the high supply temperature is 70 °C. The system employs individual heat storage tanks for supplying domestic hot water of the end-users when working on low-temperature supply mode [22]. Moallemi et al. [23] presented their own version of this concept using decentralized neighborhood-scale heat storage units for a case study in Brazil and found their simulation results interesting if not considering the effects of the dynamic operation and the thermal inertia of the pipes. In another work, Arabkoohsar [24] investigated the use of decentralized heat pumps in a VTDH system to keep the transmission pipeline always at an ultralow-temperature level and bring the temperature variation in neighborhood-scale pipes.

In this study, the compatibility of twin-pipes, which are commonly used in the existing district heating systems, with the three discussed potential future district heating schemes is thoroughly investigated. Indeed, the objective is to see if the proposed future district heating systems are realistically feasible with the existing twin-pipes or there should be a revision of the pipes to make them practically implementable. For this, the thermal-hydraulic performance of such pipes is investigated via computational fluid dynamic methods (finite volume approach) in ANSYS FLUENT software.

2. Twin-Pipes in District Heating Systems

Twin-pipes came into service when district heating technology jumped into its third generation design from the second generation concept. This concept change comprised the extremely lower temperatures of supply and return and a fundamental revision in the components of the system, including the pipeline structure [25]. Figure 1 shows the schematic of a piece of a twin-pipe with its various components marked on it. As the figure shows, in a twin-pipe, both of the supply and return lines come into a single outer casing while efficient insulation (usually polyurethane foam—PUR) prevents them exchanging heat to each other and the ambient.

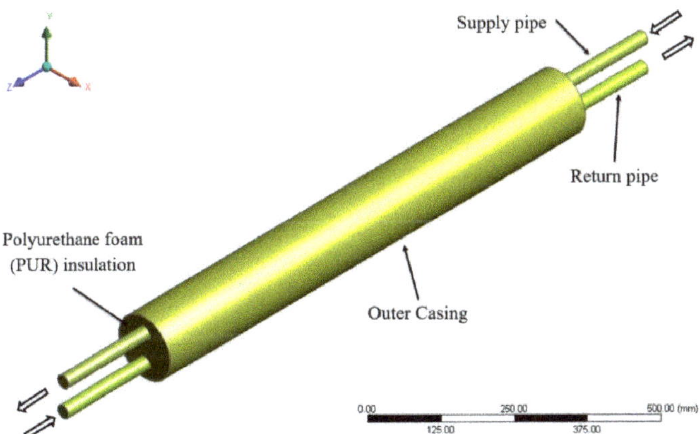

Figure 1. Schematic of a piece of a twin-pipe.

In this work, a typical twin-pipe is considered to fulfill the investigations. Table 1 presents information about the dimensions of different typical twin-pipes usually seen in various sections of a district heating pipeline including the transmission part, the street branch pipes, and the house connection pipes. As the focus of this study is on the transmission section of the pipeline, pipe type 5 is the one investigated in this work.

Table 1. Dimensions of various twin-pipes [26].

Pipe Type	Pipe			Casing	
	Nominal Diameter (mm)	Outer Diameter (mm)	Thickness (mm)	Outer Diameter (mm)	Thickness (mm)
Type 1	20	26.9	2.6	125	3
Type 2	20	26.9	2.6	140	3
Type 3	65	76.1	2.9	225	3.5
Type 4	65	76.1	2.9	250	3.5
Type 5	**200**	**219.1**	**4.5**	**560**	**6**
Type 6	200	219.1	4.5	630	6.6

Table 2. Further information about the characteristics of the pipes and district heating medium (i.e., water).

Table 2. Characteristics of twin-pipes and district heating flow through the pipes [26].

Component	Material	Density (kg/m^3)	Specific Heat (J/kg·K)	Thermal Conductivity (W/m·K)	Viscosity (kg/m·s)
Heat Carrier	Water	982	4136.5	0.65	0.001
Insulation	PUR	30	133	0.021	-
Pipe	Steel	8030	502.5	16.27	-

Note that the thermal conductivity and density of water may change as the temperature varies, however, this change is so small that it is considered negligible in this study.

Table 3 presents information about the three different district heating systems under evaluation in this work.

Table 3. Characteristics of the three district heating schemes.

District Heating Type	Supply Temperature (°C)	Return Temperature (°C)	Pressure (MPa)
ULTDH	45 (all the time)	25 (all the time)	1.2
LTDH	55 (all the time)	30 (all the time)	1.2
VTDH	70 (2 h a day)/45 (22 h a day)	35 (2 h a day)/25 (22 h a day)	1.2

3. Governing Equations and the Solution Method

In this section, the governing equations on the thermal and hydraulic behavior of twin-pipes and the heat carrier (the solid and the fluid sides) are presented. Then, the numerical solution method is explained in details.

3.1. Governing Equations

The governing equations on the motion of an incompressible flow are Reynold-Averaged Navier–Stokes, which can be written in the following format in a Cartesian coordinate [27]:

$$\frac{\partial \rho u_i}{\partial x_i} = 0, \tag{1}$$

$$\frac{\partial u_i}{\partial t} + \frac{\partial u_i u_j}{\partial x_j} = -\frac{1}{\rho}\frac{\partial p}{\partial x_i} + \frac{\partial}{\partial x_j}\left[v\left(\frac{\partial u_i}{\partial x_j} + \frac{\partial u_j}{\partial x_i} - \frac{2}{3}\delta_{ij}\frac{\partial u_l}{\partial x_l}\right) \right] + \frac{\partial}{\partial x_j}\left(-\overline{u_i' u_j'}\right), \tag{2}$$

In which, u_i, u_j, and u_l are the velocity components in the directions of i, j, and l, respectively. x_i, x_j, and x_l are the Cartesian coordinates, δ_{ij} is the Kronecker delta function, v and μ are the kinematic and dynamic viscosities, and ρ is density.

The Reynolds stresses term $(-\overline{u_i'u_j'})$ is evaluated using Boussinesq assumption as follow:

$$-\overline{u_i'u_j'} = v_t \left(\frac{\partial \overline{u}_i}{\partial x_j} + \frac{\partial \overline{u}_j}{\partial x_i} \right) - \frac{2}{3} k \delta_{ij}, \tag{3}$$

where, v_t is turbulent viscosity. For considering the effect of turbulence the Realizable k-ε model is used in this work. This model is based on the solution of the two heat transfer equations of turbulent kinetic energy (k) and dissipation of turbulence energy (ε) as below [28]:

$$\frac{\partial k}{\partial t} + \frac{\partial k u_j}{\partial x_j} = \frac{\partial}{\partial x_j}\left[\left(v + \frac{v_t}{\sigma_k} \right) \frac{\partial k}{\partial x_j} \right] + G_k + G_b - \varepsilon, \tag{4}$$

$$\frac{\partial \varepsilon}{\partial t} + \frac{\partial \varepsilon u_j}{\partial x_j} = \frac{\partial}{\partial x_j}\left[\left(v + \frac{v_t}{\sigma_\varepsilon} \right) \frac{\partial \varepsilon}{\partial x_j} \right] + C_1 S\varepsilon + C_2 \frac{\varepsilon^2}{k + \sqrt{v\varepsilon}}, \tag{5}$$

where:

$$C_1 = \max\left[0.43, \frac{\eta}{\eta + 5} \right], \; \eta = S\frac{k}{\varepsilon}, S = \sqrt{2 S_{ij} S_{ij}}. \tag{6}$$

In these equations, C_2 is a constant, σ_k and σ_ε are turbulent Prandtl numbers for k and ε. Also, G_k and G_b are the productions of turbulence kinetic energy due to the average velocity gradients and buoyancy, which can be given by the following equations, respectively.

$$G_k = -\overline{u_i'u_j'}\frac{\partial u_j}{\partial x_i} = v_t S^2 \tag{7}$$

$$G_b = \beta g_i \frac{v_t}{Pr_t} \frac{\partial T}{\partial x_i} \tag{8}$$

where, β and g_i are the coefficient of thermal expansion and the gravity acceleration component in each direction. Pr_t is the turbulence Prandtl number for energy, considered as 0.85 here. As such, S is the modulus of the mean rate-of-strain tensor, defined as:

$$S \equiv \sqrt{2 S_{ij} S_{ij}}, \tag{9}$$

$$S_{ij} = \frac{1}{2}\left(\frac{\partial u_j}{\partial x_i} + \frac{\partial u_i}{\partial x_j} \right), \tag{10}$$

When the terms k and ε and calculated, the turbulent viscosity may be calculated by the following relations:

$$v_t = C_\mu \frac{k^2}{\varepsilon}, \tag{11}$$

$$C_\mu = \frac{1}{A_0 + A_s \frac{kU^*}{\varepsilon}}, \tag{12}$$

$$U^* \equiv \sqrt{S_{ij} S_{ij} + \widetilde{\Omega}_{ij}\widetilde{\Omega}_{ij}}, \tag{13}$$

$$\widetilde{\Omega}_{ij} = \Omega_{ij} - 2\varepsilon_{ijk}\omega_k, \tag{14}$$

$$\Omega_{ij} = \overline{\Omega_{ij}} - \varepsilon_{ijk}\omega_k, \tag{15}$$

where, $\overline{\Omega_{ij}}$ is the mean rate of rotation observed in the moving reference frame with the angular velocity of ω_k. The rest of the parameters in the above formulation are given as follow:

$$A_0 = 4.04, \ A_s = \sqrt{6}\cos\phi, \tag{16}$$

$$\phi = \frac{1}{3}\cos^{-1}\left(\sqrt{6}W\right), W = \frac{S_{ij}S_{jk}S_{ki}}{\tilde{S}^3}, \tilde{S} = \sqrt{S_{ij}S_{ij}}, S_{ij} = \frac{1}{2}\left(\frac{\partial u_j}{\partial x_i} + \frac{\partial u_i}{\partial x_j}\right), \tag{17}$$

$$C_{1\varepsilon} = 1.44, \ C_2 = 1.9, \ \sigma_k = 1.0, \ \sigma_\varepsilon = 1.2, \tag{18}$$

The energy equation is used for calculating the rate of heat transfer as:

$$\frac{\partial \rho E}{\partial t} + \frac{\partial}{\partial x_i}[u_i(\rho E + p)] = \frac{\partial}{\partial x_j}\left(k_{eff}\frac{\partial T}{\partial x_j}\right), \tag{19}$$

In which, E is the total energy, p is pressure, T is temperature and k_{eff} is the effective thermal conductivity term given by:

$$k_{eff} = k + \frac{c_p \mu_t}{Pr_t}, \tag{20}$$

where, c_p is the specific thermal capacity.

Then, the energy flow for the solid side of the problem (the pipe body) can be written as:

$$\frac{\partial}{\partial t}(\rho h) + \nabla.(\vec{v}\rho h) = \nabla.(k\nabla T) + S_h, \tag{21}$$

where, h is sensible enthalpy and the second term on the left-hand side represents convective energy transfer due to rotational or translational motion of the solids. The terms on the right-hand side are the heat flux due to conduction and volumetric heat sources within the solid (S_h), respectively.

The value of heat transfer between the wall surface and the fluid for a unit area of the wall is calculated as:

$$q = h\left(T_f - T_w\right), \tag{22}$$

in which, h is the local convective heat transfer coefficient, T_f is the local fluid temperature and T_w is the local wall temperature. The heat transfer rate for the boundary of a solid cell of the pipe is calculated as below:

$$q = \frac{k_s}{\Delta n}(T_w - T_s), \tag{23}$$

where, k_s is the thermal conductivity of the solid cell, T_s is the local temperature of the solid cell, and Δn is the distance of the center of the solid cell and the wall surface.

3.2. Solution Method, Boundary Conditions, and Validation

For the simulation of the heat transfer process in this work, first, the computational space is discretized by octagonal 3-D elements in a non-uniform grid. According to the importance of the heat distribution along the radius of the pipe, the meshing has been done with very small elements. Figure 2 shows the mesh grid made on a cross section of the pipe.

The boundary condition of uniform flow for the inlets and the boundary condition of static pressure for the outlets are used. The no-slip boundary condition for the walls is also considered here. Note that for the simulation of the flow behavior close to the wall, the standard wall function is used which is broadly used for industrial flows.

Due to the large geometry of the problem and the high cost of the computations, in this work, the momentum equations (Equation (2)) are considered in the governing equations only for analyzes where the effects of turbulence are important (e.g., for calculating the average Nusselt number and pressure drop). For the rest of the computations, energy and continuity equations assuming constant and uniform velocity are used only. This assumption eliminates the need for the momentum equation and makes the effects of the turbulent fluctuations and secondary flows negligible.

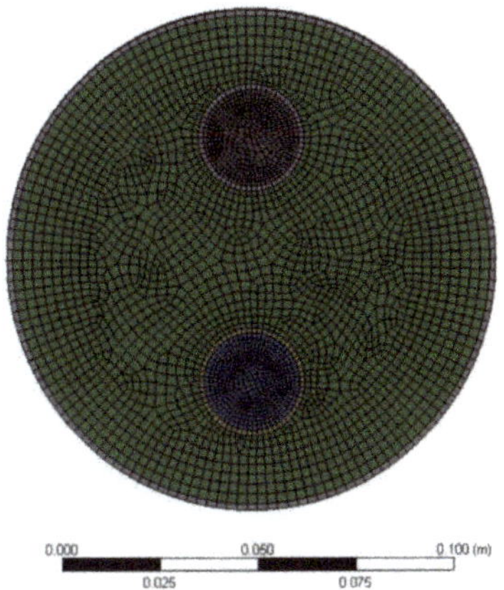

Figure 2. The mesh grid on the pipe cross-section.

The simulations are accomplished based on the finite volume method within the computational fluid dynamic model of ANSYS FLUENT.

Validation and grid sensitivity analysis:

Normally, in order to use a numerical model, one needs to first make sure about the validity of the developed model and accomplish a mesh grid sensitivity analysis. Table 4 gives information about the three different mesh grids considered in the simulations.

Table 4. Dimensions of mesh cells in different mesh grids.

Grid Type	Average Mesh Size (mm)	
	Radial Direction	Longitudinal Direction
Grid A	0.02	0.08
Grid B	0.01	0.04
Grid C	0.005	0.02

Figure 3 compares the values obtained from the simulations with various mesh grid structures (A, B, and C) for the rate of heat transfer from the twin pipes (with different nominal diameters) and those reported by the pipe manufacturer. Therefore, not only does this figure present the results of the sensitivity analysis on the mesh grids, but it also is a reference for the validity of the results of the simulations. As seen, there is a very good agreement between the results obtained from the simulations (in all the mesh grid styles) and those reported by Logstor [26]. In this figure, the graph 'CFD results: Grid C + Simplified Model', which lies almost on the graph 'CFD results: Grid C' refers to the case in which the momentum equation is not included in the simulations for the sake of simplification. As seen, this simplification does not affect the accuracy of the simulations either.

Note that for all of the pipes with different diameters, their corresponding characteristics are considered just similar to the reference cases. The supply and return temperatures are 120 °C and 70 °C, respectively.

As the pipeline length is to be quite long (10 km, here), for reducing the cost of computations, a griding towards the length direction is required to do the computations based on that. Figure 4 shows the effect of considering longer segments of the pipe in the calculations on the accuracy of the results. Here, the considered parameter is the overall heat loss rate from the pipe when used for a supply temperature of 70 °C and the return temperature is 30 °C, the pipe is made of steel, the insulation is Series 1, the nominal diameter of the pipe is 200 mm and the ambient temperature is 10 °C.

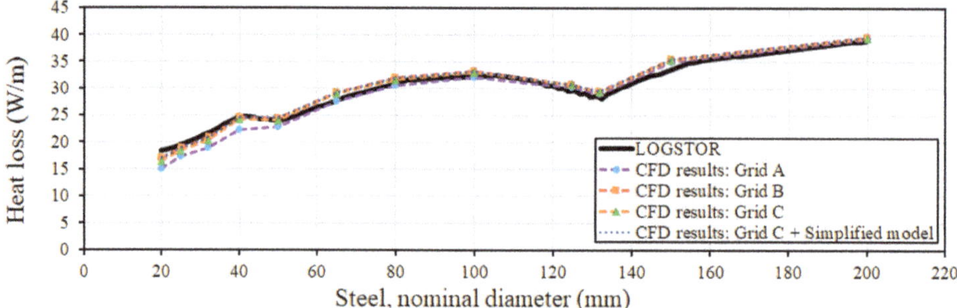

Figure 3. The mesh grid sensitivity analysis and validation of the CFD results.

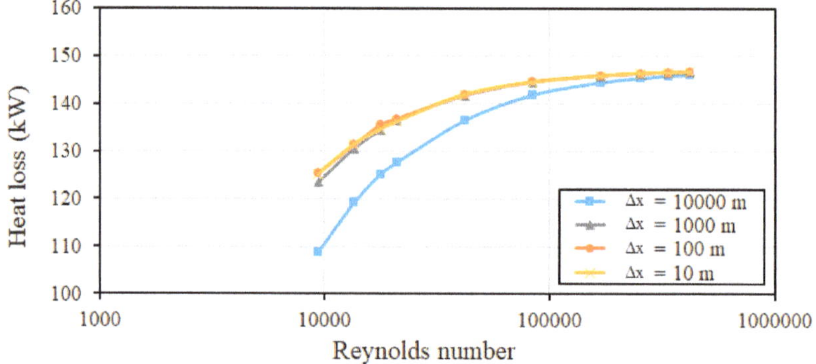

Figure 4. The effect of the grid size on length direction on the accuracy of the results.

According to the figure, as the segment length increases the accuracy of the calculations drops, however, segments up to 100 m do not affect the accuracy significantly. Here, at lower Reynolds numbers, where the deviation is larger, the error is just about 1% (from 125.1 kW to 123.5 kW) and this error decreases when Reynolds number goes to higher values (though super high Reynolds values will not happen in district heating applications as the maximum velocity through the pipes is limited to 2 m/s). Thus, for reducing the computations cost, the segment length of 100 m is considered for meshing toward length direction of the pipe in the numerical simulations.

4. Results and Discussion

In this section, the results of the simulations carried out on the pipe in various scenarios are presented and discussed.

Figure 5 presents the contours of temperature distribution in the radial direction of the insulations in each of the three cases. The legends of the contours give an indication of the temperatures that the colors represent for each case.

Figure 6 presents the profile of temperature drop in the axial direction along the supply and return lines of the pipeline (up to 10 km) for the three different scenarios. For making these graphs, the same ambient temperature (5 °C) is applied for all the three cases. As such, in order to make a fair comparison, the same heat delivery rate (5 MW), instead of the same mass flow rate, is applied. Naturally, for delivering the same amount of heat to the end-users, the mass flow rate of different cases should be different (as a direct function of the temperature difference of the supply and return lines).

Thus, the ULTDH system will result in the highest mass flow rate while the VTDH system when working on the high-temperature supply mode makes the lowest mass flow rate. It is reminded that the VTDH system has two operational modes. When it is on the low-temperature supply mode, it makes a profile just the same as the ULTDH case. As seen, the highest level of temperature drop is for the VTDH system while working on the high-temperature supply mode. It will be just about 1 °C and 0.23 °C for the supply and return lines, respectively. However, one should note that, based on the design of the VTDH system, this only applies for a short period of time a day, e.g., 2 h, and the rest of the day, the temperature drop profile will be just similar to the ULTDH system. In this case, the temperature drop values for the supply and return lines over a 10 km pipeline will be 0.33 °C and 0.07 °C, respectively. The LTDH system presents the temperature drop values of 0.52 °C and 0.14 °C as moderate temperature drop values among the three cases.

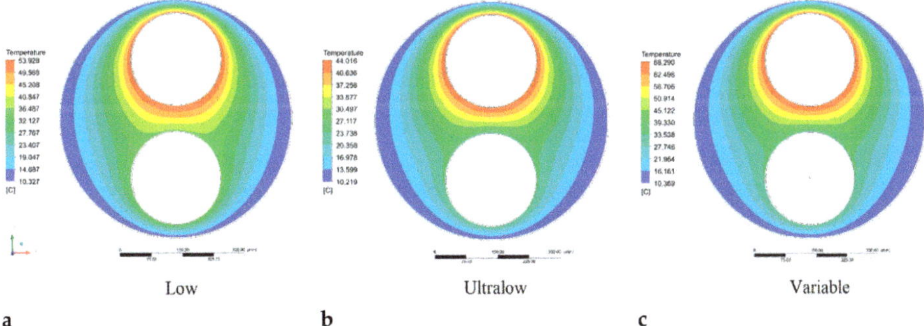

Figure 5. Contours of radial temperature distribution for the pipes through the insulations. (**a**) Low-Temperature DH systems; (**b**) Ultralow-Temperature DH system; (**c**) Variable-Temperature DH system.

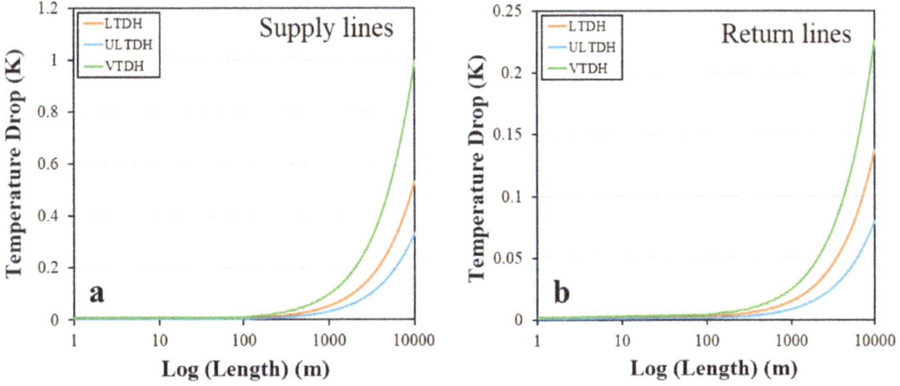

Figure 6. The temperature profile of the supply and return lines for different cases. (**a**) supply lines; (**b**) return lines.

Figure 7 investigates the rate of temperature drop along a 10 km pipeline for the three different cases if the same mass flow rate is applied through the pipe. Here again, the twin-pipe type 5 is considered for all the cases, the supply/return temperatures are 45/25 °C, 55/30 °C, and 70/35 °C for the ULTDH, LTDH, and VTDH systems, respectively, and the surrounding temperature is 5 °C. Clearly, when the same mass flow rate is passing through a specific pipe, a higher temperature will make a larger heat loss, and subsequently, a larger temperature drop. Thus, expectedly, the VTDH system (in high-temperature supply mode) makes a higher temperature drop compared to the others. Naturally, when VTDH system goes to ultralow supply mode, the same trend as the ULTDH system will be observed. The LTDH system results in a moderate temperature drop among the three cases again. A similar trend is observed for the return lines as well. Comparing the values of losses in this figure and Figure 6, one could see that the level of temperature drop in this case is lower. This is mainly due to the fact that the maximum velocity of 2 m/s has been considered for all the three cases, which gives a shorter time to the district heating water to exchange heat with the surrounding.

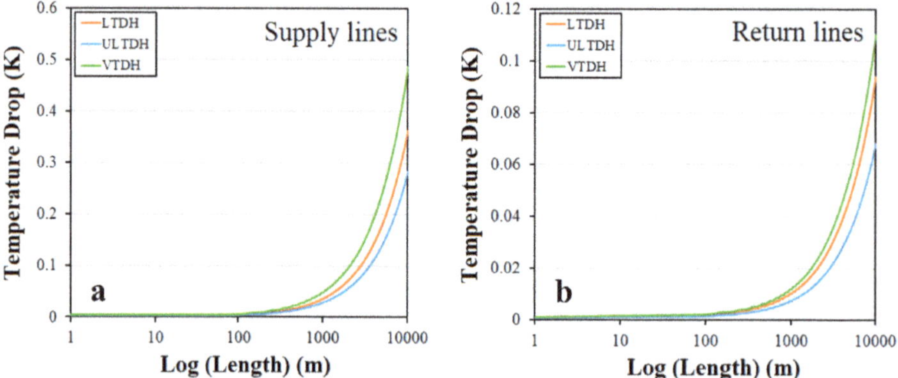

Figure 7. The rate of temperature fall along the pipeline for the three scenarios when the same mass flow rate is applied. (**a**) supply lines (**b**) return lines.

Figure 8 investigates the effect of Reynolds number on the level of total heat loss through the pipes in different cases. This includes the heat loss from the supply lines (the left panel), and the return lines (the right panel). As expected, regardless of the Reynolds number, for the same mass flow rate, the VTDH system while working in the high-temperature supply mode makes the higher rate of loss while the ULTDH system results in the lowest rate of heat loss. This is true for both the supply and return lines. On the other hand, the increase in the Reynolds number makes a growth in the rate of loss, though this effect is not that significant.

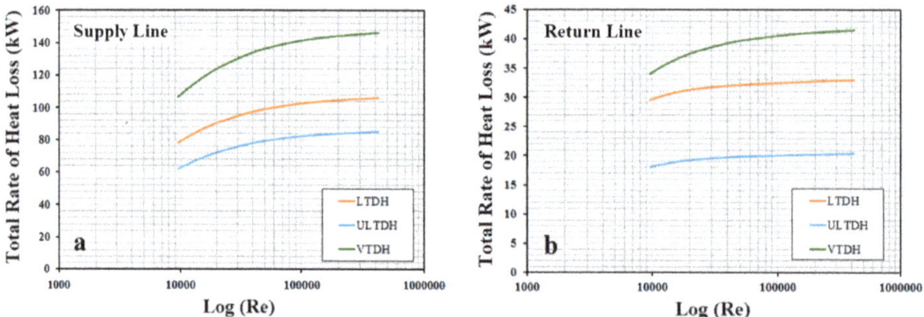

Figure 8. The effect of Reynolds number on the level of heat loss through the supply and return lines in each case. (**a**) supply lines; (**b**) return lines.

Figure 9 shows the total rate of heat loss from both of the supply and return lines along the 10 km pipelines. This figure, indeed, gives an indication of the summation of the losses reported in the above figure for each case. Naturally, the trends remain the same and the ULTDH system results in the lowest rate of loss while the LTDH system presents an average level of loss and the VTDH system (when in high-temperature supply mode) gives the highest rate of total heat loss.

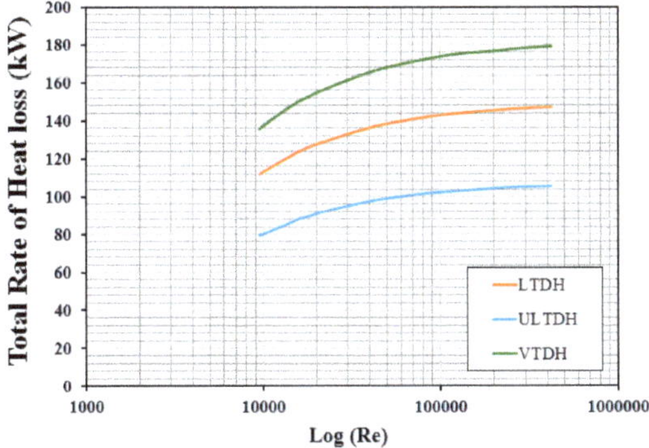

Figure 9. The effect of Reynolds number on the level of total rate of heat losses over the entire pipeline in each case.

Figure 10 illustrates the effect of Reynolds number on the level of temperature drop for the supply and return lines over the entire pipeline for each of the three scenarios. Again, the VTDH system while working at the high-temperature supply mode makes the largest value of temperature drop in the supply line while the ULTDH system makes the lowest value of temperature drop through that.

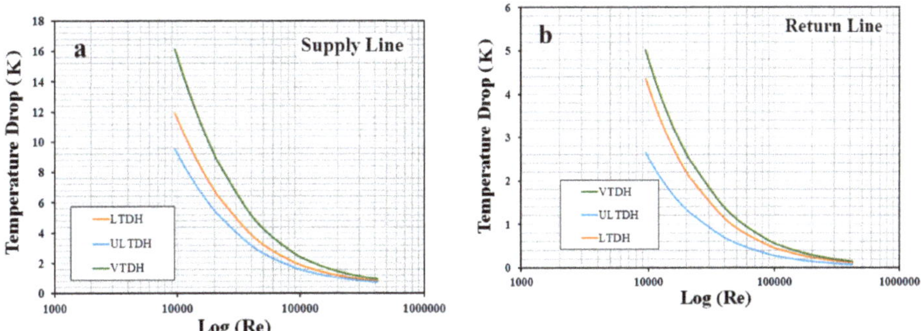

Figure 10. The effect of Reynolds number on the level of temperature drop over the entire pipeline in each case. (**a**) supply lines; (**b**) return lines.

Figure 11 illustrates the profile of pressure for a unit meter of the pipeline for the three cases. Here again, to make a fair comparison, the same amount of heat delivery (rather than a similar mass flow rate) is applied to the pipes in different cases. The presented pressure drop profiles are related to the supply lines only, as almost the same values as the supply lines are obtained for the return lines. Note that the value of pressure drop increases proportionally as the length of the pipe increases. The pressure drop values are important because they can represent the amount of work required for running the booster pumps along the paths for compensating the pressure losses through the pipes.

Figure 12 shows the variation of Nusselt number for each of the three cases as a function of Reynolds number. As seen, there is not a significant difference between different cases, neither in the supply lines nor through the return lines.

Figure 13 presents the profile of the heat loss from the supply and return lines of each case as a function of the angle. The right panel in the figure shows how the angle is considered, where the upper tube is the supply line and the lower tube is the return line. According to the figure, at angle 0°, where the supply line is closer to the surrounding soil, the loss is at its maximum level while it gets minimal when the angle is close to 180°. For the return line, the trend is revers because the supply line at angle 0° heats it up, and the losses get maximal when the angel goes toward 180°. The noteworthy point is that the heat loss from the supply line to the return line is not considered as a loss for the whole pipe because it remains in the system.

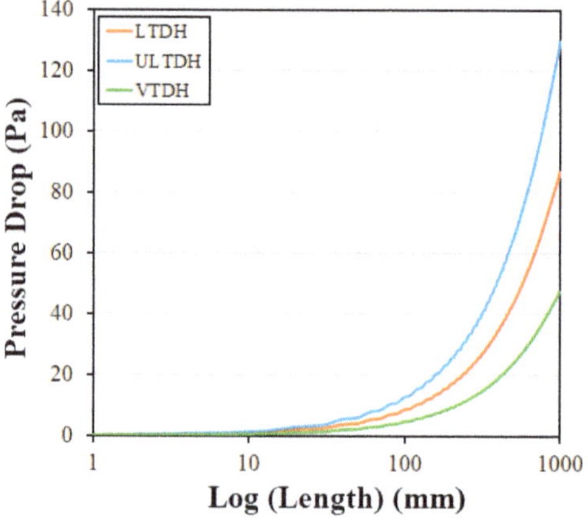

Figure 11. The level of pressure drop through the supply lines in each case.

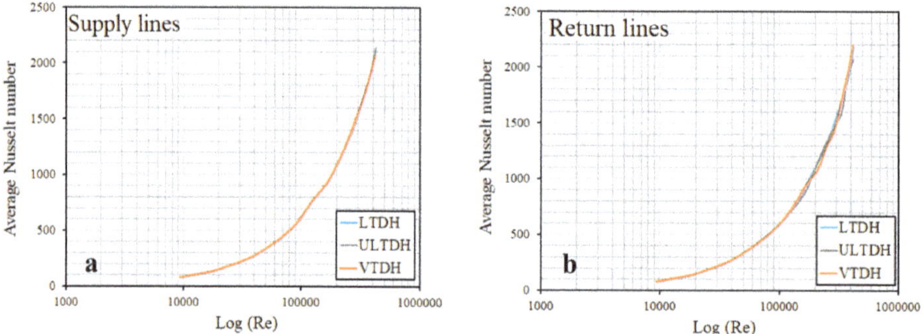

Figure 12. The effect of Reynolds number on the Nusselt number through the pipes in each case. (**a**) supply lines; (**b**) return lines.

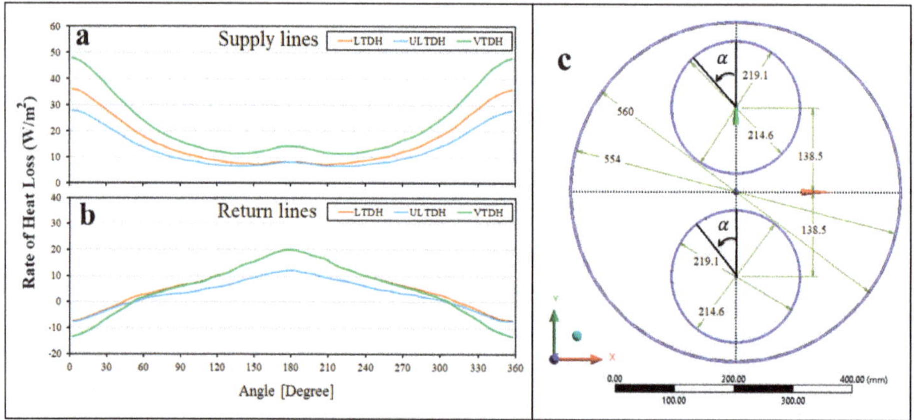

Figure 13. The rate of heat loss from the supply and return pipes as a function of angle in each case. (**a**) supply lines; (**b**) return lines; (**c**) pipe dimensions and angles.

Figure 14 shows the effect of variation in the insulation thickness of each case on the rate of heat losses. The considered range for the thickness of the insulation come in the form of the outer diameter of the casing just above the outer diameter of the pipes up to an outer diameter of 870 mm. According to the figure, although the effect of increasing the thickness of the insulation gets milder as the thickness grows, a significant reduction in the rate of loss is seen in the lower ranges of the outer diameter. For example, an increase of the outer diameter from 554 mm to 590 mm results in about 20%, 18%, and 14% less heat loss from the supply lines of the VTDH system (from 150 kW to 123 kW), the LTDH system (from 110 kW to 90 kW) and ULTDH (from 85 kW to 72 kW), respectively. The rates of improvement for the return lines are even more impressive, where for the VTDH, LTDH, and ULTDH systems the saving rates of respectively 33%, 32%, and 30% are observed for the same amount of insulation strengthen. The heat loss values (before and after increasing the insulation thickness) for the whole pipe, including both supply and return lines, are 180 kW to 143 kW, 138 kW to 110 kW, and 105 kW to 83 kW, respectively. Since the main objective of lowering the operating temperatures in district heating systems is getting a lower rate of loss, the reinforcement of the insulations in the transmission pipelines compared to the existing standard pipes could be highly helpful. Of course, this needs an optimization based on techno-economic considerations to see how much it would cost to make such a reinforcement of the insulation per meter of the pipe and how much benefit it would make. This will, naturally, be different for various district heating schemes.

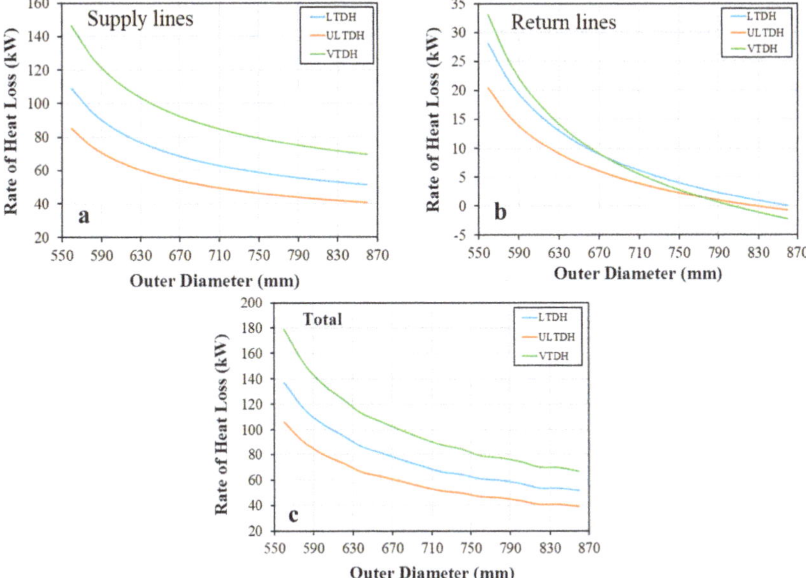

Figure 14. The effect of strengthening the insulation of pipes for various cases. (**a**) supply lines; (**b**) return lines; (**c**) total.

Figure 15 shows the variation of the outlet temperature of the supply and return lines of the VTDH system over time. The main objective is to investigate the effect of the thermal inertia of the pipe to see if this parameter allows the system to operate as the proposers [22] and investigators [23] of the system expect. For this, it is assumed that the system is in its ultralow-temperature supply mode (45/25 °C), and then suddenly, the high-temperature supply mode starts (70/35 °C). Naturally, as the operation mode changes, due to the long length of the pipe (10 km) and the limited velocity of the flow through the pipes (2 m/s here), even if there is no external parameter affecting the flow, it would take time to reach the outlet point. In addition, it can be seen that even if the effects of the thermal inertia of the pipe is neglected, the external effects cause the outlet temperature to gradually approach the supply temperature. This will even take longer (almost double) when the effects of the thermal inertia of the pipe are taken into account. A similar trend is observed for the return line as well. Therefore, it is clearly seen that a VT supply could not be effectively taken for a pipeline as long as 10 km, which is a very regular long of transmission pipeline in a district heating system.

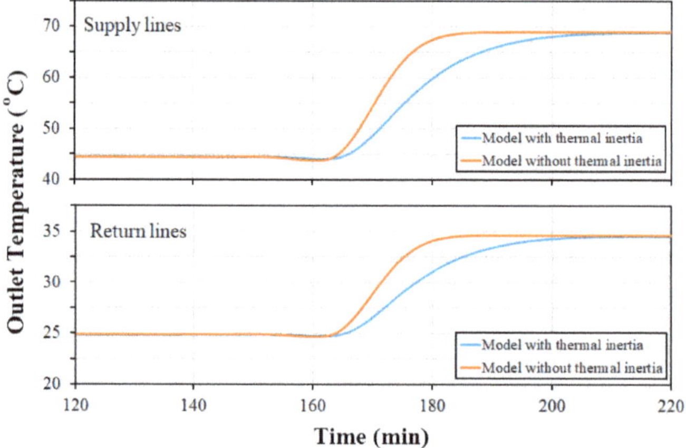

Figure 15. The effect of thermal inertia on the outlet temperature of the Variable-temperature district heating (VTDH) system.

Figure 16 compares the trend of variation of the outlet temperature of a VTDH system when the length of the pipeline is 10 km and 1 km. The former is as the previous case and the latter is based on what Arabkoohsar [24] proposed for the local use of decentralized heat pumps for increasing the temperature from 45 °C to 70 °C, which makes the length of pipe associated with the variation of the temperature below 1 km. For this, a complete high-temperature, ultralow-temperature supply process, based on the given definition for this system, is accomplished. That is, the pipeline is first at 45 °C, then suddenly two hours of 70 °C supply, followed by 2 h of 45 °C supply, is applied.

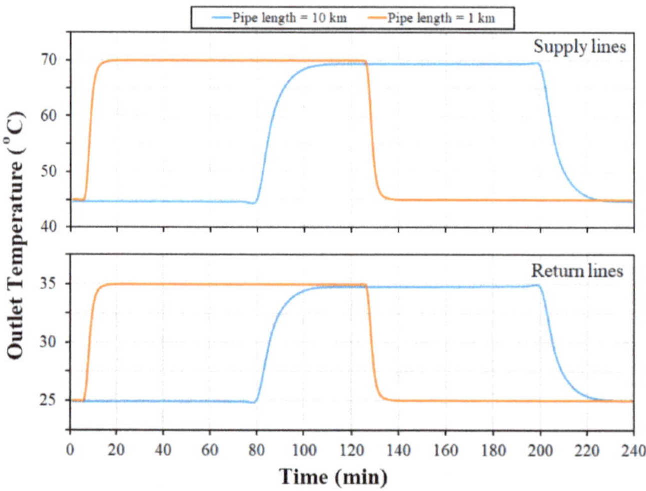

Figure 16. The trend of outlet temperature in a VTDH system with different pipe lengths.

According to the figure, the outlet temperature of the long pipe gets quite disordered and kind of unexpected while the small length of the pipe for the second case makes the desired outlet temperature achievable after a short delay. This results in a favorable outlet temperature profile for the VTDH system with a short length of 1 km (e.g., a street branch pipe).

5. Conclusions

This study presents a thermal-hydraulic performance analysis of regular twin-pipes in the third generation district heating systems to investigate how they fit into the future district heating schemes. After presenting the general thermal and hydraulic performance assessment results of such pipes for all the three considered cases, i.e., the ULTDH, the LTDH and the VTDH systems, the effect of changing the insulation thickness and the thermal inertia of the pipe on the performance of each of the cases were investigated.

The results show that, expectedly, the ULTDH system shows the lowest rate of loss and temperature drop along the pipe, where the VTDH system results in a higher rate of loss and temperature drop when working on a high-temperature supply mode. The results show that the twin-pipe (as they are) may fit the ULTDH and LTDH systems while there seems to be space for strengthening the insulation of the pipes for a further reduction of the rate of losses in a cost-effective way. This, however, requires an economic trade-off to see how much would be gained for a certain amount of reinforcement of the insulation and how much this will increase the cost of the pipes.

On the other hand, it was shown that the concept VTDH system, as proposed in [22,23] would never be practically applicable. This is mainly because of the effect of the thermal inertia of the pipe. The other scheme of the VTDH system, i.e., the one proposed by Arabkoohsar [24], however, does not suffer from this problem as this concept brings the dynamic supply temperature to the pipes close to the end-users (below 1 km distance). Although the thermal inertia of the pipe affects the practically achievable supply temperature of the end-users in this system as well, the effect is so small that the system can reliably be employed by the regular existing twin-pipes.

Author Contributions: Simulations, validation of results and software tasks of this research work was carried out by M.K. A.A. contribution was supervision of the work, interpretation of the results and writing of the manuscript.

Funding: This research received no external funding.

Conflicts of Interest: The authors declare no conflict of interest.

Nomenclature

Parameter	Explanation
$A_o, C, \sigma_k, \sigma_\varepsilon$	Turbulence model constant
g_i	Gravity acceleration component
h	Convective heat transfer coefficient/enthalpy
k	Turbulent kinetic energy
k_{eff}	Effective thermal diffusivity (heat transfer coefficient)
k_s	Thermal conductivity of solid cell
LTDH	Low-temperature district heating
Pr	Prandtl number
Pr_t	Turbulent Prandtl number
q	Rate of heat transfer/loss
T_f	Local fluid temperature
T_w	Local wall temperature
u_i, u_j, u_l	Velocity components in the i, j and l direction
ULTDH	Ultralow- temperature district heating
VTDH	Variable-temperature district heating
x_i, x_j, x_l	Cartesian coordinates in the i, j and l direction

Greek Symbols

v	Kinematic viscosity
μ	Dynamic viscosity
v_t	Turbulent viscosity
ρ	Density
δ_{ij}	Kronecker delta function
ε	Dissipation of turbulence energy
β	Coefficient of thermal expansion
ω	Angular velocity

References

1. Dincer, I.; Acar, C. Smart energy systems for a sustainable future. *Appl. Energy* **2017**, *194*, 225–235. [CrossRef]
2. Mathiesen, B.V.; Lund, H.; Connolly, D.; Wenzel, H.; Østergaard, P.A.; Möller, B.; Nielsen, S.; Ridjan, I.; Karnøe, P.; Sperling, K.; et al. Smart Energy Systems for coherent 100% renewable energy and transport solutions. *Appl. Energy* **2015**, *145*, 139–154. [CrossRef]
3. Lund, H.; Duic, N.; Østergaard, P.A.; Mathiesen, B.V. Smart energy systems and 4th generation district heating. *Energy* **2016**, *110*, 1–4. [CrossRef]
4. Werner, S. District heating and cooling in Sweden. *Energy* **2017**, *126*, 419–429. [CrossRef]
5. Werner, S. International review of district heating and cooling. *Energy* **2017**, *137*, 617–631. [CrossRef]
6. Mazhar, A.R.; Liu, S.; Shukla, A. A state of art review on the district heating systems. *Renew. Sustain. Energy Rev.* **2018**, *96*, 420–439. [CrossRef]
7. Ziemele, J.; Gravelsins, A.; Blumberga, A.; Vigants, G.; Blumberga, D. System dynamics model analysis of pathway to 4th generation district heating in Latvia. *Energy* **2016**, *110*, 85–94. [CrossRef]
8. Nord, N.; Nielsen, E.K.L.; Kauko, H.; Tereshchenko, T. Challenges and potentials for low-temperature district heating implementation in Norway. *Energy* **2018**, *151*, 889–902. [CrossRef]
9. Gadd, H.; Werner, S. Achieving low return temperatures from district heating substations. *Appl. Energy* **2014**, *136*, 59–67. [CrossRef]
10. Im, Y.-H.; Liu, J. Feasibility study on the low temperature district heating and cooling system with bi-lateral heat trades model. *Energy* **2018**, *153*, 988–999. [CrossRef]
11. Cai, H.; You, S.; Wang, J.; Bindner, H.W.; Klyapovskiy, S. Technical assessment of electric heat boosters in low-temperature district heating based on combined heat and power analysis. *Energy* **2018**, *150*, 938–949. [CrossRef]
12. Averfalk, H.; Werner, S. Novel low temperature heat distribution technology. *Energy* **2018**, *145*, 526–539. [CrossRef]
13. Li, Y.; Xia, J.; Su, Y.; Jiang, Y. Systematic optimization for the utilization of low-temperature industrial excess heat for district heating. *Energy* **2018**, *144*, 984–991. [CrossRef]
14. Yang, X.; Svendsen, S. Achieving low return temperature for domestic hot water preparation by ultra-low-temperature district heating. *Energy Procedia* **2017**, *116*, 426–437. [CrossRef]
15. Yang, X.; Svendsen, S. Ultra-low temperature district heating system with central heat pump and local boosters for low-heat-density area: Analyses on a real case in Denmark. *Energy* **2018**, *159*, 243–251. [CrossRef]
16. Vivian, J.; Emmi, G.; Zarrella, A.; Jobard, X.; Pietruschka, D.; de Carli, M. Evaluating the cost of heat for end users in ultra low temperature district heating networks with booster heat pumps. *Energy* **2018**, *153*, 788–800. [CrossRef]
17. Zühlsdorf, B.; Meesenburg, W.; Ommen, T.S.; Thorsen, J.E.; Markussen, W.B.; Elmegaard, B. Improving the performance of booster heat pumps using zeotropic mixtures. *Energy* **2018**, *154*, 390–402. [CrossRef]
18. Ommen, T.; Thorsen, J.E.; Markussen, W.B.; Elmegaard, B. Performance of ultra low temperature district heating systems with utility plant and booster heat pumps. *Energy* **2017**, *137*, 544–555. [CrossRef]
19. Thorsen, J.E.; Ommen, T. Field experience with ULTDH substation for multifamily building. *Energy Procedia* **2018**, *149*, 197–205. [CrossRef]
20. Knudsen, M.D.; Petersen, S. Model predictive control for demand response of domestic hot water preparation in ultra-low temperature district heating systems. *Energy Build.* **2017**, *146*, 55–64. [CrossRef]
21. Johnson, R.C. *Variable Temperature Water District Heating*; ResearchGate: Berlin, Germany, 1986.

22. Rindt, C. Optimization of a Variable Temperature 4th Generation District Heating Network Using the CONICO Thermo-Differential Valve Technology. Available online: https://www.researchgate.net/project/Optimization-of-a-variable-temperature-4th-generation-district-heating-network-using-the-CONICO-thermo-differential-valve-technology (accessed on 1 April 2019).
23. Moallemi, A.; Arabkoohsar, A.; Pujatti, F.J.P.; Valle, R.M.; Ismail, K.A.R. Non-uniform temperature district heating system with decentralized heat storage units, a reliable solution for heat supply. *Energy* **2018**, *167*, 80–91. [CrossRef]
24. Arabkoohsar, A. Non-uniform temperature district heating system with decentralized heat pumps and standalone storage tanks. *Energy* **2019**, *170*, 931–941. [CrossRef]
25. Lund, H.; Werner, S.; Wiltshire, R.; Svendsen, S.; Thorsen, J.E.; Hvelplund, F.; Mathiesen, B.V. 4th Generation District Heating (4GDH): Integrating smart thermal grids into future sustainable energy systems. *Energy* **2014**, *68*, 1–11. [CrossRef]
26. Logstor, D.H. *Pipeline Developer*. Available online: https://www.logstor.com/ (accessed on 3 April 2019).
27. White, F.M. *Fluid Mechanics*; Mc Graw Hill Book Company: New York, NY, USA, 1986.
28. Tennekes, H.; Lumley, J.L. *A First Course in Turbulence*, 6th ed.; MIT Press: Cambridge, MA, USA, 1972.

Article

Analyzing the Performance and Control of a Hydronic Pavement System in a District Heating Network

Stefan Blomqvist [1,*], Shahnaz Amiri [1,2], Patrik Rohdin [1] and Louise Ödlund [1,†]

[1]   Division of Energy Systems, Department of Management and Engineering, Linköping University, SE-581 83 Linköping, Sweden; shahnaz.amiri@liu.se (S.A.); patrik.rohdin@liu.se (P.R.); louise.odlund@liu.se (L.Ö.)

[2]   Division of Building, Energy and Environment Technology, Department of Technology and Environment, University of Gävle, SE-801 76 Gävle, Sweden

*   Correspondence: stefan.blomqvist@liu.se; Tel.: +46-132-88-932

†   Louise Ödlund (former Trygg).

Received: 29 April 2019; Accepted: 27 May 2019; Published: 30 May 2019

Abstract: A hydronic pavement system (HPS) is an alternative method to clear snow and ice, which avoids the use of salt, sand, and fossil fuel in conventional snow clearance, and minimizes the risk of accidents. The aim is to analyze the performance of different control strategies for a 35,000 m^2 HPS utilizing heat from a district heating and cooling (DHC) system. The key performance indicators are (1) energy performance of the HPS, and (2) primary energy use, (3) electricity production and (4) greenhouse gas (GHG) emissions from the DHC system. The methodology uses a simulation model of the HPS and an optimization model of the DHC system. Three operational strategies are analyzed: A reference scenario based on the current control strategy, and scenarios where the HPS is shut down at temperatures below −10 °C and −5 °C. The study shows that the DHC return temperature is suitable for use. By operational strategies, use during peak demand in the DHC system can be avoided, resulting in reduced use of fossil fuel. Moreover, the energy use of the HPS could be reduced by 10% and the local GHG emissions by 25%. The study emphasizes that the HPS may have positive effects on global GHG emissions, as it enables electricity production from renewable resources.

Keywords: hydronic pavement system; district heating; primary energy use; energy system modeling; greenhouse gas emissions

---

1. Introduction

In 1976, Sweden established the following definition for the heating of ground surfaces such as pavement areas: "Ground heat refers to devices for raising the surface temperature in order to avoid slipping, keeping the surface free of snow and ice or prolonging the vegetation period" [1]. Moreover, a system of 2000 m^2 was installed in downtown Klamath Falls (Oregon, USA) in 1948 [2]. It operated for 50 years before being replaced due to external corrosion of the iron pipes [3]. The systems of today make use of electrical, infrared or, most commonly, hydronic techniques [4,5]. A hydronic pavement system (HPS) is a technique in which heat is transported in pipes embedded in the pavement structure using circulating water or other liquid heat media.

HPS can utilize different energy sources, such as geothermal energy, district heating or solar, and in some cases, it is also combined with thermal storage [6]. A review conducted by Lund and Boyd [7] states that an area of 2,500,000 m^2 world-wide is heated by HPS utilizing geothermal energy, with required power of 130–180 W/m^2 [7]. In addition, standalone solutions can be found, particularly in airports [8–12], and bridge decks [13,14]. In Sweden, there are 400 systems utilizing district heating as a heat source with annual use of 150–200 GWh [15]. The technique is increasingly common in for example Iceland, as Gunnlaugsson et al. [16] mention that systems such as HPS may utilize the low-grade heat of district heating return water.

Pan et al. [6] state that the technology of HPS is not only suitable for melting snow and ice. Pavements are exposed to a large amount of solar radiation, causing the pavement to reach temperatures of 55–75 °C [17]. This can impair the performance of the material, as well as causing urban heat islands [6,18]. A possible solution is to use pavements for energy harvesting [19], with pipes embedded in the structure channeling away the heat for immediate use elsewhere or for seasonal energy storage, to be used by a DHC system, for electrical purposes or recharging [6].

When compared to conventional snow clearance, the most common arguments in favor of using an HPS are the avoidance of sand, of potential material damage from conventional snow clearance, and of salt with its negative local environmental effects [20–26]. Moreover, conventional snow clearance contributes to greater greenhouse gas (GHG) emissions than HPS [11,27], since HPS enables the use of renewable energy, compared with the use of fossil fuel to operate heavy machinery as snow plows. Furthermore, it is desirable to avoid heavy machinery in crowded areas, which also reduces the risk of material damage to the pavement structure or its surroundings [1,28,29]. Crowded areas, such as commercial streets, squares, entrances, stairs or other areas with intensive use, are suitable areas for HPS [1]. Figure 1 displays such areas, a square and walkway, in the central parts of Linköping, Sweden, which is the location of this study.

(a)

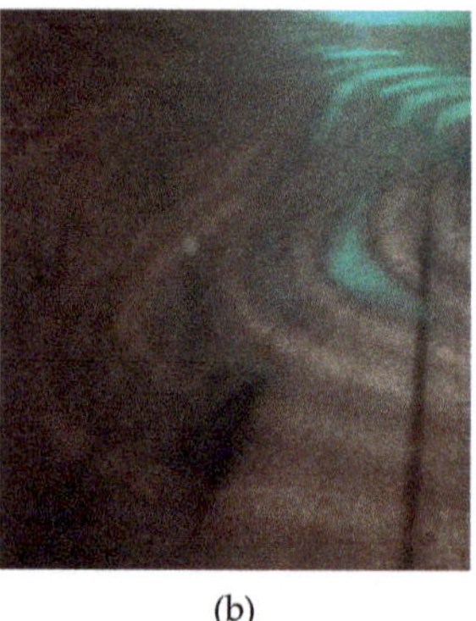

(b)

Figure 1. Pictures from the center of Linköping, Sweden. (**a**) A square at an outdoor temperature of −4 °C with an active hydronic pavement system (HPS) in the outer perimeter, plus conventional snow clearance and use of sand in the middle. (**b**) A walkway shortly after precipitation at 0 °C. The positioning of the HPSs embedded pipes can be discerned when the surface starts to dry up.

The most common cause of accidents involving pedestrians, often older people, in Sweden is slipping, with 74% caused by snow and ice formation [30]. The accidents tend to occur in central areas [31,32] that are suitable for HPS. A study conducted in Sweden by Carlsson et al. [31] indicates that 80% of slipping accidents could be prevented by HPS. Moreover, studies indicate that the cost of injured pedestrians is more than four times higher than the winter maintenance. Accordingly, it has been shown to be cost-efficient to invest more in winter maintenance in pedestrian areas from a national perspective [32,33]. A study made by Nevalainen Henaes [27] in 2018 in Lund, Sweden, put the estimated construction cost of an HPS at 100 EUR/m^2, whilst the maintenance cost is the same as conventional snow removal at 3 EUR/m^2. The case study concludes the annual cost of slipping accidents to be 90 EUR/m^2.

A control strategy for the HPS may be used in order to balance performance criteria, environmental criteria, and energy use [34]. Other studies have proposed ideas to minimize energy use, e.g., by stop heating during cold sub-zero temperatures and, thus, reduce the energy use [34,35]. Moreover, there is extensive research done to study strategies by steady-state models, as well as transient models, assessing energy use for different specific weather phenomena and locations [34–36], and many studies have been done in connection to geothermal heating [5]. However, this research intends to study a control strategy on an HPS utilizing district heating during a whole heating season using a transient model.

The national targets of Sweden, originating from the targets of the European Council [37], state that by 2020 GHG emissions shall be decreased by 40% (relative to 1990) and no net emissions shall occur by 2045. By 2020, the share of renewable energy shall be 50% of total energy use, and by 2040 electricity production shall be from 100% renewable resources [38,39]. The Swedish government [38] also stated in 2008 that fossil fuel is not to be used for heating purposes. This has led energy companies operating DHC system to phase out fossil fuel from their production by 2030 or by 2025, as some companies have committed to [40]. District heating utilizing the technology of combined heat and power (CHP) plants is said to provide an opportunity to make use of energy that would otherwise be wasted [38,39].

Studies emphasize the role of the CHP technique in the energy transition from fossil fuel to renewable resources [41–44]. Furthermore, considering the work of reducing GHG emissions, the potential electricity output from European CHP plants could be more than doubled [45]. On a global level, there is a low utilization and low awareness of the benefits of DHC and there is a potential for DHC systems to be a viable option of the future energy system [46]. However, studies also point to the unclear role of a DHC system in a future energy system, where questions regarding surplus electricity from intermittent sources and future access to conventional fuel as waste and biofuel are unexplored [47]. Stankeviciute et al. [45] argue that competition regarding biomass between the transport sector and other sectors will act as a limitation on the potential of CHP. The potential for a reduction in global GHG emissions highly depends on whether biofuel is seen as a limited or unlimited resource and on the alternative use of biofuel [48]. Studies also highlight potential issues where CHP plants are unprofitable in the future Nordic market, with a prevailing trend of heat-only boilers replacing CHP plants in DHC production [49].

Achieving low return temperatures is an important factor in obtaining an efficient DHC system, as it may increase the heat recovery from flue gas condensation and electricity generation in the CHP plants, as well as increasing the potential use of excess heat from industrial processes [50,51]. Moreover, the economic value of a reduced return temperature varies between 0.05 and 0.5 EUR/MWh, °C [52]. The average return temperature in Swedish DHC systems is 47.2 °C [50]. In order to have an efficient system, it is necessary to have as large difference as possible in the supply and return temperature [50], and in a future fourth generation of DHC systems, the supply temperature will most likely be reduced [53]. It will, therefore, be even more important to achieve low return temperatures, and in this work, an HPS utilizing the return temperatures of the DHC system can make a useful contribution.

The aim of this paper is to analyze the consequences of different control strategies for an HPS in a DHC network. The key performance indicators are (1) energy performance of the HPS, (2) the resulting primary energy use, (3) electricity production and (4) greenhouse gas emissions from the DHC system. The study was conducted by using a unique transient model and simulating three scenarios for operations of the HPS. Thereafter, the results were analyzed in the DHC system settings using an optimization model.

The main contribution of this work is the study of the environmental effects of an HPS, in terms of local GHG emissions and the effects on a global level. Also, as a contribution, the unique simulation model created in ANSYS software utilizing weather data of outdoor temperature, precipitation, wind, and solar irradiance enables detailed studies on how operational strategies may contribute to more efficient energy use.

2. Method

A framework including the development of a numerical simulation model and an optimization model was designed for this study, presented in Figure 2. A transient simulation model of the HPS system was created using the software ANSYS®Workbench™ and the patch ANSYS®CFX®Release 18.0 [54]. Three scenarios are analyzed in this setup. Data collection of material and weather properties plus boundary equations are inputs to the transient simulation model of the HPS. The simulation results are then scaled up to a larger system area, and the time step is converted into a flexible time division suitable for larger energy systems by using the software Converter [55]. The scenarios are then analyzed within the scope of the DHC system using the linear optimization program MODEST [56]. Sections 2.1 and 2.2 present the software tools used in the study, while Sections 3 and 4 describes the design of the scenario study as well as design and use of the software tools.

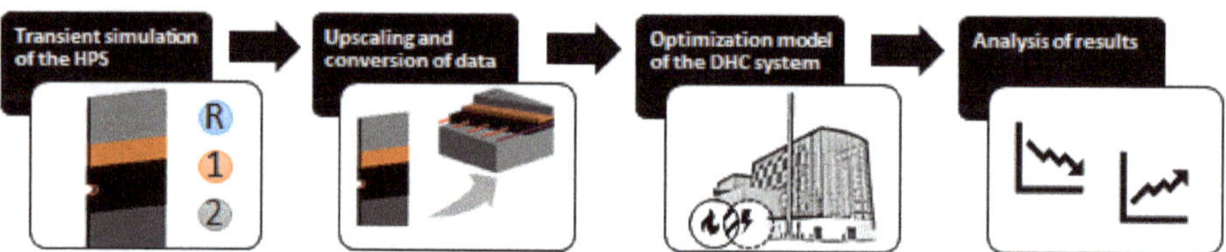

Figure 2. A course illustration of the framework of this study, which includes development of a transient simulation model of the HPS, upscaling, and use of an optimization model of the district heating and cooling (DHC) system. A general description of the software tools used is presented in Sections 2.1 and 2.2. The design of the scenario study, inputs, and use of the software tools are presented in Sections 3 and 4.

2.1. The Simulation Software ANSYS

The ANSYS software is an engineering simulation platform used to analyze how product designs will behave in an operative environment. The tool ANSYS CFX used in this study is general purpose computational fluid dynamics software capable of modeling, e.g., transient flows, heat transfer, and thermal radiation. The governing equations in ANSYS CFX are the unsteady Navier–Stokes equations [57] in their conservation form, which describe momentum, heat, and mass transfer. ANSYS uses the finite element method to reach a numerical solution, by iteratively solving the equations for each element and in so doing deriving a full picture of the flow in the model [58].

2.2. The Optimization Software MODEST

MODEST, short for "model for optimization of dynamic energy systems with time-dependent components and boundary conditions", is a optimization software utilizing linear programming which was developed at Linköping University [59]. MODEST is structured according to energy flows, starting with fuel that, via conversion and distribution, serves a demand, making it suitable for analysis of large energy systems. The model's objective is to minimize the system cost to supply demand [59]. Hence, the results from MODEST will represent a cost-effective production mix, which will be used to analyze the key performance of this paper. Other results to analyze are the system's GHG emissions, expressed as CO_2 equivalents, peak power, and primary energy use in the production mix. The strength of MODEST is the scope for arbitrary prerequisites regarding geographical, sectoral and temporal conditions, and energy carrier [60]. MODEST may be used to analyze different energy systems and components, both on a local and national level. The software was mainly developed for studies regarding DHC [59,61–63], but other studies using MODEST include the national electricity grid [64], utilizing waste heat from industries [65,66], introducing large-scale heat pumps in DH [67] and biogas systems [68].

MODEST has a flexible time division to depict fluctuations. A full year is depicted as several periods reflecting seasonal, weekly, and daily dependencies [60]. The seasonal climate changes are represented in the used time division, presented in Table 1. The months of spring, summer and autumn are divided into periods of night and daytime to cope with the variations. Meanwhile, the high-use winter months are analyzed more closely, by selecting the peak power of each time period in each month.

Table 1. A course description of the time division of a full normal year (8760 h) in model for optimization of dynamic energy systems with time-dependent components and boundary conditions (MODEST), with respect to seasonal, weekly, and daily dependencies.

Seasons	Months	Days and Hours	Analyzed Peak Hours
Winter	Jan–Mar, Nov–Dec	Mon–Fri 06:00–07:00	Peak day 06:00–07:00
		Mon–Fri 07:00–08:00	Peak day 07:00–08:00
		Mon–Fri 08:00–16:00	Peak day 08:00–16:00
		Mon–Fri 16:00–22:00	Peak day 16:00–22:00
		Mon–Fri 22:00–06:00	Peak day 22:00–06:00
		Sat, Sun 06:00–22:00	
		Sat, Sun 22:00–06:00	
Spring, summer, and autumn	Apr–Oct	Mon–Fri 06:00–22:00	
		Mon–Fri 22:00–06:00	
		Sat, Sun 06:00–22:00	
		Sat, Sun 22:00–06:00	

The software Converter is used to convert the data from ANSYS into the MODEST time division in Table 1. The software was developed at the Division of Energy Systems at Linköping University.

3. The Studied Scenarios

In this study, different scenarios are analyzed with a system perspective approach. The location is Linköping, Sweden, which has 160,000 residents and is located 200 km southwest of Stockholm. The aim is to analyze the consequences of different control strategies for an HPS utilizing heat from a DHC system, as visualized in Figure 3. Key performance indicators are performance and energy use of the HPS, and the resulting primary energy use, electricity production and the GHG emissions of the DHC system. Three scenarios for different control strategies for the HPS are analyzed, as presented in Section 3.

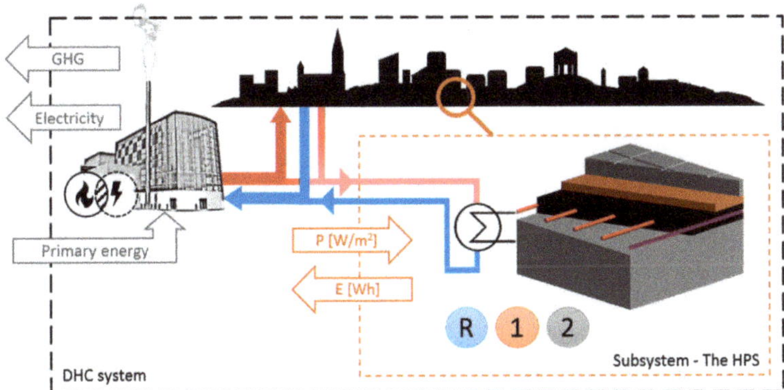

Figure 3. A visualization of the scenario study with the studied HPS as a subsystem of the DHC system. The return water of the DHC system can be utilized as a heat source for the HPS, and the hot water as back up. Three scenarios, called R, 1 and 2, regarding control strategies of the HPS which generate different power demands and thus energy use, are studied. The effects on the DHC system are analyzed and the key performance indicators are input of primary energy as fuel and output of electricity and greenhouse gas (GHG) emissions at local and global levels. Image used courtesy of ANSYS, Inc.

The DHC system in Linköping is the third largest high-temperature system in Sweden. The majority of the heat, cooling, and electricity production comes from CHP plants, mainly using the fuels household waste, biomass, coal, and oil. The demand in a normal year amounts to 1700 GWh heat, 60 GWh cooling, and 400 GWh electricity.

The HPS is concentrated in the central parts of Linkoping and has a total area of 35,000 m², divided into nine subsystems. Each subsystem uses heat from the DHC system and is controlled by a substation. The HPS is active during the months of Jan–Apr and Oct–Dec and is operated when the outdoor temperature is below 4 °C.

Scenarios and Specifics

Three scenarios regarding the HPS system are analyzed. In order to quantify the scenarios, they are analyzed relative to a situation with no HPS. The scenarios are:

- Scenario R: Reference scenario, business as usual
- Scenario 1: The HPS system shuts down at temperatures lower than −10 °C
- Scenario 2: The HPS system shuts down at temperatures lower than −5 °C

Scenario R is used as a reference scenario. The control strategy is to keep the temperature of the ground surface at 2 °C during periods of no precipitation and at 5 °C in presence of precipitation, which is done in order to minimize energy use and has previously been studied for this system [69]. This solution is intended to use weather forecast to function properly. Scenario R is also used to compare the simulation model to statistical data. The idea of scenarios 1 and 2 is to examine control strategies which include shutdown periods at sub-zero temperatures, whilst still maintaining the performance of the HPS in keeping the surface dry and non-slippery. Snowfall is most common at temperatures around 0 °C, with the frequency decreasing as the outdoor temperature falls below −5 °C to −10 °C, due to the reduced moisture content in the air as the temperature drops. Consequently, the risk of slipperiness due to snowfall decreases as the temperature drops. This can potentially be an efficient way of minimizing the energy needed in an HPS. However, the pickup loads to restore the surface temperatures after a shutdown period during cold temperatures must be analyzed.

The simulation model uses hourly data on temperature, wind, solar radiation, and precipitation collected from the Swedish Meteorological and Hydrological Institute's (SMHI) weather station Malmslätt in Linköping, Sweden [70]. A course compilation of the data in monthly values is presented in Table 2.

Table 2. Weather parameters of temperature and precipitation for the studied year of 2016 and of a normal year. The analyzed time period is the cold months of January-April and October-December, a total of 213 days for 2016. The data is collected from Swedish Meteorological and Hydrological Institute's (SMHIs) weather station Malmslätt in Linköping, Sweden [70].

Temperature	Unit	Jan	Feb	Mar	Apr	Oct	Nov	Dec	Total
Average temp.	°C	−5	−0.2	2.7	5.7	6.8	1.8	2.0	-
Temperature in a normal year	°C	−2.8	−3.0	0.5	5.2	7.6	2.4	−1.3	-
Hours < 4 °C	h	652	583	504	217	69	530	486	3041
Hours < 0 °C	h	494	346	144	51	3	223	230	1491
Precipitation									
Amount of precipitation	mm	26.7	11.8	20	27.6	70.8	37.7	25.6	220.2
−with temp < 0 °C	mm	16.7	4.2	2.2	0	0	4.8	5	32.9
Precipitation in a normal year	mm	41.0	27.1	33.6	34.2	46.0	52.2	46.8	280.2

4. Computational Setup and Numerical Procedure

The computational work in this study comprises two parts: the simulation model of the HPS and the optimization model of the DHC system.

4.1. Parameters and Properties of the HPS Simulation Model

The HPS is simulated as a transient model. A time step of 15 min is used, in order to capture the fluctuant weather, as well as capturing the response time needed in an HPS to ensure good performance. Fifteen minutes is also an approximate value of the HPSs cycle time. A residual error of $1.0e^{-4}$ (RMS level) is used in this study. The model includes the parameters:

- Weather properties regarding temperature, precipitation, solar irradiation, and wind.
- Material properties regarding density, thermal conductivity, and specific heat capacity.
- Heat transfer equations regarding conductivity, convection, and radiation.
- Operational parameters.

The construction of the HPS model is illustrated in Table 3 and, along with the material properties of the layers. Table 3 includes intervals for each value, which have been found in related literature. The surface layer A consists of the pavement which could be asphalt or, as in this study, paving stone. Layer B is a layer of rammed sand. The properties of layer C depend on the context of use. If the surface is regularly exposed to heavy loads of traffic, the normal material would be asphalt, as in this study. If the area serves as a walkway with minor loads, layer C could consist of rammed sand with embedded plastic PEX-pipes. Lastly, in layer E, there is a bearing layer of gravel or similar. In this study, plastic PEX-pipes have a spacing of 0.25 m. As a final note on the construction, Adl-Zarrabi et al. [28] conclude that the thermal properties of the material and spacing of the pipes have a large influence on the system's performance, and the buried depth of the pipes are of less importance.

Table 3. Material properties and model setup of the studied HPS, here visualized as 1 m². The construction consists of five layers of different materials, where density, thermal conductivity, and specific heat capacity are used by the software ANSYS to calculate the internal conduction occurring in the ground layers. The values in parenthesis are intervals found in the literature. Image used courtesy of ANSYS, Inc.

	Layer	Material	ρ (kg/m³)	λ (W/m·°C)	C_p (J/kg·°C)
	A	Pavement	2300 (1906–2450) [71,72]	2 (0.5–3.2) [71,73]	840 (767–2000) [72]
	B	Sand	1700 (1677–1771) [74,75]	1 (0.25–3) [75,76]	1000 (919–1117) [75]
	C	Asphalt	2100 (1906–2450) [77]	0.75 (0.74–2.9) [71,77]	920 (800–1853) [77]
	D	PEX-plastic	925 [78]	0.35 [78]	2300 [78]
	E	Gravel	2100 (1928–2129) [75]	1.5 (0.51–1.77) [75]	1150 (1088–1307) [75]

The final simulation model of the HPS is visualized in Figure 4. Symmetry, adiabatic conditions, and areal conversion equations are used to reduce the model size, which thus requires less computational power. The unstructured mesh grid used is based on proximity and curvature, containing 3936 elements. The simulation will predict heat transfer in the X- and Y-direction, as illustrated in Figure 4. Therefore, the boundary condition of the two surfaces in the Z-direction is set to be adiabatic. Also, symmetry is used on the surfaces in the X-direction. The bottom boundary condition is set to an average temperature of 9 °C, to reflect lower ground level in the urban area. The depth of the model is generous enough to not influence the result. The boundary conditions at the top surface of the model are presented in Equation (1), which includes convection, radiation, solar irradiance, and effects of precipitation. The heat transfer occurring by conduction is calculated in the model using the properties presented in Table 3.

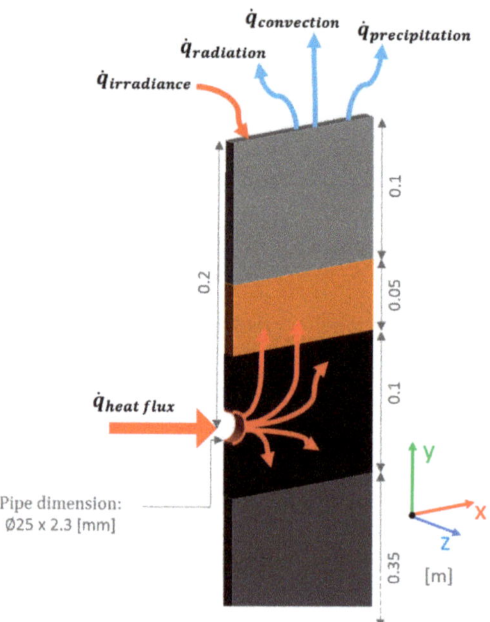

Figure 4. The final simulation model is a slip and cross-sectional part of the ground. Compared to the illustration in Table 3, this model utilizes symmetry by cutting through the plastic pipes, and between two pipes at the other end. The surfaces in the Z-direction are adiabatic. Also illustrated are the boundary conditions on the top surface summarized in Equation (1), and the power applied in the pipes, expressed as $\dot{q}_{heat\ flux}$. Image used courtesy of ANSYS, Inc.

The control strategy depends on sensors registering the temperature of the surface, $T_{surface}$, and controlling the power accordingly. As long as the outdoor temperature is below the starting temperature, the surface temperature, $T_{surface}$, is regulated relative to a set point temperature, $T_{set\ point}$. The set point temperature is either 2 °C as a standby mode during no precipitation, or 5 °C in an active mode in presence of precipitation. For example, the outdoor temperature may be −3 °C and precipitation is due in a couple of hours. Then the set point temperature will be set to 5°C well before the upcoming precipitation and then melt the snow fall, thus, preventing snow and ice formation on the pavement. The active mode refers to a preheat period of 4 h before precipitation and to when the surface is dry again after the precipitation. The control strategy, analyzing the difference between the surface temperature, $T_{surface}$, and the desired set point temperature, $T_{set\ point}$, determines the power applied by the HPS. The heat transfer occurring between the heat medium and the embedded pipes is expressed as power levels in the model, seen as $\dot{q}_{heat\ flux}$ in Figure 4. The levels range from 20 W/m^2 to 160 W/m^2, which corresponds to a range of 11 °C to 33 °C of the heat medium in the pipes. The control strategy is illustrated in Figure 5, both for times with no precipitation and in presence of precipitation. The HPS is only active at outdoor temperatures below 4 °C and the operational formula aims at keeping the temperature of the surface, $T_{surface}$, above the set point temperature by a proactive approach to slow down the temperature drop as it approaches the set point temperature.

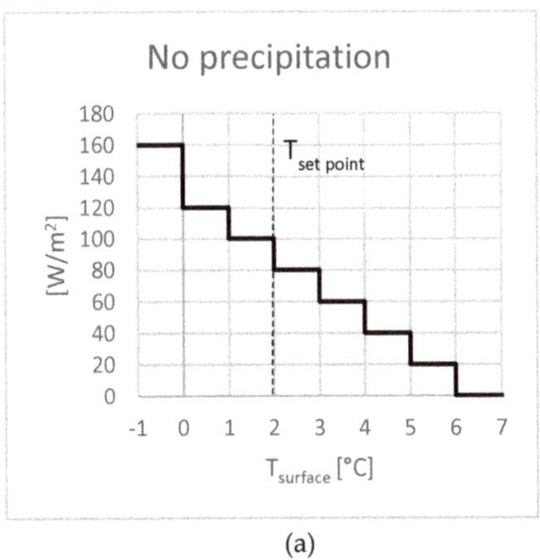

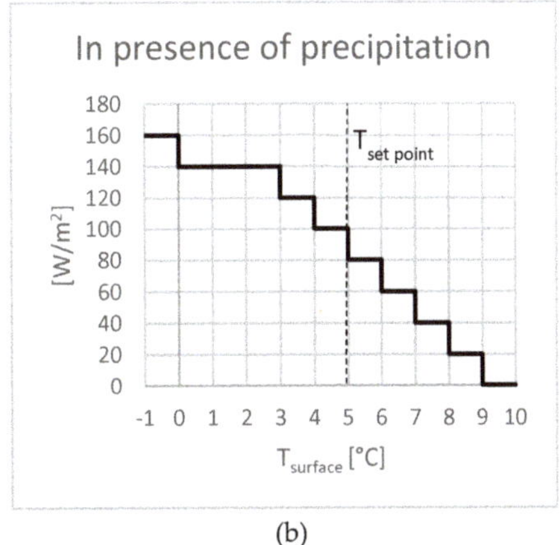

(a)   (b)

Figure 5. Illustration of the control strategy used in this study. (**a**) The control strategy of the standby mode, during periods with no precipitation. (**b**) The control strategy of the active mode, in the presence of precipitation. This mode activates four hours before precipitation.

Equation (1) is the governing boundary condition on the top surface, including convection radiation, precipitation, and irradiance, as illustrated in Figure 4.

$$\dot{q}_{surface}(t) = \dot{q}_{irradiance} - \dot{q}_{radiation} - \dot{q}_{convection} - \dot{q}_{precipitation}\left[W/m^2\right], \tag{1}$$

where the four parameters of the equation are described in more detail in the following Equation (2)–(5). Equation (2) regards solar irradiance acting on the top surface.

$$\dot{q}_{irradiance}(t) = \alpha \cdot I(t)\left[W/m^2\right], \tag{2}$$

where $I(t)$ is the solar irradiance which are collected for Linköping from the data model STRÅNG, which is a mesoscale model for solar radiation [79]. α is the absorptivity of the surface, which is set to 0.2 in accordance with the literature presenting a range of 0.05 to 0.35 [80].

Equation (3) regards the radiation acting on the surface.

$$\dot{q}_{radiation}(t) = \sigma \cdot \varepsilon \cdot \left(T^4_{surface}(t) - T^4_{ambient\ air}(t)\right), \tag{3}$$

where σ is the Stefan-Boltzmann constant. ε is the emissivity, which is set to 0.9 in accordance with the literature of urban surfaces ranging from 0.71 to 0.95 [81]. $T_{surface}$, is the temperature of the surface, which is measured in the simulation model. $T_{ambient\ air}$, is assumed to be the outdoor temperature, which is collected from the SMHIs open data and the weather station Malmslätt in Linköping, Sweden [70].

$$\dot{q}_{convection}(t) = h(t) \cdot \left(T_{surface}(t) - T_{ambient\ air}(t)\right), \tag{4}$$

where h is the heat transfer coefficient including Reynolds and Prandtl numbers and calculated in accordance with Çengel et al. [82], Holman [83], and Storck et al. [84]. The heat transfer coefficient is calculated at each time step of the simulation model and is dependent on the wind speed and temperature measurements collected from SMHIs open data and the weather station Malmslätt in Linköping, Sweden [70].

Equation (5) regards the precipitation, and if snowfall occurs, the snow's sensible heat is first calculated. Secondly, the heat required to melt the snow is calculated and lastly, a calculation is made of the latent heat required to keep the ground surface at a constant temperature while the remaining precipitation is evaporated.

$$\dot{q}_{precipitation}(t) = \rho_w \cdot n_s(t) \cdot \left(c_{p,s} \cdot \left| 0\,°\text{C} - T_{air}(t) \right| + c_m + (1 - \varphi) \cdot c_e \right) / t, \tag{5}$$

where ρ_w is the density of water, which is set to be 1000 kg/m^3 [82], n_s is the amount of precipitation collected from SMHI open data [70], $c_{p,s}$ is the specific heat capacity of snow and ice and set to 2110 J/kgK [82], c_m is the melt enthalpy of water and is set to 333,700 J/kg [82], φ is the surface runoff factor and is set to 0.9, which is in range of the literature ranging between 0.7 and 0.95 for urban surfaces [85]. c_e is the evaporation enthalpy of water, which is set to 2,500,000 J/kg [82]. t, is the time step of the model, which is 900 s (15 min) in this study.

Limitations and Comments on the Model

The model does not regard the choice of heat medium and electricity use of, e.g., pumps, since construction and use may differ significantly from one system to another. The properties stated in Table 3 may vary depending on moisture, porosity, potential air voids, and temperature of the material [76]. If the materials in the ground are at sub-zero temperatures, it can affect the thermal properties and increase energy use due to moisture transport in the ground, as analyzed by Xu and Tan [34]. However, as the HPS studied aims at drying the surface after precipitation and thereby minimize moisture, as well as keeping the ground layers above sub-zero temperatures, these issues are neglected in this study.

Regarding the boundary condition of the bottom surface, the surrounding ground often contains nearby infrastructure, such as storm drains, sewers or district heating pipes, which causes the average temperature of the ground to be higher than in rural areas. In addition to this, the HPS itself is an influencing factor when it turns on at 4 °C in the autumn, whereby normal cooling can never occur. The inertia of the ground causes the temperature to be higher lower into the ground than at the surface.

Moreover, Equations (1)–(5) represent parameters with the largest impact on HPS performance. Other parameters, such as dew or moisture on the surface, are neglected since the HPS aims to keep the surface warmer than the ambient air. Regarding Equation (2), the α-value of the pavement has a large impact on the affecting solar irradiance. In urban areas the surroundings will disturb irradiance, causing urban canyons [80] and making it difficult to assign a general α-value. When regarding radiation in Equation (3), the ambient temperature is considered to be that of the urban environment, as opposed to the sky temperature commonly used. However, during fully cloudy conditions the sky temperature may also be considered as the ambient temperature [35]. When snowfall occurs, the wind in Equation (4) acts on a surface film temperature of 0 °C regardless of the outdoor temperature. Moreover, during rainfall, the film temperature is assumed to be the same as the ambient temperature. Regarding Equation (5), the ground layers are considered to be impervious as it is desirable to keep the surface and ground layers dry. Moreover, urban areas often suffer heavy compaction due to demands for bearing capacity, which results in a reduced infiltration capacity [86]. Also, the temperature of the precipitation is assumed to be the same as the ambient temperature as it hits the top surface. The characteristics of snow vary, depending especially on whether the snow is uncompressed or compressed, with the latter requiring more energy to melt [87]. As the HPS starts to heat the ground four hours before predicted precipitation and aims to melt the snowflakes as they fall to the ground, the snow is considered to be uncompressed.

According to [1] the HPS should be designed to be able to keep the ground surface temperature at +5 °C when the air temperature is 5 °C higher than the design outdoor temperature, which is −16.6 °C for Linköping, Sweden [88]. Statistically, this means that a temporary snow cover will remain for no longer than two hours on five different occasions over a ten-year period. Statistically, snowfall of such intensity that snow cover will remain for more than eight hours occurs once every ten years.

4.2. Parameters and Properties of the DHC System's Optimization Model

The DHC production in Linköping is based on the incineration plants: The Gärstad waste-fueled CHP plant, located in the northern part of Linköping, and a mixed-fueled CHP plant located in the central part. The system is complemented with a biomass-fueled CHP plant and a heat-only boiler (HOB) using biomass in the nearby town of Mjölby. As a backup, there are also heat-only boilers using oil and electricity to cover peak demand. The model of the studied DHC system is visualized in Figure 6, and the basic input data for the utilities is presented in Table 4. The DHC system and optimization model has been studied previously by Blomqvist et al. [89].

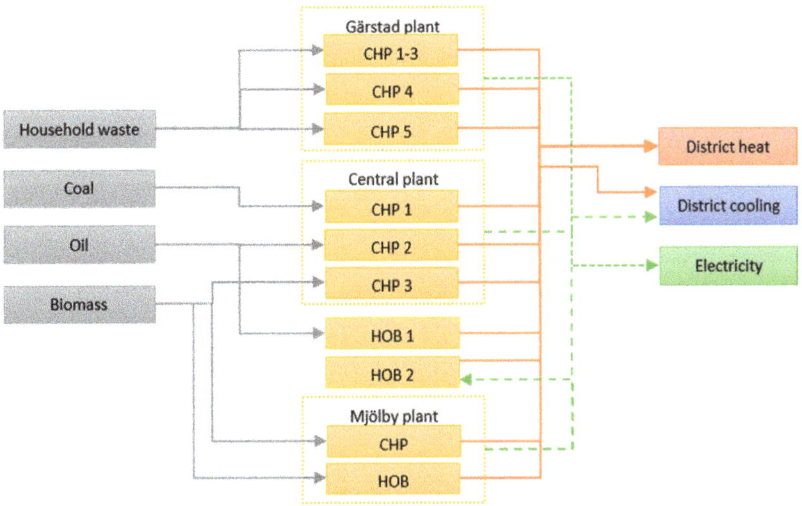

Figure 6. Schematic view of the optimization model created in MODEST representing the studied DHC system. The model consists of fuel that is converted using combined heat and power (CHP) and heat-only boilers (HOBs) to serve a demand of district heat, cooling, and electricity. The production units are based on the plants Gärstad, Central and Mjölby, and standalone HOBs. Table 4 presents technical data of the production units. The district cooling is produced by an absorption plant of 12 MW and a compression plant of 6 MW.

The Gärstad CHP plant consists of three waste incineration boilers, illustrated as CHP 1–3, 4, and 5 in Figure 6. They are hybrid systems, with flue gas condensing and steam turbine through a gas turbine heat recovery steam generator. CHP 1–3 has a maximum capacity of 75 MW heat, with an additional 15 MW heat from flue gas condensing and 10 MW electricity. CHP 4 has a steam turbine capacity of 68 MW heat, and an additional 15 MW heat from flue gas condensing and 19 MW electricity. CHP 5 has a steam turbine capacity of 84 MW, and an additional 12 MW heat from flue gas condensing and 21 MW electricity. The central CHP plant consists of three boilers and three steam turbines, where two are backpressure turbines, and one combined condensing and backpressure turbine. The first boiler is fueled with coal, with fractions of rubber. The second boiler uses heating oil. The third boiler, with flue gas condensing, is fueled with biomass made from wood products and fractions of plastics. The central plant can produce electricity and heat or use a direct condenser for the sole production of heat. Cooling of the condensing turbine is achieved using water from the nearby river Stångån, resulting in 50 GWh heat potentially being wasted in Stångån annually. The technical input data may be seen in Table 4.

Table 4. Production capacity and fuel use of the units in the DHC system presented in Figure 6, including fuel, heat, and power production capacity, as well as the capacity of the flue gas condensation.

Unit		Fuel	Heat [1] (MW)	Power (MW)	Heat from Flue Gas Condensation (MW)
Gärstad	CHP 1–3	Household waste [2]	75	10	15
	CHP 4	Household waste [2]	68	19	15
	CHP 5	Household waste [2]	84	21	12
Central	CHP 1	Coal [3]	83	31	-
	CHP 2	Oil	154	41	-
	CHP 3	Wood [4]	78	32 or 22 [5]	20
Standalone	HOB 1	Oil	144	-	-
	HOB 2	Electricity	25	-	-
Mjölby	CHP	Wood	33	10	-
	HOB	Wood	32.5	-	-

[1] Heat from steam production. [2] The annual use of household waste is limited to 1781 GWh. [3] Fuel also contains fractions of rubber. [4] Fuel also contains fractions of plastics. [5] 22 MW back-pressure power or 32 MW condensing power.

The district cooling in the system is produced in an absorption plant with a capacity of 12 MW cooling using heat from the DHC system and an electricity-powered compression-cooling plant with a capacity of 6 MW cooling.

In order to calculate the local GHG emissions, the factors presented in Table 5 are used. The locally emitted GHG is a result of the fuel use in the DHC system. The biomass consists of primary and secondary wood fuels. The majority of household waste is organic and comes from the surrounding region. The subsequent effects on the global GHG emissions are calculated and analyzed by using three different factors, also presented in Table 5. The factors are of Swedish electricity mix, Nordic electricity mix, and electricity produced by coal condensing plants. The effects on the global GHG emissions are caused by the local changes in electricity production from the CHP plants and the use of biomass, which is seen as a scarce resource in this study.

Table 5. Local GHG emission factors for the fuel used in the model of the DHC system. The factors include incineration, production, and transportation. Also presented is the global GHG emission factors used to analyze the global effects generated by changes in the local DHC system.

Local Emission	GHG Emission Factor [90] (g CO_2eq/kWh)	Global Emission	GHG Emission Factor [91] (g CO_2eq/kWh)
Household waste	143	Swedish electricity mix	36.4
Wood [1]	14.5	Nordic electricity mix	97.3
Oil	297	Coal condensing production	968.6
Coal [1]	340		
Electricity (internal)	0		
Flue gas cond.	0		

[1] Emission factors are weighted in order to reflect a fuel mixture used in the central plant CHP 1 (coal with fractions of rubber) and CHP 3 (primary and secondary wood fuels with fractions of plastics).

5. Results

The results in the paper include three sections, (1) where the two models of the HPS and DHC system will be compared to actual system performance from 2016, and (2) where the overall performance of the HPS will be analyzed. In the third (3) section the impact of the HPS on the DHC system will be analyzed.

5.1. Comparing Models of the HPS and DHC System to Actual System Performance

Simulation of the HPS model is compared to annual and monthly statistical data for the winter months, Jan–Apr and Oct–Dec 2016. In Figure 7, the result of the reference Scenario R is presented together with the statistical data. The simulation model generates an annual difference of 2%, and a monthly maximum difference of 10%, except during the low use months of April and October. Moreover, the operational hours of the reference scenario are 3055 h, compared with 3041 h at temperatures below the starting temperature of 4 °C, as presented in Table 2. The small difference is due to modeling issues. The model time step of 15 min has a higher resolution than the hourly statistical data and causes parts of an hour to count when the outdoor temperature fluctuates around the starting temperature of 4 °C.

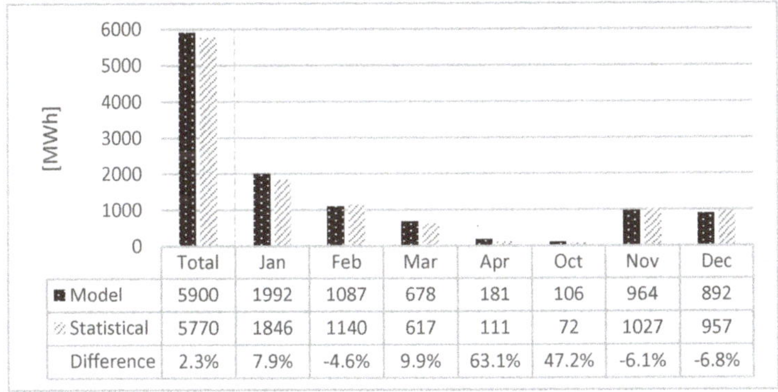

	Total	Jan	Feb	Mar	Apr	Oct	Nov	Dec
Model	5900	1992	1087	678	181	106	964	892
Statistical	5770	1846	1140	617	111	72	1027	957
Difference	2.3%	7.9%	-4.6%	9.9%	63.1%	47.2%	-6.1%	-6.8%

Figure 7. Results of Scenario R compared to the statistics for the HPS from 2016. To the left, the annual values are compared, and the active months of Jan-Apr and Oct-Dec follows. The annual value differs by 2% and the monthly values differ within 10% except for the low use months of April and October.

The fuel use of reference Scenario R from the optimization model in MODEST is compared to statistics as visualized in Figure 8. The total annual fuel use differs by 5%, with the optimization model resulting in energy use of 2440 GWh and the statistics showing 2323 GWh.

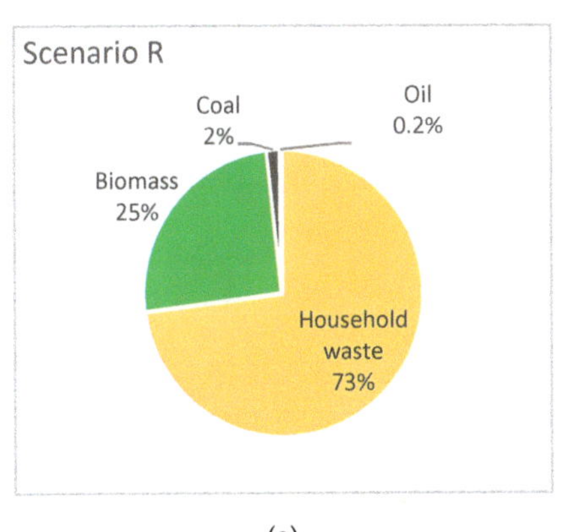

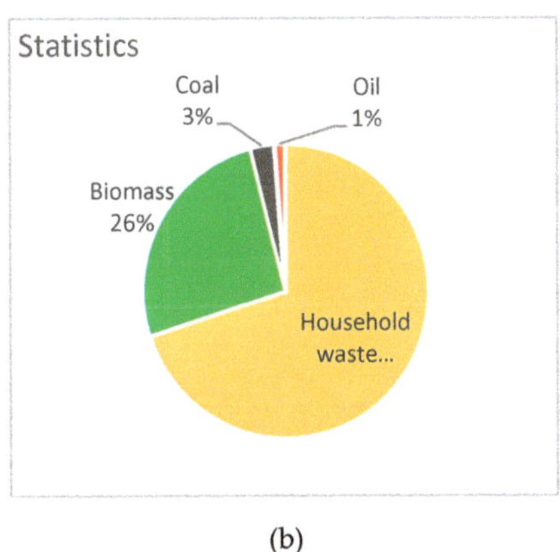

(a)   (b)

Figure 8. Results of the reference scenario R (**a**) compared to the statistics for the system from 2016 (**b**). The annual fuel use differs 5% between the optimization model (2440 GWh) and the statistics (2323 GWh).

5.2. Performance of the HPS and Energy Use of the Studied Scenarios

Figure 9 shows the performance of the HPS in the studied scenarios during January and February of 2016. The temperature of the top surface in the scenarios is displayed along with outdoor temperature and precipitation. A rapid decrease in the surface temperature indicates precipitation, as the surface in the simulation receives large negative power, in accordance with boundary Equation (5).

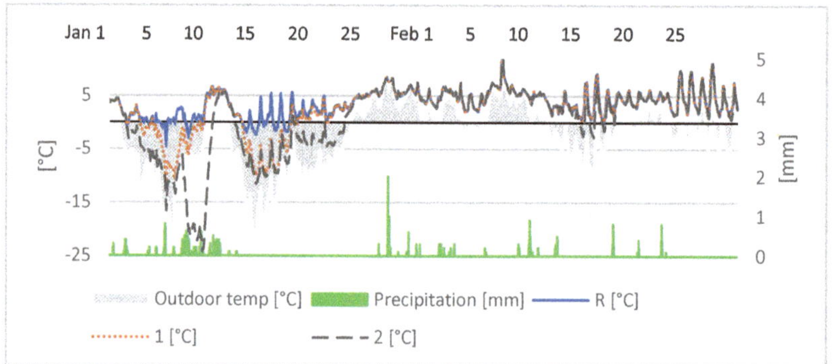

Figure 9. Performance, in terms of temperature at the top surface, of the HPS and the simulated scenarios R, 1, and 2 during January and February. The outdoor temperature is presented in gray background, and the precipitation is visualized at the secondary axis.

The energy use of scenarios R, 1 and 2 is presented in Figure 10. The annual energy use of the studied scenarios amounts to 6.3 GWh for Scenario R, 5.6 GWh for Scenario 1, and 4.5 GWh for Scenario 2, generating a potential energy saving of 28%. Moreover, the results correspond to an average annual use 180 kWh/m^2 for Scenario R, 161 kWh/m^2 for Scenario 1, and 129 kWh/m^2 for Scenario 2.

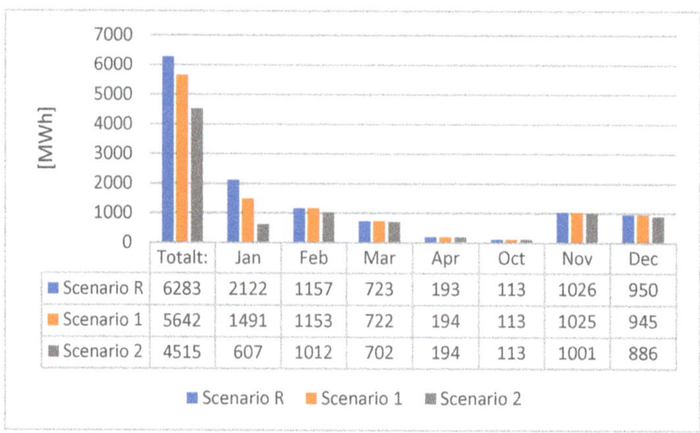

	Totalt:	Jan	Feb	Mar	Apr	Oct	Nov	Dec
Scenario R	6283	2122	1157	723	193	113	1026	950
Scenario 1	5642	1491	1153	722	194	113	1025	945
Scenario 2	4515	607	1012	702	194	113	1001	886

Figure 10. Results presenting the total energy use in 2016 of the HPS simulation for scenarios R, 1, and 2. The annual values of each scenario are presented to the left and the active months of Jan–Apr and Oct–Dec follow. The largest changes are seen in January and February.

A duration diagram of the HPS is presented in Figure 11. Savings are made in all the power steps, and the average energy use ranges from 57.5 W/m^2 to 49.7 W/m^2, corresponding to a potential energy saving of 14% between the scenarios. The operational hours of the scenarios differ by 500 h, ranging from 3055 h to 2525 h.

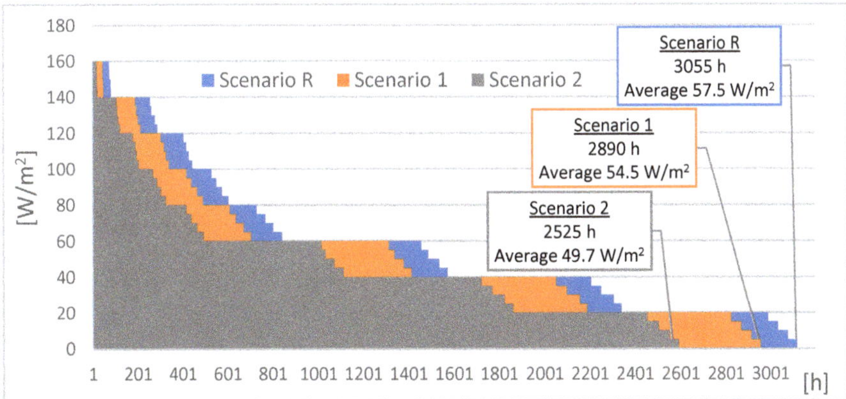

Figure 11. Duration diagram of the simulated scenarios R, 1, and 2. The HPS is active for 3055 h in Scenario R with average energy use of 57.5 W/m^2, 2890 h with average energy use of 54.5 W/m^2 for Scenario 1, and 2525 h, with average energy use of 49.7 W/m^2 for Scenario 2.

5.3. Evaluation of the Impact the HPS Has on the DHC System

The primary energy use of the DHC system is presented in Figure 12. Household waste is the main source, followed by biomass. Coal is needed in the peak 1000 h, and at a production level of 375 MW the HOB using biomass is needed, while oil is required once the production level exceeds 400 MW.

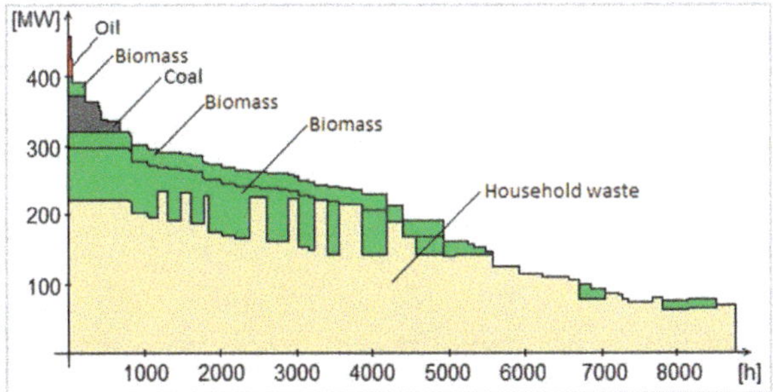

Figure 12. Duration diagram of the fuel use for production in Linköping's DHC system, reference Scenario R. Results of the optimization model in MODEST showing the fuel use of the DHC system with household waste as a base, biomass, and peak fuels of coal and use of the HOBs using biomass and oil to cover the final peak hours. Image used courtesy of MODEST.

Figure 13 shows the time of operation of the HPS and how it coincides with the demand in the DHC system. For the reference Scenario R, the time of operation coincides with the peak demand in the DHC system. Scenario 1 is excluded from the top days of the peak demand, approximately 50 h. Furthermore, Scenario 2 is excluded from the top weeks, approximately 300 h.

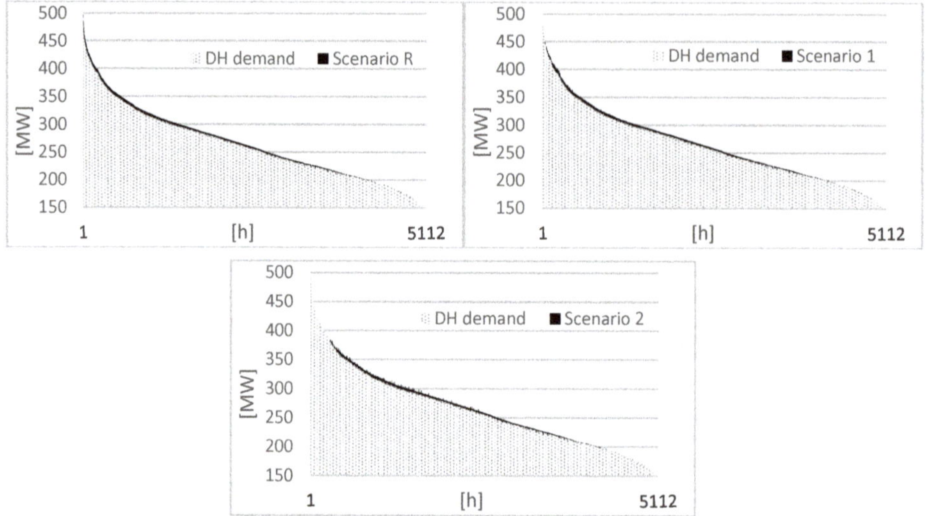

Figure 13. Duration diagram of the heat demand in Linköping during the studied time period of 5112 h in Jan-Apr and Oct-Dec, with the demand of the HPS, visualized for scenarios R, 1, and 2. Scenario 1 is excluded from the peak 50 h, and scenario 2 from the peak 300 h. Please note that the Y-axis is truncated and starts at 150 MW.

Figure 14a presents the primary energy savings of each scenario, while Figure 14b shows the increase in electricity production due to increased production at the CHP plants. It is worth noting that mostly renewable resources such as biomass are used for the HPS and that electricity production increases as the use of HPS increases.

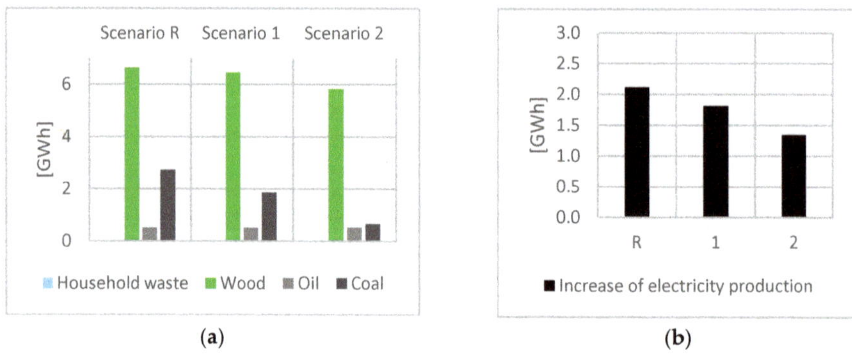

Figure 14. (**a**) Results showing the fuel use of the DHC system productions units in order to satisfy the demand of the HPS for each scenario. Results are presented relative to a scenario were no HPS are used and conventional snow clearance must be used. (**b**) Results showing the increase in electricity production at the CHP plants for each scenario.

Figure 15 presents the increase in local emissions and decrease in global emissions. The local emissions increase in all scenarios, as a result of increased production in the DHC system. The local emissions amount to 34 $kgCO_{2,\,eq}/m^2$ for Scenario R, 25 $kgCO_{2,\,eq}/m^2$ for Scenario 1, and 13 $kgCO_{2,\,eq}/m^2$ for Scenario 2. Including the operational hours of the studied system, the value could be expressed as 11 $gCO_{2,\,eq}/m^2$, h for Scenario R, 9 $gCO_{2,\,eq}/m^2$, h for Scenario 1, and 5 $gCO_{2,\,eq}/m^2$, h for Scenario 2.

As presented in Figure 15, global emissions are reduced in all scenarios, mostly due to the increased production of electricity in the CHP plants, as presented in Figure 14b. Moreover, reduced use of the scarce resource of biomass has a positive impact on global GHG emissions.

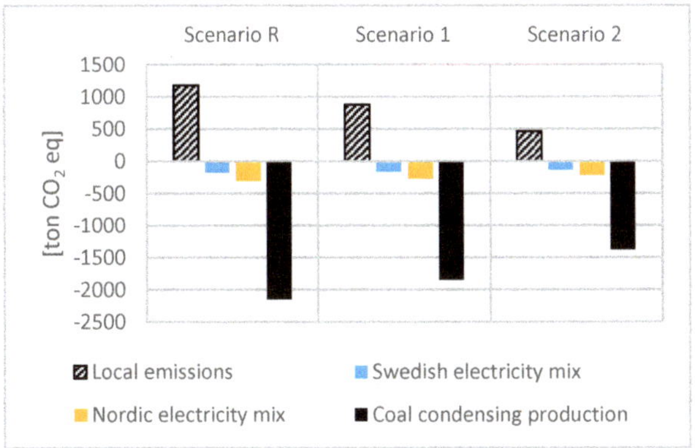

Figure 15. Results for scenarios R, 1, and 2 showing the increase in local GHG emissions due to increased production in the DHC systems production units, as presented in Figure 14a. Also presented, is the positive effect and decrease of global GHG emissions, mostly due to increased electricity production in the CHP plants, as presented in Figure 14b. The three emission factors used are presented in Table 5. Results are presented relative to a scenario where no HPS is used and conventional snow clearance must be used.

6. Discussion

6.1. The Scenarios and Modeling of the HPS

For scenarios 1 and 2, the heat medium should have freeze protection, or the HPS should have an idling mode where the medium is kept above freezing and circulating in the pipes. The model is constrained to a maximum power of 160 W/m^2, which is lower than the theoretical design power for the geographical location stated in 1976 [1], but in line with other more recent literature [7,78]. The maximum power corresponds to 33 °C, which enables the use of the return water in the DHC system. Besides being able to operate an HPS, the opportunity to decrease the return temperature can result in a more efficient DHC system, as it increases the output of the flue gas condensation, as argued in the introduction.

It should be noted that, overall, the studied year was warmer than a normal year. However, January of 2016 was colder, as were October and November. For 3041 h during the studied months, the temperature dropped below 4°C, which is the temperature at which the HPS becomes active. In terms of precipitation, the studied year was dryer than the normalized year. October is the only month when the precipitation was above normal. The cold month of January, in particular, had less precipitation than the normalized data. This is in line with the ideas of the scenario, with shutdown periods at lower temperatures, since the air's saturation level decreases at lower temperatures, which, in turn, reduces the occurrence of precipitation.

The simulation model of the HPS shows good agreement with measurements. The difference between the annual averaged model prediction and the actual performance was 2%, and the monthly difference was below 10% for the cold winter months. A higher relative average was only found for April and October when the system was only scarcely used. The comparison was applied to an HPS area of 33,000m^2, due to imperfect data from one subsystem at the time of the study. This is the reason why Figure 7 shows lower energy use than reference Scenario R in Figure 10. The optimization model of the DHC system in MODEST resulted in an annual difference of 5% and good agreement in terms of allocation of the primary energy sources.

6.2. Performance of the HPS

In order to analyze the performance of the HPS, it is stated that the HPS should be designed to keep the surface temperature at +5 °C when the air temperature is 5 °C higher than the design outdoor temperature [1]. In Figure 9, reference Scenario R indicates that it is possible to achieve this, as the surface temperature is above 0 °C when the outdoor temperature is −11 °C. The temperature set point in the model is 2 °C, but as the model includes cooling effects from wind, radiation, and precipitation, it is possible to keep a lower temperature setting than 5 °C. Moreover, the air is dry at these low temperatures and in combination with the surface temperature always being higher than the that of the air, the risk of a slippery surface is minimized. It comes down to a choice between keeping the surface above freezing at all times or settling for minimizing the risk of slipping, in relation to efficient energy use.

In Figure 9, which presents the performance of the HPS, a rapid decrease in the surface temperature indicates precipitation, as the surface in the simulation receives negative power derived from Equation (5). The temperature increases quickly as an indication of the HPS reacting to the precipitation. Looking more closely at Scenario 2 in January, one can see a large drop in temperature. The drop is explained by that the system is turned off and prolonged precipitation, in the form of snow, falls when the outdoor temperature increases from low levels. This indicates that the snow will remain on the surface, whereas the snow will be melted in Scenario 1, as the temperature climbs above 0 °C. This poses a design question of how many hours the HPS is allowed to fail and still be seen as acceptable, which can be decided upon the desired function of the system and location. Moreover, during late February, the large temperature variation is explained by the spring sun.

6.3. The impact of the HPS on the DHC System

The largest change in energy use between the scenarios is seen in the cold month of January, as shown in Figure 10. In connection to the duration diagram in Figure 11, where the operational hours of the peak power 160 W/m^2 are decreasing, it is concluded that the pickup loads after shutdown periods in scenarios 1 and 2 do not result in the HPS demanding more peak power than the reference Scenario R. However, as presented in Figure 9, on the days around 10 January, the performance of the scenario may enable snowfall to remain on the surface, resulting in poor performance of the HPS in Scenario 2. This indicates that an HPS that shuts down at a temperature of −5 °C or lower performs poorly but presents good energy performance values. However, Scenario 1, when the HPS shuts down at a temperature of −10 °C or lower, presents good performance and also reduced the energy use of 3 W/m^2 (5%) for average use or 640 MWh (10%) for annual energy use, compared to the reference Scenario R. Moreover, Scenario 1 is excluded from the top 50 h regarding peak demand in the DHC system, as presented in Figure 13, which in turn reduces the use of fossil fuel, as seen in Figure 14a. In addition, Scenario 1 reduces local GHG emissions by 25% relative to the reference Scenario R. However, Scenario 1 also results in less electricity generated at the CHP plants relative to Scenario R, leading to less available electricity on the market. This, in turn, leads to a reduced positive effect on global GHG emissions.

The local emissions in this study only include the use of HPS and not the alternative use of conventional snow clearance using heavy machinery. However, the literature points to the HPS being the more GHG efficient alternative, indicating reduced local emissions if a comparison were to be made. The HPS generates a low value of 11 gCO$_2$, eq/m^2, h for the reference Scenario R. In a future fossil-free DHC system, the local GHG emissions from the HPS will decrease even more.

The factor to analyze at implementation is the time it takes for the HPS to produce a sufficient temperature at the surface to minimize the risk of slipperiness, at times when the temperature rises from below −10 °C in conjunction with precipitation shortly thereafter.

7. Conclusions

- The study indicates that the HPS is suitable for the use of return temperatures in a DHC system. An HPS could further decrease the return temperature, thereby potentially increasing the efficiency of the DHC system. Furthermore, in a future DHC system with lower supply temperatures, it is also desirable to achieve lower return temperatures to maintain an efficient DHC system.

- A control strategy that shuts down the HPS at temperatures below $-10\,°C$ results in a 10% energy saving, avoidance of use during the top 50 h of peak demand in the DHC system, reduced use of fossil fuel and a 25% reduction in local GHG emissions, whilst maintaining sufficient performance of the HPS.

- Utilizing HPS connected to a DHC system which has CHP can potentially result in increased electricity production. This generates a positive effect and reduction on the global GHG emissions if a coal condensing power plant is regarded as the marginal production unit in the European electricity market. In a future fossil-free production of DHC, the generated electricity will fully derive from renewable resources. This will improve the impact an HPS has on the GHG emissions even further.

Author Contributions: Conceptualization, S.B., P.R. and L.Ö.; data curation, S.B.; formal analysis, S.B.; funding acquisition, L.Ö.; investigation, S.B.; methodology, S.B.; project administration, S.B. and P.R.; software, S.B. and S.A.; supervision, P.R. and L.Ö.; validation, S.B.; visualization, S.B.; writing—original draft, S.B., S.A. and P.R.; writing—review & editing, S.B.

Funding: This study has been financed by the Swedish Agency for Economic and Regional Growth and the energy companies Tekniska verken AB and E.ON Sweden AB, as well as the housing companies Stångåstaden AB, AB Lejonfastigheter, Fastighets AB L E Lundberg, and Akademiska Hus AB.

Acknowledgments: The authors gratefully acknowledge Tekniska verken AB, for its insights and information regarding the HPS and DHC system.

Conflicts of Interest: The authors declare no conflict of interest.

References

1. Byggnadsstyrelsen. *Markvärme*; KBS Rapport; Byggnadsstyrelsen: Stockholm, Sweden, 1976.
2. Lund, J.W.; Boyd, T.L. Geothermal direct-use in the United States update: 1995–1999. In Proceedings of the Proceedings World Geothermal Congress 2000, Tohoku, Japan, 28 May–10 June 2000; Geo-Heat Center, Oregon Institute of Technology Klamath Falls: Kyushu, Japan, 2000; pp. 297–305.
3. Lund, J.W. Reconstruction of a pavement geothermal deicing system. *Geo-Heat Cent. Q. Bull.* **1999**, *20*, 14–17.
4. American Society of Heating, Refrigerating and Air-Conditioning Engineers. *2011 ASHRAE Handbook: Heating, Ventilating, and Air-Conditioning Applications*; ASHRAE: Atlanta, GA, USA, 2011; ISBN 9781936504077.
5. Mensah, K.; Choi, J.M. Review of technologies for snow melting systems. *J. Mech. Sci. Technol.* **2015**, *29*, 5507–5521. [CrossRef]
6. Pan, P.; Wu, S.; Xiao, Y.; Liu, G. A review on hydronic asphalt pavement for energy harvesting and snow melting. *Renew. Sustain. Energy Rev.* **2015**, *48*, 624–634. [CrossRef]
7. Lund, J.W.; Boyd, T.L. Direct utilization of geothermal energy 2015 worldwide review. *Geothermics* **2016**, *60*, 66–93. [CrossRef]
8. Zwarycz, K. *Snow Melting and Heating Systems Based on Geothermal Heat Pumps at Goleniow Airport, Poland*; The United Nations University, Geothermal training programme: Reykjavik, Iceland, 2002.
9. Shen, W.; Ceylan, H.; Gopalakrishnan, K.; Kim, S.; Taylor, P.C.; Rehmann, C.R. Life cycle assessment of heated apron pavement system operations. *Transp. Res. Part D Transp. Environ.* **2016**, *48*, 316–331. [CrossRef]
10. Shen, W.; Gopalakrishnan, K.; Kim, S.; Ceylan, H. Assessment of Greenhouse Gas Emissions from Geothermal Heated Airport Pavement System. *ISSN Int. J. Pavement Res. Technol. Int. J. Pavement Res. Technol.* **1997**, *88*, 233–242.
11. Shen, W.; Ceylan, H.; Gopalakrishnan, K.; Kim, S.; Nahvi, A. *Sustainability Assessment of Alternative Snow-Removal Methods for Airport Apron Paved Surfaces*; Federal Aviation Administration: Atlantic City, NJ, USA, 2017.

12. Chi, Z.; Yiqiu, T.; Fengchen, C.; Qing, Y.; Huining, X. Long-term thermal analysis of an airfield-runway snow-melting system utilizing heat-pipe technology. *Energy Convers. Manag.* **2019**, *186*, 473–486. [CrossRef]

13. Zhang, N.; Yu, X.; Li, T. Numerical Simulation of Geothermal Heated Bridge Deck. In Proceedings of the International Conference on Transportation Infrastructure and Materials (ICTIM 2017), Qingdao, China, 9–12 June 2017.

14. Yu, X.; Zhang, N.; Pradhan, A.; Puppala, A.J. Geothermal Energy for Bridge Deck and Pavement Deicing—A Brief Review. In Proceedings of the Geo-Chicago 2016, Chicago, IL, USA, 14–18 August 2016; American Society of Civil Engineers: Reston, VA, USA, 2016; pp. 598–609.

15. Swedish Energy Agency Annual Energy Statistics (Electricity, Gas and District Heating). Available online: http://www.scb.se/en0105-en (accessed on 18 April 2019).

16. Gunnlaugsson, E.; Reykjavikur Baejarhals, O. District heating in Reykjavík; past–present–future. In Proceedings of the 30th Anniversary Workshop, Reykjavík, Iceland, 26 August 2008.

17. Solaimanian, M.; Kennedy, T.W. Predicting Maximum Pavement Surface Temperature Using Maximum Air Temperature and Hourly Solar Radiation. *Transp. Res. Rec.* **1993**, *1417*, 1–11.

18. Gustavsson, T.; Bogren, J.; Green, C. Road Climate in Cities:A Study of the Stockholm Area, South-East Sweden. *Meteorol. Appl.* **2001**, *8*, 481–489. [CrossRef]

19. Andriopoulou, S. *A Review on Energy Harvesting from Roads*; [TSC-MT 12-017]; KTH Royal Institute of Technology: Stockholm, Sweden, 2012.

20. National Research Council (US). *Comparing Salt and Calcium Magnesium Acetate*; National Research Council: Washington, DC, USA, 1991.

21. Amrhein, C.; Strong, J.E.; Mosher, P.A. Effect of deicing salts on metal and organic matter mobilization in roadside soils. *Environ. Sci. Technol.* **1992**, *26*, 703–709. [CrossRef]

22. Bäckström, M.; Karlsson, S.; Bäckman, L.; Folkeson, L.; Lind, B. Mobilisation of heavy metals by deicing salts in a roadside environment. *Water Res.* **2004**, *38*, 720–732. [CrossRef] [PubMed]

23. Fischel, M. *Evaluation of Selected Deicers Based on a Review of the Literature*; Colorado Department of Transportation: Denver, CO, USA, 2001.

24. Howard, K.W.F.; Haynes, J. Groundwater Contamination due to Road De-icing Chemicals - Salt Balance Implications. *Geosci. Can.* **1993**, *20*, 1–8.

25. Kroening, S.; Ferrey, M. *The Condition of Minnesota's Groundwater, 2007–2011*; Minnesota Pollution Control Agency: Saint Paul, MI, USA, 2013.

26. Fay, L.; Shi, X. Environmental Impacts of Chemicals for Snow and Ice Control: State of the Knowledge. *Water Air Soil Pollut.* **2012**, *223*, 2751–2770. [CrossRef]

27. Nevalainen Henaes, H. *Impact on Society from Conventional Winter Road Maintenance and Ground Heating (sv: Samhällspåverkan vid Konventionell Vinterväghållning och Markvärme)*; Lunds University—Faculty of Engineering: Lund, Sweden, 2018.

28. Adl-Zarrabi, B.; Mirzanamadi, R.; Johnsson, J. Hydronic pavement heating for sustainable ice-free roads. *Transp. Res. Procedia* **2016**, *14*, 704–713. [CrossRef]

29. Nixon, W.A. *Improved Cutting Edges for Ice Removal*; Strategic Highway Research Program: Washington, DC, USA, 1993.

30. Schyllander, J. *Fotgängarolyckor: Statistik och Analys*; Swedish Civil Contingencies Agency: Karlstad, Sweden, 2014.

31. Carlsson, A.; Sawaya, B.; Kovaceva, J.; Andersson, M. *Skadereducerande Effekt av Uppvärmda Trottoarer, Gång- och Cykelstråk (TRV 2016/19843)*; Chalmers University of Technology: Gothenburg, Sweden, 2018.

32. Öberg, G.; Arvidsson, A.K. *Injured Pedestrians: The Cost of Pedestrian Injuries Compared to Winter Maintenance Costs*; Swedish National Road and Transport Research Institute: Linköping, Sweden, 2012.

33. Mattsson, A. *Samhällsekonomiska Effekter av VinterväGhåLlning för GåEnde.: En Kostnads–Nyttoanalys av VinterväGhåLlning och GåNgtrafikanters Singelolyckor i Stockholms Stad*; Linköping University: Linköping, Sweden, 2017.

34. Xu, H.; Tan, Y. Modeling and operation strategy of pavement snow melting systems utilizing low-temperature heating fluids. *Energy* **2015**, *80*, 666–676. [CrossRef]

35. Mirzanamadi, R. *Ice Free Roads Using Hydronic Heating Pavement with Low Temperature Thermal Properties of Asphalt Concretes and Numerical Simulations*; Chalmers University of Technology: Gothenburg, Sweden, 2017.

36. Rees, S.J.; Spitler, J.D.; Xiao, X. Transient analysis of snow-melting system performance. *ASHRAE Trans.* **2002**, *108 Pt 2*, 406–423.

37. Commission of the European communities. *20 20 by 2020 Europe's Climate Change Opportunity*; Commission of the European Communities: Brussels, Belgium, 2008.

38. Reinfeldt, F.; Olofsson, M. *Prop. 2008/09:163: En Sammanhållen Klimat- och Energipolitik*; Government Office: Stockholm, Sweden, 2008.

39. Löfven, S.; Baylan, I. *Prop. 2017/18:228: Energipolitikens inriktning*; Government Office: Stockholm, Sweden, 2018.

40. Fossil free Sweden (sv: Fossilfritt Sverige). *Roadmaps for Fossil Free Competitiveness—The Heating Sector (sv: Färdplan för Fossilfri Konkurrenskraft—Uppvärmningsbranschen)*; Fossil Free Sweden: Stockholm, Sweden, 2018.

41. Kim, H.-J.; Yu, J.-J.; Yoo, S.-H.; Kim, H.-J.; Yu, J.-J.; Yoo, S.-H. Does Combined Heat and Power Play the Role of a Bridge in Energy Transition? Evidence from a Cross-Country Analysis. *Sustainability* **2019**, *11*, 1035. [CrossRef]

42. Lake, A.; Rezaie, B.; Beyerlein, S. Review of district heating and cooling systems for a sustainable future. *Renew. Sustain. Energy Rev.* **2017**, *67*, 417–425. [CrossRef]

43. Sayegh, M.A.; Danielewicz, J.; Nannou, T.; Miniewicz, M.; Jadwiszczak, P.; Piekarska, K.; Jouhara, H. Trends of European research and development in district heating technologies. *Renew. Sustain. Energy Rev.* **2016**, *68*, 1183–1192. [CrossRef]

44. Lund, R.; Mathiesen, B.V. Large combined heat and power plants in sustainable energy systems. *Appl. Energy* **2015**, *142*, 389–395. [CrossRef]

45. Stankeviciute, L.; Krook Riekkola, A. Assessing the development of combined heat and power generation in the EU. *Int. J. Energy Sect. Manag.* **2014**, *8*, 76–99. [CrossRef]

46. Werner, S. International review of district heating and cooling. *Energy* **2017**, *137*, 617–631. [CrossRef]

47. Sernhed, K.; Lygnerud, K.; Werner, S. Synthesis of recent Swedish district heating research. *Energy* **2018**, *151*, 126–132. [CrossRef]

48. Djuric Ilic, D.; Dotzauer, E.; Trygg, L.; Broman, G. Introduction of large-scale biofuel production in a district heating system e an opportunity for reduction of global greenhouse gas emissions. *J. Clean. Prod.* **2014**, *64*, 552–561. [CrossRef]

49. Helin, K.; Zakeri, B.; Syri, S.; Helin, K.; Zakeri, B.; Syri, S. Is District Heating Combined Heat and Power at Risk in the Nordic Area?—An Electricity Market Perspective. *Energies* **2018**, *11*, 1256. [CrossRef]

50. Gadd, H.; Werner, S. Achieving low return temperatures from district heating substations. *Appl. Energy* **2014**, *136*, 59–67. [CrossRef]

51. Lauenburg, P. Temperature optimization in district heating systems. *Adv. Dist. Heat. Cool. Syst.* **2016**, 223–240. [CrossRef]

52. Frederiksen, S.; Werner, S. *District Heating and Cooling*, 1st ed.; Studentlitteratur: Lund, Sweden, 2013; ISBN 9789144085302.

53. Lund, H.; Werner, S.; Wiltshire, R.; Svendsen, S.; Thorsen, J.E.; Hvelplund, F.; Mathiesen, B.V. 4th Generation District Heating (4GDH): Integrating smart thermal grids into future sustainable energy systems. *Energy* **2014**, *68*, 1–11. [CrossRef]

54. ANSYS Inc. *ANSYS® Workbench^{TM}, Release 18.0.*; ANSYS®CFX®, ANSYS Inc.: Canonsburg, PA, USA, 2016.

55. Division of Energy System at Linköping University. *Converter*; Linköping University: Linköping, Sweden, 2016.

56. Henning, D. *MODEST for Windows*; IEI Energy Systems; Linköping University: Linköping, Sweden, 2014.

57. Temam, R. *Navier-Stokes Equations: Theory and Numerical Analysis*; AMS Chelsea Pub: Amsterdam, The Netherlands, 2001; ISBN 0821827375.

58. ANSYS Inc. *ANSYS Documentation Release 18.0*; ANSYS Inc.: Canonsburg, PA, USA, 2016.

59. Henning, D. MODEST—An energy-system optimisation model applicable to local utilities and countries. *Energy* **1997**, *22*, 1135–1150. [CrossRef]

60. Henning, D. *Optimisation of Local and National Energy Systems: Development and Use of the MODEST Model*; Division of Energy Systems, Department of Mechanical Engineering, Linköpings University: Linköping, Sweden, 1999; ISBN 9172193921.

61. Åberg, M.; Widén, J.; Henning, D. Sensitivity of district heating system operation to heat demand reductions and electricity price variations: A Swedish example. *Energy* **2012**, *41*, 525–540. [CrossRef]

62. Åberg, M.; Henning, D. Optimisation of a Swedish district heating system with reduced heat demand due to energy efficiency measures in residential buildings. *Energy Policy* **2011**, *39*, 7839–7852. [CrossRef]

63. Lidberg, T.; Olofsson, T.; Trygg, L. System impact of energy efficient building refurbishment within a district heated region. *Energy* **2016**, *106*, 45–53. [CrossRef]

64. Henning, D.; Trygg, L. Reduction of electricity use in Swedish industry and its impact on national power supply and European CO_2 emissions. *Energy Policy* **2008**, *36*, 2330–2350. [CrossRef]

65. Sundberg, G.; Orgen, J.; Odin, S. Project financing consequences on cogeneration: Industrial plant and municipal utility co-operation in Sweden. *Energy Policy* **2003**, *31*, 491–503. [CrossRef]

66. Gebremedhin, A. The role of a paper mill in a merged district heating system. *Appl. Therm. Eng.* **2003**, *23*, 769–778. [CrossRef]

67. Lund, R.; Ilic, D.D.; Trygg, L. Socioeconomic potential for introducing large-scale heat pumps in district heating in Denmark. *J. Clean. Prod.* **2016**, *139*, 219–229. [CrossRef]

68. Amiri, S.; Henning, D.; Karlsson, B.G. Simulation and introduction of a CHP plant in a Swedish biogas system. *Renew. Energy* **2013**, *49*, 242–249. [CrossRef]

69. Blomqvist, S.; Nyberg, S.; Energisystem, A. *Modellering och Energieffektivisering av Befintligt Markvärmesystem*; Linköping University: Linköping, Sweden, 2014.

70. Swedish Meteorological and Hydrological Institutes SMHI's Open Data. Available online: https://www.smhi.se/en/services/open-data/search-smhi-s-open-data-1.81004 (accessed on 14 March 2019).

71. Hassn, A.; Chiarelli, A.; Dawson, A.; Garcia, A. Thermal properties of asphalt pavements under dry and wet conditions. *Mater. Des.* **2016**, *91*, 432–439. [CrossRef]

72. Li, H.; Harvey, J.; Jones, D. Multi-dimensional transient temperature simulation and back-calculation for thermal properties of building materials. *Build. Environ.* **2013**, *59*, 501–516. [CrossRef]

73. Shin, A.H.; Kodide, U. Thermal conductivity of ternary mixtures for concrete pavements. *Cem. Concr. Compos.* **2012**, *34*, 575–582. [CrossRef]

74. Ohemeng, E.A.; Yalley, P.P.-K. Models for predicting the density and compressive strength of rubberized concrete pavement blocks. *Constr. Build. Mater.* **2013**, *47*, 656–661. [CrossRef]

75. Ižvolt, L.; Dobeš, P. Test Procedure Impact for the Values of Specific Heat Capacity and Thermal Conductivity Coefficient. *Procedia Eng.* **2014**, *91*, 453–458. [CrossRef]

76. Sundberg, J. *Thermal Properties of Soil and Rock (sv: Termiska Egenskaper i jord och Berg)*; Swedish Geotechnical Institute (SGI): Linköping, Sweden, 1991; Volume 12.

77. Luca, J.; Mrawira, D. New Measurement of Thermal Properties of Superpave Asphalt Concrete. *J. Mater. Civ. Eng.* **2005**, *17*, 72–79. [CrossRef]

78. Uponor Uponor surface heating system (sv: Uponor ytvärmesystem). *Heating, Ventilation and Sanitation Manual*; Uponor AB: Västerås, Sweden, 2018; pp. 547–565.

79. Swedish Meteorological and Hydrological Institute; Swedish Radiation Safety Authority; Swedish Environmental Protection Agency STRÅNG. A Mesoscale Model for Solar Radiation. Available online: http://strang.smhi.se/ (accessed on 26 April 2019).

80. Santamouris, M. *(Matheos) Energy and Climate in the Urban Built Environment—Chapter 11 Appropiate Materials for the Urban Environment*; Santamouris, M., Ed.; James & James: Athens, Greece, 1999; ISBN 9781873936900.

81. Brown, R.D.; Gillespie, T.J. *Microclimatic Landscape Design: Creating Thermal Comfort and Energy Efficiency*; J. Wiley & Sons: Guelph, ON, Canada, 1995; ISBN 0471056677.

82. Çengel, Y.A.; Turner, R.H.; Cimbala, J.M. *Fundamentals of Thermal-Fluid Sciences*; McGraw Hill Higher Education: New York, NY, USA, 2008; ISBN 9780071266314.

83. Holman, J.P.; Jack, P. *Heat Transfer*; McGraw Hill Higher Education: New York, NY, USA, 2010; ISBN 9780073529363.

84. Storck, K.; Karlsson, M.; Andersson, I.; Renner, J.; Lloyd, D. *Formelsamling i Termo- och Fluiddynamik*; Linkoping University: Linköping, Sweden, 2009.

85. Lindeburg, M.R. *Civil Engineering Reference Manual for the PE Exam, Appendices A-45*, 9th ed.; Professional Publications: Belmont, CA, USA, 2003; ISBN 9781888577952.

86. Dobre, R.-G.; Gaitanaru, D.S. Snowmelt modelling aspects in urban areas. *Procedia Eng.* **2017**, *209*, 127–134. [CrossRef]

87. Nuijten, A.D.W.; Høyland, K.V. Comparison of melting processes of dry uncompressed and compressed snow on heated pavements. *Cold Reg. Sci. Technol.* **2016**, *129*, 69–76. [CrossRef]

88. Boverket Öppna Data—Dimensionerande Vinterutetemperatur (DVUT 1981-2010) för 310 orter i Sverige. Available online: https://www.boverket.se/sv/om-boverket/publicerat-av-boverket/oppna-data/dimensionerande-vinterutetemperatur-dvut-1981-2010/ (accessed on 28 Junuary 2018).

89. Blomqvist, S.; La Fleur, L.; Amiri, S.; Rohdin, P.; Ödlund, L. The Impact on System Performance When Renovating a Multifamily Building Stock in a District Heated Region. *Sustainability* **2019**, *11*, 2199. [CrossRef]

90. Swedenergy (sv. Energiföretagen Sverige). *Överenskommelse i Värmemarknadskommittén 2018*; Swedenergy: Stockholm, Sweden, 2018.

91. Gode, J.; Martinsson, F.; Hagberg, L.; Öman, A.; Höglund, J.; Palm, D. *Miljöfaktaboken 2011—Estimated Emission Factors for Fuels, Electricity, Heat and Transport in Sweden (Sv: Uppskattade Emissionsfaktorer för Bränslen, el, värme och Transporter)*; Värmeforsk Service AB: Stockholm, Sweden, 2011; ISBN 1653-1248.

Article

An Investigation into the Limitations of Low Temperature District Heating on Traditional Tenement Buildings in Scotland

Michael-Allan Millar *, Neil Burnside and Zhibin Yu

James Watt School of Engineering, Systems, Power and Energy division, University of Glasgow, Glasgow G12 8QQ, UK

* Correspondence: michael.millar@glasgow.ac.uk; Tel.: +44-(0)-141-330-2000

Received: 31 May 2019; Accepted: 4 July 2019; Published: 6 July 2019

Abstract: Domestic heating accounts for 64% of domestic energy usage in the UK, yet there are currently very few viable options for low carbon residential heating. The government's carbon plan commits to improving the uptake of district heating connections in new build dwellings, but the greatest carbon saving can be made through targeting traditional housing stock. This paper aims to quantify the potential carbon and energy savings that can be made by connecting a traditional tenement building to a district heating scheme. The study uses a transient system simulation tool (TRNSYS) model to simulate the radiator system in a tenement block and shows that a significant benefit can be achieved by reducing the supply temperature; however, the minimum supply temperature is drastically limited by the building condition. Therefore, the study also critically compares the benefits of a lower supply temperature against minor refurbishments. It was found that improving building conditions alone could offer a 30% reduction in space heating energy consumption, while building improvements and integration of a river source heat pump could offer almost a 70% reduction. It is the recommendation of this study that a dwelling be improved as much as economically possible to achieve the greatest carbon and energetic savings.

Keywords: district heating; residential; domestic; Scotland; TRNSYS; retrofit

---

1. Introduction

Residential energy use has changed significantly from the 19th century until now, moving from solid fuel combustion (e.g., coal/wood stoves) to a predominantly gas heating market, which totals 64% of domestic energy usage in the UK in 2017 [1]. The Scottish government has set ambitious targets to provide 11% of non-electrical heat demand by renewable sources by 2020, and for 35% of domestic heat to be provided by renewable sources by 2032 [2]. The government also aims for all Scottish homes to have an energy performance certificate (EPC) of at least band C by 2040, where "technically feasible and cost effective" [2]. In Scotland, the greatest number of dwellings by type is tenement flats, and more than 74% of housing stock was built pre-1982, as shown in Figure 1 [3]. This suggests, in order to achieve Scottish government targets, the greatest focus must be on modernizing existing housing stock, rather than new housing.

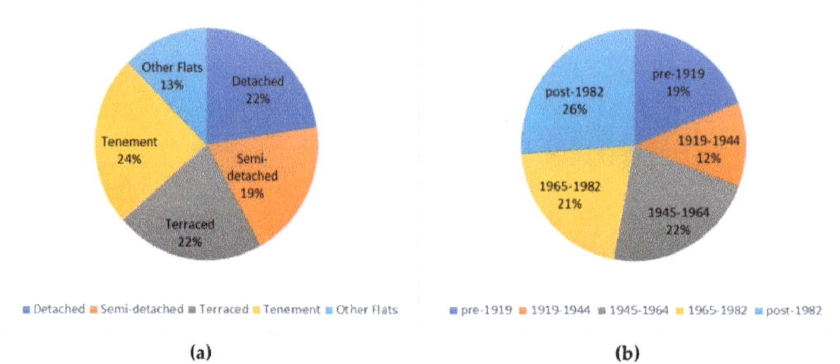

Figure 1. Breakdown of Scottish dwelling type by (**a**) number and (**b**) year built, in 2017 [3].

Limited work has been completed to the assess the suitability of existing residential buildings to connect to a district heating network (DHN); examples shown in [4–11]. Brand and Svendsen [6] discuss the necessary upgrades to existing stock in order to integrate a low temperature district heating network (LTDHN). They show that for a typical single-family, Danish house from the 1970s, small refurbishment can allow the district heating supply temperature to drop from 78 °C to 67 °C, and below 60 °C for 98% of the year. This study uses a home already connected to a DHN and compares the energy demand between the traditional DHN and a LTDHN. The paper shows promising results with minor investment. However, while the paper acknowledges the need for water disinfection for supply temperatures below 60 °C, it is difficult to know how this will influence the overall efficiency, carbon savings or cost. Østergaard and Svendsen [8] provide an investigation into the use of LTDHNs from the 1930s in single family houses, a similar study to Brand and Svendsen [6]. This study considers the influence of replacing critical radiators, and found that while 50% of the case studies could be converted to LTDH with minor renovations, 50% would require substantial work. This work is not directly transferable to the UK due to different building styles and weather patterns, but does, however, show the first steps in considering options for existing housing stock.

Wang and Holmberg [11] discuss retrofitting Swedish multi-family buildings from 1965–1975 with low temperature heating and a heat recovery ventilation system (FTX ventilation). While this discussion is limited, it does show that savings could be made on space heating—albeit with significant renovations, which would likely be out-weighed by the significant cost of improving/installing the DHN substation, installing the ventilation system, and improving the air tightness, as is recommended by the paper.

Burzynski et al. [12] provide a valuable insight to space heating and domestic hot water demands from newer tenement flats (built 2007–2010) connected to district heating schemes in the UK. The study makes use of metered data provided by one of the big six energy providers, Scottish and Southern Energy (SSE), to present floor area normalized energy usages for space heating and domestic hot water. A standard heat interface unit in a UK dwelling with district heating will only measure the total heat supplied to the property and will give no indication of the split between space heating or hot water. To find this split, Burzynski et al. [12] first applied a regression analysis to estimate a base temperature for heating degree days. The heat supplied on the calculated non-heating days was then assumed to be only for hot water, giving a baseline usage which can be subtracted from the total heat for the rest of the year to differentiate between space heating and hot water heating. The results of Burzynski et al. [12] do not correlate with SAP 2005 or SAP 2009; this could be due to an underestimation of heating in the methodology of the authors, which differs from the SAP method for estimating energy consumption, however this performance gap has been well documented elsewhere [13–17]. This is a significant piece of work for the UK district heating market, but will have limited applications to a significant majority of housing, which does not follow the Building Regulations part L or Section 6 (Scotland). The flats in this study have a district heating supply, but it is unclear if this has been from build or retrofitted later (although likely from build, due to the age of the dwellings).

Ovchinnikov et al. [18] give a comparative review of low temperature heating systems with a focus on the practicalities of the Russian building sector. In this paper, the authors discuss the merits of being able to use smaller radiators with a higher supply temperature, before going on to discuss the low energy efficiency of this approach. The authors mention the priority of addressing consumer awareness of energy usage. The authors discuss the challenges and obstruction of 4G heat networks by obsolete 3G networks. This is an interesting insight into the contrast between the challenges of heat network integration in the UK and abroad. While the UK is installing new networks, many other countries must consider how to best improve existing networks. The paper concludes that low temperature heating can be used in existing Russian housing, however, significant energy efficiency can only be achieved with vast refurbishment and building improvement. In a further paper, using an IDA Indoor Climate and Energy (IDA ICE) tool, Ovchinnikov et al. [19] provide a dynamic model and assessment of Russian building regulations and the feasibility of low-temperature heating for residential buildings. The study investigates four hydronic space-heating configurations with either a high temperature supply (75 °C) or low temperature supply (45 °C). The paper concluded that a heat pump supply could offer good energy savings for many of the case studies and operating conditions.

Peeters et al. [20] assess heating control in residential buildings for a Belgian case study. The study describes the current heating practice in Flanders by first summarizing previous housing surveys and boiler conditions. This data is then used in a TRNSYS model to evaluate the efficiency of gas boiler systems with varying levels of insulation. The case study models a terraced house with a multizone thermostat and night set back and concludes that optimal efficiency can be achieved when a flexible heating design is used, which is able to cope with large variations in heating load. A very similar study was performed by Liao et al. [21] in a UK context, however this focused on non-domestic users and no new information is provided for UK domestic dwellings.

On considering the current state of the literature, we present in this study the transient system simulation tool (TRNSYS) models, where we consider the necessary building improvements for a typical Scottish tenement flat to be connected to a district heating network or a low temperature district heating network. Lowering the supply temperature of a heating system requires careful consideration to the building condition. Therefore, we first consider and discuss the minimum supply temperature achievable to maintain a reasonable thermal comfort level at different levels of building renovation. The calculated minimum supply temperature is then used as the set point for the LTDH river source heat pump loop. A parametric analysis is provided, showing the energy and carbon savings achievable from district heating in each case study. The aims of this study are to:

1. Through dynamic computational modelling, assess the minimum radiator supply temperature which can maintain a reasonable thermal comfort in a Scottish/UK domestic dwelling, under various building conditions.
2. Assess the potential energy and therefore carbon reduction of implementing the minimum chosen supply temperature.
3. Qualitatively assess the feasibility of a river source heat pump to meet the demand of domestic heating.

The modelling tool chosen is TRNSYS. TRNSYS is a simulation environment which can be used to extensively model HVAC and building systems, amongst other things. The user can select from a range of pre-installed "types", which computationally represent physical components. At each time-step, the TRNSYS kernel feeds inputs to the different types that produce the outputs. The process is described in further detail in the TRNSYS documentation [22,23].

2. Methodology

The methodology is as follows:

1. Define the case study.
2. Develop a building model.
3. Assess minimum supply temperature for each case using a TRNSYS model.
4. Use chosen supply temperature to assess operability and control operation of river source heat pump.
5. Assess energetic and carbon benefits.

2.1. Case Studies

There are many choices available to improve the energy efficiency of a dwelling. For this study, two of the most common home improvements were chosen for consideration—double glazing and wall insulation. The case studies are summarized in Table 1.

Table 1. Summary of case studies.

Case Number	Single Glazing	Double Glazing	Insulation	No Insulation
1	X			X
2		X		X
3	X		X	
4		X	X	

The chosen case study is a traditional sandstone tenement flat, a common building type in Scotland. Tenement walls are typically solid wall, with no cavity. This makes insulation difficult, as it must be either internal or external. External insulation is not a recommended choice as it will inevitably change the appearance of the building. Internal insulation is possible, however, it will remove a small amount of internal space. Internal insulation is the only feasible option and therefore is the only one considered here. The supply temperature is varied from 60–100 °C to mimic a broad range of typical DHN supply temperatures.

2.2. Building Modelling

To be of any significance, the building choice must be typical tenement housing stock; unfortunately, due to the age of the buildings, accurate and updated plans are not publicly available. The building layout was chosen from available plans of a typical tenement and is therefore not specific to any site (however, many tenement buildings will follow this structure). As the plans are not updated, they do not include any consideration to building modifications or renovations; however, as the modelled dwellings are less than 150 m^2 floor area, they can each be modelled as a single thermal zone. This makes any error due to un-accounted for renovations likely to be insignificant.

The building geometry was produced from building plans of a typical 20th century Glasgow tenements, shown in Figure 2. The geometry was created in Sketchup (previously Google Sketchup), shown in Figure 3, and TRNSYS3d. TNSYS3d is a Sketchup extension which allows the construction types to be defined in Sketchup (e.g., external wall, window, roof etc.) and then exported as a *.idf file, which is then imported to TRNBuild, where the thermal properties of the building can be implemented. TRNBuild produces a *.b18 file which can then be used in the TRNSYS simulation studio with Type56 multizone modelling component.

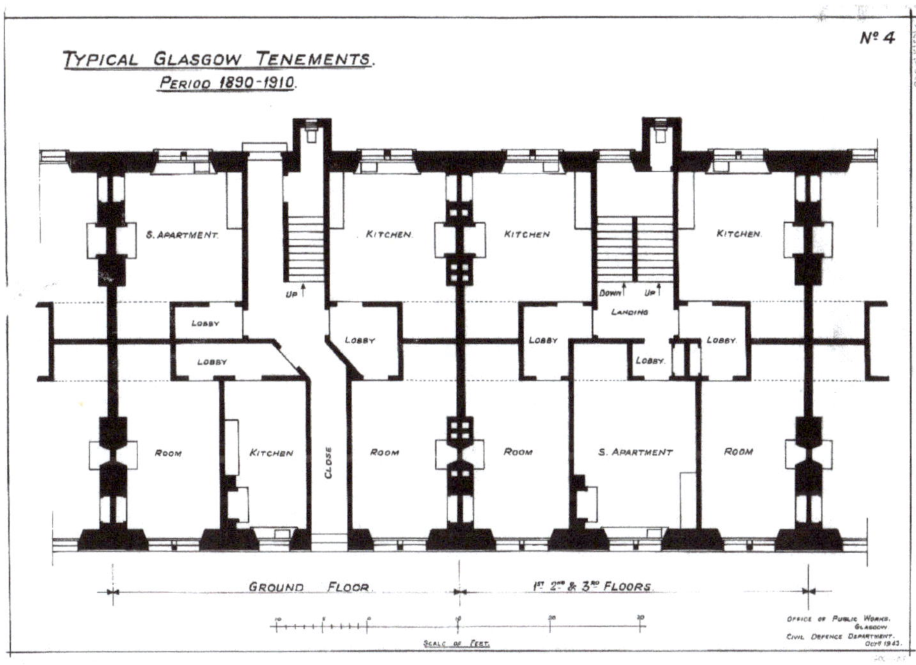

Figure 2. Typical Glasgow tenement plans. Reproduced with permission from Glasgow City Archives.

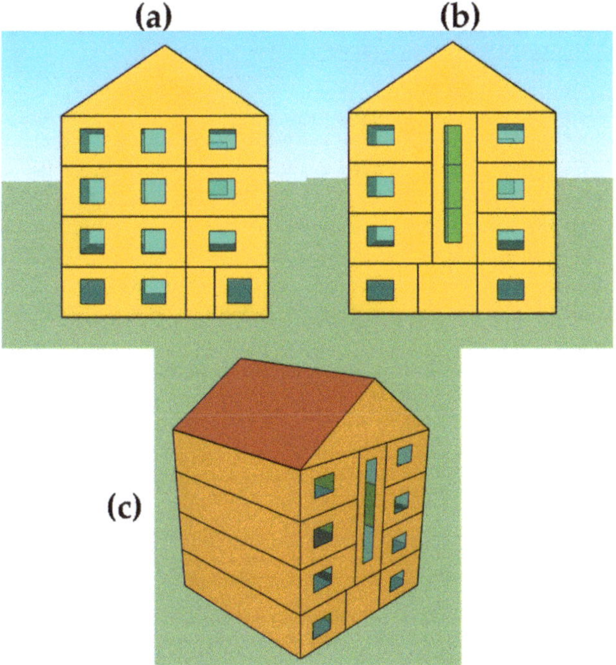

Figure 3. (**a**) Front, (**b**) back, and (**c**) isometric view of modelled tenement.

The single close of flats contains eight dwellings. Only one close is shown, although it is typical for tenement blocks to have 20 to 30 closes.

The initial TRNBuild construction types chosen are shown in Table 2.

Table 2. Initial heat transfer co-efficient of TRNBuild constructions.

Construction Type	Materials	U Value (W/m^2 K)
EXT_WALL	Plasterboard Sandstone	1.0
EXT_ROOF	Plasterboard Slate	2.5
ADJ_WALL	Plasterboard Brick	2.4
ADJ_CEILING	OAK	2.4
GROUND_FLOOR		0.78

Data is not available to consider how many properties exist with original fixtures and structures, however, the considered modifications are shown in Table 3 with thermal conductivity (U) values [24].

Table 3. Heat Transfer co-efficient for building materials.

Building Component.	Thermal Conductivity (W/m^{2}K)
Single-glazed wooden windows	5.8
Double-glazed PVC windows	1.2
Solid wall—no Insulation	1.0
Solid wall—insulated	0.18

It is assumed that the roof has been replaced since initial construction, however, since this is not part of the upper dwellings, it is not considered with renovations.

2.3. Minimum Supply Temperature

For each dwelling, there is a minimum supply temperature of space heating, dependent on the dwelling's ability to retain heat and the radiator capacity. Using TRNSYS, this is determined for each building construction case, as shown in Table 1. These temperatures are then used as a basis for the following sections. The TRNSYS model used to determine the minimum supply temperature is shown in Figures 4 and 5. The expanded macro shown in Figure 5 is the same for all "Flat X" macros. Radiators in the UK are typically designed for an 82 °C supply and 71 °C return temperature and are supplied by gas boilers. Energy and cost savings are therefore calculated against this as the base case [25].

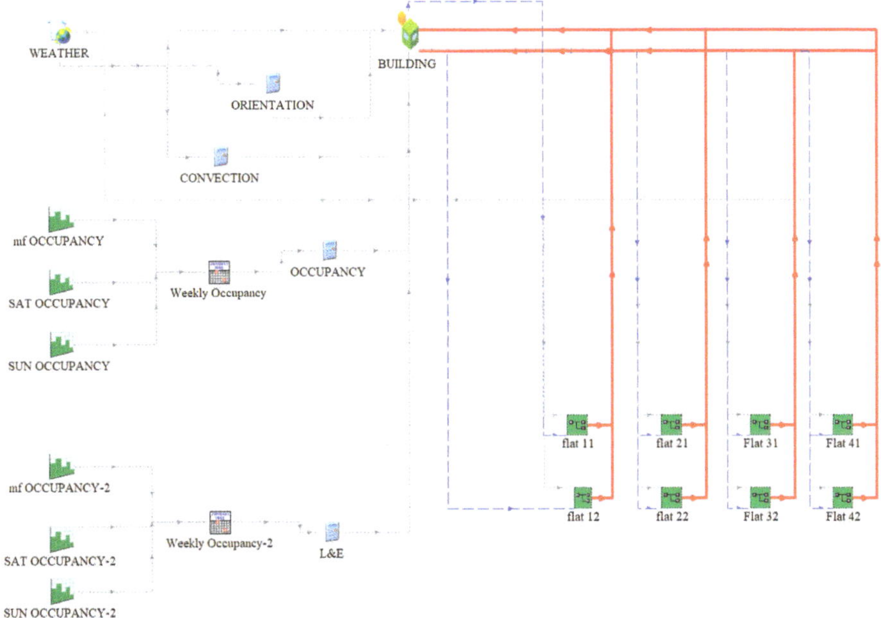

Figure 4. TRNSYS Simulation Model. Blue lines show cold streams, red shows hot streams and grey shows auxiliary streams (occupancy schedules, control signals etc.).

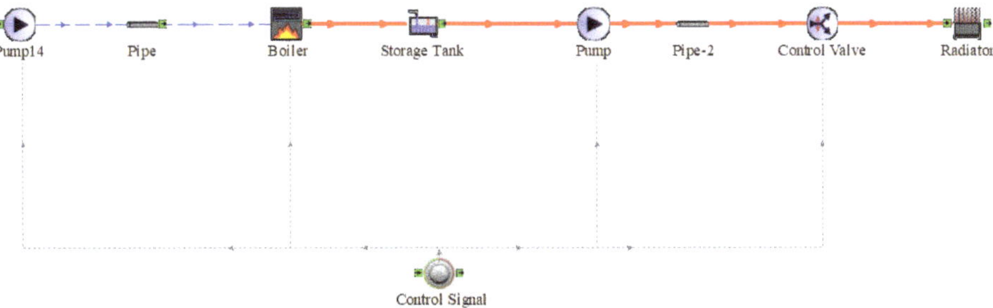

Figure 5. Expanded TRNSYS Model macro.

The air set point temperature of the dwellings is chosen as 20 °C during occupied periods, based on a generic occupancy schedule [6,26]. To maintain the set point temperature, the radiators must balance the thermal losses from each dwelling. The energy balance used in the building model is given in Equations (1)–(3) [27].

$$\dot{Q}_{ConGain} = \dot{Q}_{surface} + \dot{Q}_{infil} + \dot{Q}_{vent} + \dot{Q}_{ICG} + \dot{Q}_{CAG} + \dot{Q}_{Solwin} + \dot{Q}_{solshade} \tag{1}$$

where $\dot{Q}_{ConGain}$ is the air node convective heat gain, $\dot{Q}_{surface}$ is the convective surface gains, $\dot{Q}_{infil}$ is the infiltration gains, $\dot{Q}_{vent}$ is the ventilation gains, $\dot{Q}_{ICG}$ is the internal convective gains, $\dot{Q}_{CAG}$ is the convective air gains from other thermal zones, $\dot{Q}_{Solwin}$ the convective solar gains from external windows and $\dot{Q}_{solshade}$ is the portion of convective gains from absorbed solar radiation on shading devices. There is no mechanical ventilation and so:

$$\dot{Q}_{vent} = 0 \tag{2}$$

It is impossible to accurately determine air exchange between flats without further study, so it is assumed to be negligible for the purpose of this investigation. Therefore

$$\dot{Q}_{CAG} = 0$$

The heat addition from infiltration is given as

$$\dot{Q}_{infil} = \dot{V}\rho c_p(T_{outside\ air} - T_{inside\ air}) \tag{3}$$

where $\dot{V}$ is the volumetric flow rate of air, ρ the air density, c_p the specific heat capacity and T the temperatures of the outside and inside air.

The radiative fraction calculations are complex and explained in detail elsewhere [27–29].

2.4. Water Source Heat Pump Design

The heat supply technology chosen is a river source heat pump, due to Glasgow's large resource of river water. This is designed to operate by extracting 3 °C from the supply river water and deliver it to the main water. Although there are examples of water source heat pumps being able to condition water streams to 80 °C, WSHPs are typically only rated by manufacturers to 60/65 °C. For this reason, the heat pump is designed to condition the load stream to 60 °C. The load stream is then supplied with auxiliary heat from a gas boiler until it reaches the design supply temperature. No consideration is given to parasitic electrical load in the COP calculations (e.g., the electricity required to pump water to the heat pump). Figure 6 shows the adjusted TRNSYS model.

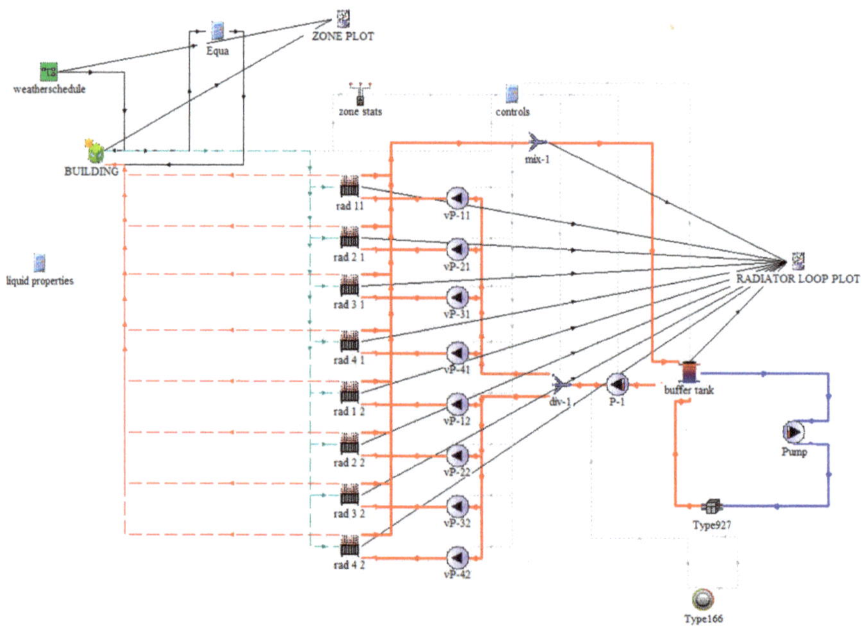

Figure 6. TRNSYS model used for heat pump supply modelling.

Figures 7 and 8 show the daily average temperature and cross-sectional flow of the River Clyde at Daldowie (NS 67154 61642). There are currently no limitations imposed by the local authority on heat extraction, however, for operational reasons the heat pump is controlled to extract 3 °C from the abstracted river flow and to switch off when the return flow to the river falls below 2 °C. This sets a lower operating temperature of 5 °C on the abstracted river stream. When the heat pump is off, the radiator loop is conditioned to the set point by the gas boilers only. A simple proportional controller is used for the purposes of this study, but a more sophisticated control system could make it possible to store heat prior to the river dropping below 5 °C.

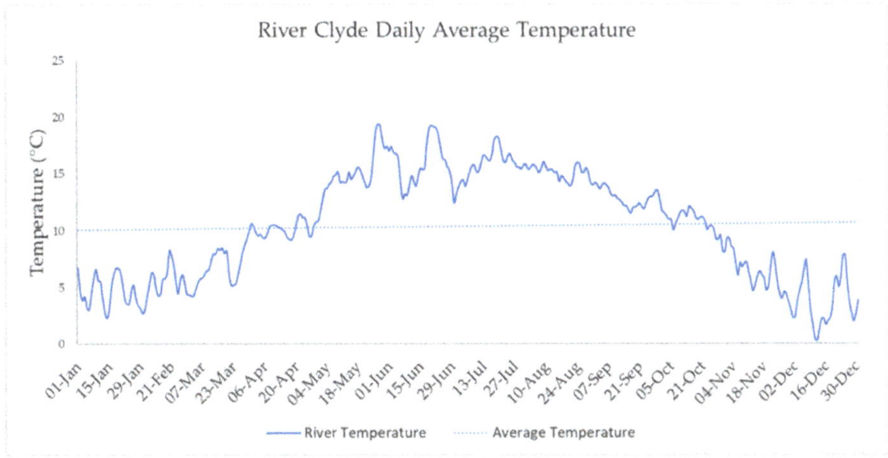

Figure 7. Average daily River Clyde water temperature at Daldowie for 2017.

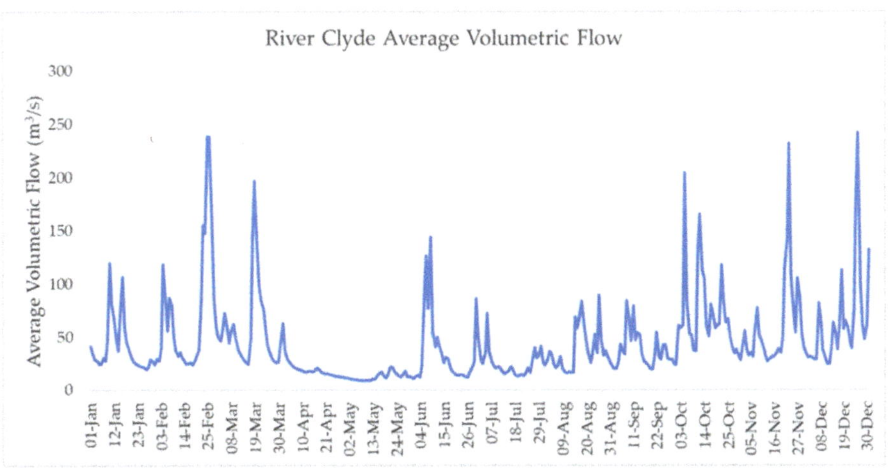

Figure 8. River Clyde average daily volumetric flow at Daldowie for 2017.

2.5. Radiator Loop

The radiator circuit is a closed loop feed, going from a small buffer tank to the radiator system and then back to the tank. The intermediate components shown in Figure 5 control the supply rates and pressure in the loop. Each dwelling is designed with 5kW of radiator capacity, which is typical of this dwelling type. From radiator sizing guidelines, this is undersized for the property—a common problem in UK housing.

2.6. Carbon Benefits

A standard carbon calculation is used to determine the carbon footprint of two energy systems. The first system is where the entire thermal load is met by a gas boiler. It is assumed that a condensing boiler is used with an efficiency of 90% [30,31]. The second system uses the heat pump to initially heat the water to 60 °C, and then uses a gas boiler to reach the set point temperature.

3. Results

3.1. Minimum Supply Temperature

Table 4 shows the average percentage of timesteps where the heating system could not maintain the set point temperature. Figures 9–12 show the percentage of timesteps where the zone air temperature fell below 19 °C. Figure 13 shows the heating power across the sample year.

Table 4. Average percent of timesteps below 19 °C across all dwellings.

	Average % of Timesteps below 19 °C			
Supply Temperature (°C)	**Case**			
	Case 1	**Case 2**	**Case 3**	**Case 4**
60	88.7	49.0	37.0	30.7
65	81.7	39.7	27.0	20.9
70	72.9	30.6	18.2	13.0
75	61.0	22.0	11.3	7.4
80	47.7	14.9	6.5	3.8
85	34.7	9.2	3.2	1.7
90	22.4	5.2	1.3	0.5
95	12.9	2.8	0.4	0.1
100	7.6	1.4	0.1	0.0

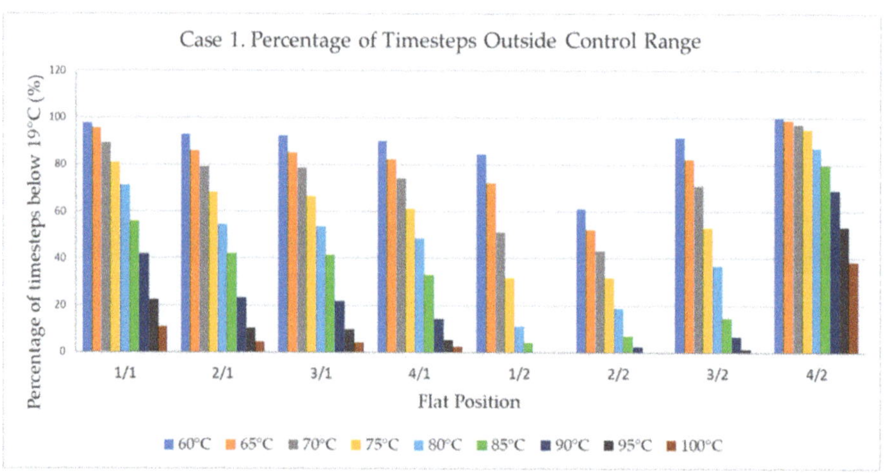

Figure 9. Case 1: no insulation and single glazing.

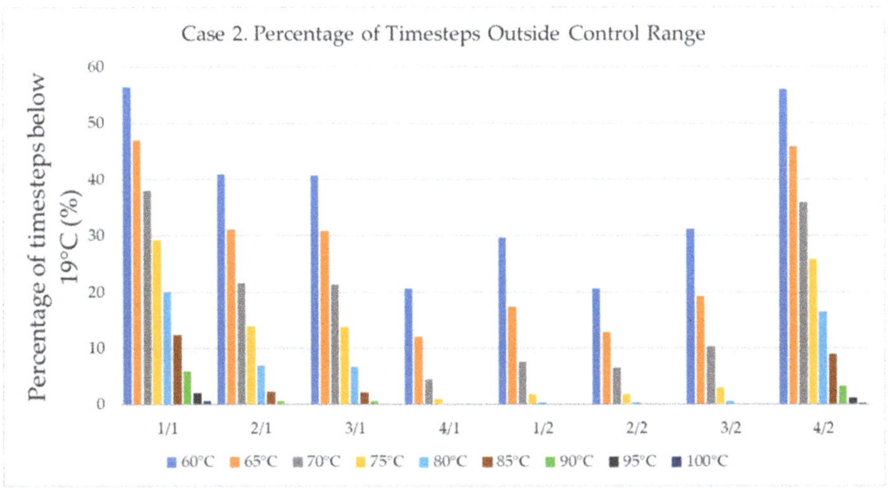

Figure 10. Case 2: no insulation and double glazing.

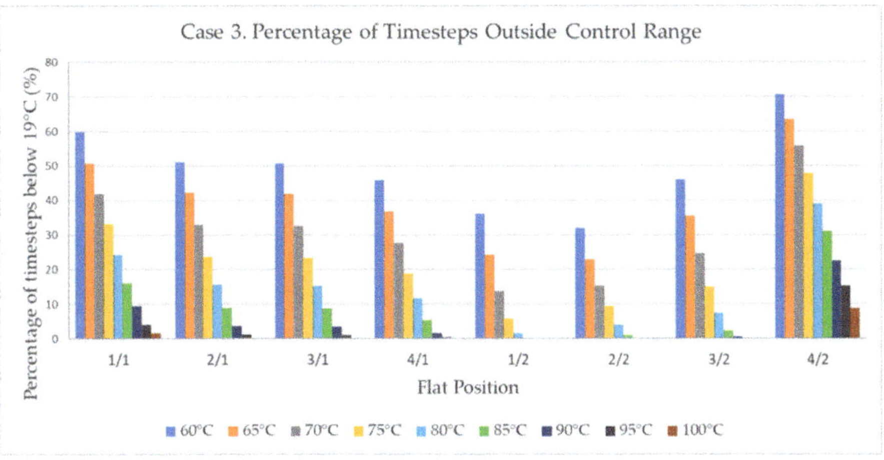

Figure 11. Case 3: insulation with single glazing.

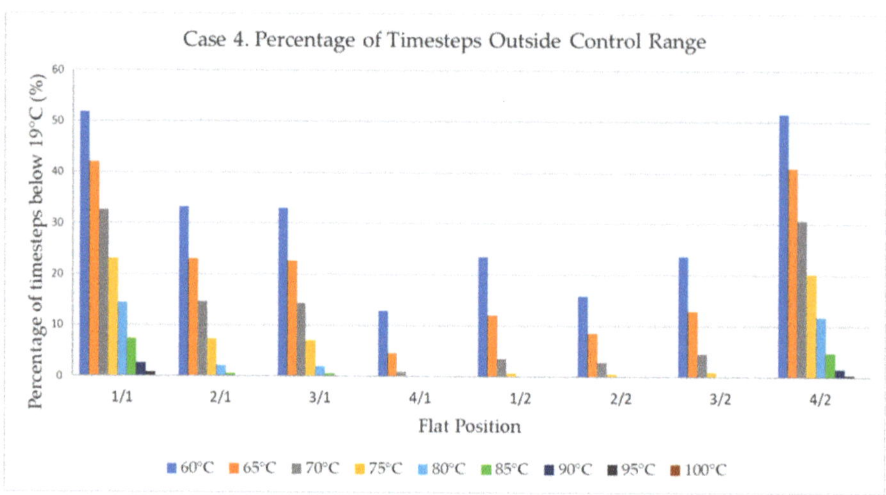

Figure 12. Case 4: insulation and double glazing.

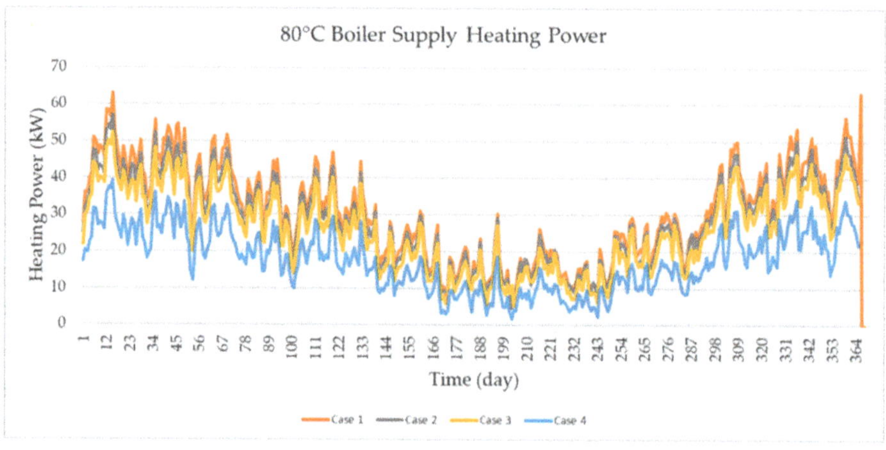

Figure 13. Case by case comparison of heating power demand at 80 °C.

3.2. Carbon and Energy Savings

The following tables show the computational results from modelling the gas boiler and heat pump energy usage at each case study. Table 5 shows the minimum achievable supply temperature chosen from Section 3.1, which is then used as the set point temperature for the modelled radiator supply. Table 6 shows the energy usage and saving when space heating is met only by the gas boiler for the base case of 80 °C supply and for the chosen minimum. Table 7 shows the electricity and gas usage when the space heating is met by the river source heat pump and supplemented by gas boilers. Table 8 shows the equivalent carbon emissions for each case. All tables show results for the full block of flats.

Table 5. Minimum supply temperature.

	Minimum Temperature
Case 1	80
Case 2	75
Case 3	70
Case 4	65

Table 6. Gas boiler energy usage.

	Gas Boiler Energy Usage (MWh)		
	80 °C	Minimum	Saving
Case 1	271	271	0
Case 2	249	234	15
Case 3	227	219	8
Case 4	154	150	4

Table 7. Boiler and heat pump energy usage.

	Boiler and Heat Pump Energy Usage (MWh)						
	80 °C Supply Temperature			Minimum Supply Temperature			Saving
	Gas	Electricity	Total	Gas	Electricity	Total	
Case 1	161	41	202	161	41	202	0
Case 2	145	37.3	182.3	133	41.7	174.7	7.6
Case 3	132	36	168	120	33	153	15
Case 4	75.4	32.5	107.9	51	36.8	87.8	20.1

Table 8. Carbon emissions.

	Carbon Emissions (ton CO_{2e})					
	Boiler			Boiler and Heat Pump		Saving
	80 °C	Minimum	Saving	80 °C	Minimum	
Case 1	55.284	55.284	0	44.447	44.447	0
Case 2	50.796	47.736	3.06	40.1359	38.9331	1.2028
Case 3	46.308	44.676	1.632	37.116	33.819	3.297
Case 4	31.416	30.6	0.816	24.5791	21.9	2.68

The carbon emissions are based on the 2018 UK government conversion factor; 1 kWh electricity is 0.283 kg CO_{2e} and 1 kWh natural gas is 0.204 kg CO_{2e} [32].

4. Discussion

4.1. Minimum Supply Temperature

Figures 9–12 show the percentage of time steps with non-zero control signal that are below 19 °C. The minimum supply temperature is chosen as the point where the system can meet approximately 80% of the demand. The additional 20% needed can be met through thermal storage; these demand side management techniques are well documented elsewhere and therefore not considered here. For dwellings with no insulation or double glazing, this does not drop significantly until the supply temperature reaches 85 °C (a typical operating temperature of domestic radiators in the UK). For smaller district heating networks, the supply temperature is often kept below 80 °C to allow the use of polyethylene or polybutylene pipes in the distribution network; these pipes can only cope with a maximum of 90 °C flow for short periods of time [25]. If thermal losses in the distribution system are considered, the temperature in the network will exceed 90 °C for a significant duration, meaning pre-insulated steel carrier pipes will likely be needed. This greatly increases the project costs. An alternative is to operate the network at a lower temperature, in order to minimize capital cost through the use of polymer piping and supplement the conditioned stream from a pre-existing heating system within the dwelling. This would add costs only to the end user, which is not preferable. Given the financial significance of dwelling connections to the network economic model, it is not recommended to adopt this approach. It can therefore be concluded that, for dwellings that are poorly insulated with low quality windows, internal improvements must be made before connection to a district heating network becomes a viable option.

When the dwelling has been fitted with insulation but no double glazing, a 70 °C flow can, on average, meet the heating demand for over 80% of the year. On addition of double glazing, a 65 °C flow can meet demand for around 79% of the year – largely similar to Case 3 with a 70 °C flow. Double glazing without insulation (Case 2) can only reach 78% of demand at 75 °C. As is to be expected, the lowest supply temperature is achievable with double glazing and insulation. The addition of insulation offers a 36.1% improvement on air temperature maintenance, while the addition of double glazing offers only 28.3% improvement in Case 1. Case 4 (both insulation and double glazing) offers a 39% improvement in Case 1, but only a 3% improvement in Case 2. It is therefore clear that, while the greatest improvement is with double glazing and insulation, the improvement by the double glazing is only marginal. The choice of double glazing should be considered based on the economic or carbon case.

4.2. Water Source Heat Pump Supply

In the sample year (2017), the average daily river Clyde temperature falls below 5 °C for 16% of the year. On these days, heating is supplied entirely from the gas boilers. For UK tenements, this necessitates a reliance on the gas boilers during this period; the boilers cannot be removed from the dwellings. While this is common in the UK, the dependence on gas can be phased out with improved thermal storage and demand side management.

For Case 1 (no insulation, single glazing), the supply temperature could not be reduced and so remained at 80 °C. When the water source heat pump is used with the gas boiler, a 25% reduction in energy and 20% carbon saving can be achieved.

From Table 7, it is clear that a reduction in total energy usage does not relate to a linear reduction in electricity to the heat pump. This is because the return temperature from the radiators is typically above the heating set point of the heat pump (60 °C), meaning that the heat pump is used to heat the radiator loop initially, but not continuously, during heating.

5. Conclusions

The UK faces challenges to decarbonize the domestic heating sector, but has few choices to do this. The best options will offer significant carbon benefits and be competitively priced to the current heating market. This work has presented a dynamic transient system simulation tool model of a typical tenement flat in the UK, one of the most common dwelling types. From this work, the following conclusions can be drawn:

The minimum supply temperature of domestic radiator systems, and therefore district heating schemes supplying tenement buildings, is strongly dependent on the building condition. Wall insulation can be difficult to install in solid wall tenement blocks but can yield a 16% energy saving on space heating per year, without lowering the supply temperature. Double glazing had less of an impact in this study but may be more significant in buildings with a greater window to wall ratio.

Tenement blocks in poor condition are unlikely to be able to connect to a district heating scheme, due to the high supply temperatures giving rise to a significant cost of carrier pipes.

In cases where no supply temperature reduction is feasible, energy and carbon savings can still be made from integrating low carbon technology. In Case 1 with heat pump supply at an 80 °C set point, energy consumption was reduced by 25% and the carbon footprint by 20%.

When building conditions permit, supply temperature can be reduced to around 65 °C and could yield almost a 70% reduction in space heating.

While there are currently no restrictions in Scotland on river heat abstraction, this is heavily dependent on the local laws.

6. Future Work

This work is presented as the start of a conversation around district heating connections for traditional housing in the UK. For this work to progress:

A substantial building condition survey of UK housing stock is needed to afford a better appreciation of the potential of low temperature heating.

Greater government incentive must be offered for privately owned dwellings to decarbonize heating.

Further minimum supply temperature studies of other dwelling types are needed, potentially offering a tool for developers to easily assess the minimum feasible supply temperature for retrofitted projects.

Author Contributions: Conceptualization, M.A.M., N.M.B. and Z.Y.; methodology, M.A.M.; investigation, M.A.M.; writing—original draft preparation, M.A.M.; writing—review and editing, M.A.M., N.M.B. and Z.Y.; supervision, N.M.B. and Z.Y.; funding acquisition, N.M.B. and Z.Y.

Funding: This research was funded by the National Productivity and Investment Fund EPSRC Doctoral Scheme (EP/R512266/1), Clyde Gateway and received support from the EPSRC "Combi-Gen" project (EP/P028829/1) and the EU H2020 LCE "DESTRESS" project (EC-691728). The research was also funded by the Newton Fund (EP/R003122/1) and research fund EP/N020472/1.

Acknowledgments: We would like to thank SEPA for supply the environmental data used in modelling. We would also like to thank Tim McDowell and David Bradley from TESS for the support received during the modelling process. Finally, we would like to thank the Glasgow City Archives for the use of building plans.

Conflicts of Interest: The authors declare no conflict of interest.

References

1. Digest of United Kingdom Energy Statistics. Department for Business, Energy and Industrial Strategy, 2018. Available online: https://www.gov.uk/government/collections/digest-of-uk-energy-statistics-dukes (accessed on 5 July 2019).

2. *Heat Policy Statement*; Scottish Government: Edinburgh, UK, 2015. Available online: https://www.gov.scot/binaries/content/documents/govscot/publications/publication/2015/06/heat-policy-statement-towards-decarbonising-heat-maximising-opportunities-scotland/documents/00478997-pdf/00478997-pdf/govscot%3Adocument (accessed on 5 July 2019).

3. Clarkson, A.; Laird, E.; Lloyd, D. Scottish House Condition Survey: 2017 Key Findings. 2018. Available online: https://www.gov.scot/publications/scottish-house-condition-survey-2017-key-findings/ (accessed on 5 July 2019).

4. Haapio, A.; Viitaniemi, P. A critical review of building environmental assessment tools. *Environ. Impact Assess. Rev.* **2008**, *28*, 469–482. [CrossRef]

5. Cetiner, I.; Edis, E. An environmental and economic sustainability assessment method for the retrofitting of residential buildings. *Energy. Build.* **2014**, *74*, 132–140. [CrossRef]

6. Brand, M.; Svendsen, S. Renewable-based low-temperature district heating for existing buildings in various stages of refurbishment. *Energy* **2013**, *62*, 311–319. [CrossRef]

7. Vallios, I.; Tsoutsos, T.; Papadakis, G. Design of biomass district heating systems. *Biomass Bioenergy* **2009**, *33*, 659–678. [CrossRef]

8. Østergaard, D.S.; Svendsen, S. Replacing critical radiators to increase the potential to use low-temperature district heating – A case study of 4 Danish single-family houses from the 1930s. *Energy* **2016**, *110*, 75–84. [CrossRef]

9. Harrestrup, M.; Svendsen, S. Changes in heat load profile of typical Danish multi-storey buildings when energy-renovated and supplied with low-temperature district heating. *Int. J. Sustain. Energy* **2015**, *34*, 232–247. [CrossRef]

10. Li, H.; Nord, N. Transition to the 4th generation district heating - possibilities, bottlenecks, and challenges. *Energy Procedia* **2018**, *149*, 483–498. [CrossRef]

11. Wang, Q.; Holmberg, S. Combined Retrofitting with Low Temperature Heating and Ventilation Energy Savings. *Energy Procedia* **2015**, *78*, 1081–1086. [CrossRef]

12. Burzynski, R.; Crane, M.; Yao, R.; Becerra, V.M. Space heating and hot water demand analysis of dwellings connected to district heating scheme in UK. *J. Cent. South Univ.* **2012**, *19*, 1629–1638. [CrossRef]

13. Allard, I.; Olofsson, T.; Nair, G. Energy evaluation of residential buildings: Performance gap analysis incorporating uncertainties in the evaluation methods. *Build. Simul.* **2018**, *11*, 725–737. [CrossRef]

14. Fedoruk, L.E.; Cole, R.J.; Robinson, J.B.; Cayuela, A. Learning from failure: Understanding the anticipated-achieved building energy performance gap. *Build. Res. Inf.* **2015**, *43*, 750–763. [CrossRef]

15. Johnston, D.; Miles-Shenton, D.; Farmer, D. Quantifying the domestic building fabric 'performance gap'. *Build. Serv. Eng. Res. Technol.* **2015**, *36*, 614–627. [CrossRef]

16. McElroy, D.J.; Rosenow, J. Policy implications for the performance gap of low-carbon building technologies. *Buil. Res. Inf.* **2019**, *47*, 611–623. [CrossRef]

17. Zou, P.X.W.; Xu, X.; Sanjayan, J.; Wang, J. Review of 10 years research on building energy performance gap: Life-cycle and stakeholder perspectives. *Energy Build.* **2018**, *178*, 165–181. [CrossRef]

18. Ovchinnikov, P.; Borodiņecs, A.; Strelets, K. Utilization potential of low temperature hydronic space heating systems: A comparative review. *Build. Environ.* **2017**, *112*, 88–98. [CrossRef]

19. Ovchinnikov, P.; Borodiņecs, A.; Millers, R. Utilization potential of low temperature hydronic space heating systems in Russia. *J. Build. Eng.* **2017**, *13*, 1–10. [CrossRef]

20. Peeters, L.; Van der Veken, J.; Hens, H.; Helsen, L.; D'haeseleer, W. Control of heating systems in residential buildings: Current practice. *Energy Build.* **2008**, *40*, 1446–1455. [CrossRef]

21. Liao, Z.; Swainson, M.; Dexter, A.L. On the control of heating systems in the UK. *Build. Environ.* **2005**, *40*, 343–351. [CrossRef]

22. Solar Energy Laboratory. *TRNSYS, a Transient Simulation Program*; Solar Energy Laboratory: Madison, WI, USA, 1975.

23. Solar Energy Laboratory. *TRNSYS18: Volume 1 Getting Started*; Solar Energy Laboratory: Madison, WI, USA, 2017.

24. Baker, P. *U-Values and Traditional Buildings*; Historic Scotland: Edinburgh, UK, 2011; Available online: https://www.historicenvironment.scot/archives-and-research/publications/publication/?publicationId=16d0f7f7-44c4-4670-a96b-a59400bcdc91 (accessed on 5 July 2019).

25. Wiltshire, R.; Williams, J.; Woods, P. *A Technical Guide to District Heating*; BRE Publications: Watford, UK, 2014; Available online: https://www.brebookshop.com/samples/327451.pdf (accessed on 5 July 2019).

26. Dansk Standard (DS). *Calculation of heat loss from buildings*; DS 418:2011; Dansk Standard (DS): København, Denmark, 2011.

27. Solar Energy Laboratory. *TRNSYS18 Volume 5: Multizone Building Modelling with Type56 and TRNBuild*; Solar Energy Laboratory: Madison, WI, USA, 2017.

28. Stephenson, D.G.; Mitalas, G.P. Calculation of heat conduction transfer functions for multi-layer slabs. *ASHRAE Trans.* **1971**, *77*, 117–126.

29. Mitalas, G.P.; Arseneault, J.G. Fortran IV Program to Calculate Z-Transfer Functions for the Calculation of Transient Heat Transfer Through Walls and Roofs. 1971. Available online: https://www.ibpsa.us/sites/default/files/publications/B10_34_1972_Jun_Fortran%20IV%20Program%20to%20Calculate%20z-Transfer%20Funtions%20for%20the%20Calculation%20of%20Transient%20Heat%20Transfer%20through%20Walls%20and%20Roofs_National%20Research%20Council%20Canada.pdf (accessed on 5 July 2019).

30. Palmer, J.; Cooper, I. *United Kingdom Housing Energy Fact Fil*; Research Consultants and Cambridge Energy: Cambridge, UK, 2013; Available online: http://www.carltd.com/sites/carwebsite/files/Housing%20Energy%20Fact%20File%202012.pdf (accessed on 5 July 2019).

31. The Building (Scotland) Regulations 2004. Available online: http://www.legislation.gov.uk/ssi/2004/406/contents/made (accessed on 5 July 2019).
32. Total greenhouse gas emission trends and projections. Available online: https://www.eea.europa.eu/data-and-maps/indicators/greenhouse-gas-emission-trends-6/assessment-1 (accessed on 5 July 2019).

Article

Exploiting Scalable Machine-Learning Distributed Frameworks to Forecast Power Consumption of Buildings

Tania Cerquitelli *, Giovanni Malnati and Daniele Apiletti

Department of Control and Computer Engineering, Politecnico di Torino, 10129 Turin, Italy
* Correspondence: tania.cerquitelli@polito.it; Tel.: +39-011-090-7178

Received: 7 June 2019; Accepted: 25 July 2019; Published: 31 July 2019

Abstract: The pervasive and increasing deployment of smart meters allows collecting a huge amount of fine-grained energy data in different urban scenarios. The analysis of such data is challenging and opening up a variety of interesting and new research issues across energy and computer science research areas. The key role of computer scientists is providing energy researchers and practitioners with cutting-edge and scalable analytics engines to effectively support their daily research activities, hence fostering and leveraging data-driven approaches. This paper presents SPEC, a scalable and distributed engine to predict building-specific power consumption. SPEC addresses the full analytic stack and exploits a data stream approach over sliding time windows to train a prediction model tailored to each building. The model allows us to predict the upcoming power consumption at a time instant in the near future. SPEC integrates different machine learning approaches, specifically ridge regression, artificial neural networks, and random forest regression, to predict fine-grained values of power consumption, and a classification model, the random forest classifier, to forecast a coarse consumption level. SPEC exploits state-of-the-art distributed computing frameworks to address the big data challenges in harvesting energy data: the current implementation runs on Apache Spark, the most widespread high-performance data-processing platform, and can natively scale to huge datasets. As a case study, SPEC has been tested on real data of an heating distribution network and power consumption data collected in a major Italian city. Experimental results demonstrate the effectiveness of SPEC to forecast both fine-grained values and coarse levels of power consumption of buildings.

Keywords: big data frameworks; data mining algorithms; machine learning; energy consumption forecast; data streams analysis

---

1. Introduction

In the last few years, an increasing number of smart meters has been deployed in smart city environments to monitor energy consumption in buildings. As a result, the collected data have increased at an exceptional rate, so that energy-related data are becoming big data. The plenitude of data provides a favorable circumstance to face valuable challenges and add intelligence in energy-related contexts. The knowledge discovery process applied to energy data can reveal hidden and actionable models and patterns, such as those characterising and predicting energy consumption, for different stakeholders, from energy managers, to analysts, and consumers.

In the last decades of the past century, data mining proved to be a valid solution for finding implicit, unknown, and useful information from very large datasets. The most popular data mining tasks include correlation analysis (e.g., association rules), prediction (e.g., classification, regression), and grouping similar data (e.g., clustering). Turning to the energy domain, the association rule mining and clustering allow unsupervised energy data exploration useful for summarizing usage patterns, while prediction algorithms enable energy consumption forecasting, which, in turn, paves the way for optimising heating distribution networks.

To effectively mine large collections of energy data, state-of-the-art data mining algorithms have often required crucial limitations to be addressed, such as those represented by computational resources. To this aim, scalable solutions have been devised in recent years, including wide-spread big data frameworks like Apache Hadoop [1], and Apache Spark [2]. However, the fast changing pace of the evolution of such distributed-computing technologies introduces two sources of problems: (i) on the one hand, they are not mature enough to be applied on a generic domain and still require some form of fine-tuning to fit the specific scenario; (ii) on the other hand, it is difficult to find suitable professional profiles trained on the latest advancements of such platforms. The proposed solution tries to accommodate the needs of energy experts and energy-provider companies of extracting data insights by exploiting machine-learning solutions which have much less stringent requirements on the professional skills of the data analysts.

The exploitation of big data platforms on energy-related data is of primary importance to extract useful, actionable, and previosuly unknown knowledge from data as well as to forecast future energy consumption. Thus, the analysis of energy data opens up a variety of interesting research issues across two research communities: energy and computer science. To design effective analytics tools, a considerable interaction between an energy scientist and a computer scientist is needed, with the former being mainly responsible for defining the end-goals and the assessment of extracted knowledge. Furthermore, a stronger involvement in the algorithm definition phase is beneficial to enrich the algorithm itself with domain-expert knowledge, such as physical laws and event models. The computer scientist tackles the task of selecting the right software platform, designing and developing efficient and effective algorithms, selecting the optimal analytic techniques to achieve the end-goals, with the right trade-off between quality of results and processing time or resources.

From the energy scientist's point of view, a lot of research efforts should be devoted to analyzing, characterizing and understanding energy-related data to effectively support different interested users in the decision-making process, from energy managers and analysts, to end-users living in buildings. Different research challenges can be addressed, whose results have a great potential to influence the overall energy balance of our communities. From the computer scientist's point of view, most of the technologies and algorithms related to big data processing and analytics have to be tailored to the specific features of the energy domain, such as heterogeneous sources and formats, variable data distributions, different abstraction levels, both fine and coarse grained, to effectively and efficiently support the knowledge extraction process.

This paper presents SPEC, a Scalable Predictor of PowEr Consumption. It provides a data mining engine for predicting the future power consumption over sliding time windows, specific to each building under exam. Different regression techniques and a classification method have been integrated into SPEC to build a model aimed at predicting the fine-grained power consumption at a time instant in the near future (i.e., with a limited time horizon): artificial neural networks, random forest regressor and ridge regression. furthermore, SPEC also includes the random forest classifier to forecast a power consumption level, instead of the fine-grained value. Each prediction model is tightly tailored to the specific building efficiency, by being trained only on power consumption historical data of the selected building. The SPEC methodology builds upon state-of-the-art distributed-computing solutions, namely, Apache Spark, which allows us to quickly analyze very large data collections. As a case study, SPEC has been validated on thermal power consumption collected in a major city in the North of Italy. Energy data have been enriched with meteorological open data. Experimental results, obtained on 12 buildings monitored every roughly five minutes for one year, demonstrate the effectiveness of the proposed methodology in predicting fine-grained power consumption with a limited average error and a good accuracy.

This paper is organized as follows. We first present the most used distributed and parallel frameworks, then the paper contribution is detailed. Next the description of the main building blocks of the SPEC engine is presented followed by the discussion of the experimental results yielded by the SPEC engine on real thermal power data. After the comparison of our approach with related works, we draw conclusions and presents future work.

2. Distributed Frameworks

The recent explosion in size of sensor-provided datasets has required the development of new distributed and parallel big data approaches, often replicated inside cloud-based services (e.g., platform-as-a-service tools) [3]. In recent years, two frameworks have emerged: MapReduce [4], as a programming paradigm, whose most popular implementation is provided by the Apache Hadoop platform, and Apache Spark [2], a more real-time data-processing solution with improved performance. Both solutions allow programmers to focus on data-processing issues, disregarding low-level details of the physical data replication and distribution over a cluster of machines, and the corresponding network coordination. The first big data approach to crunch huge datasets was the MapReduce [4] paradigm proposed by Google and then implemented in various software solutions, whose most popular is Apache Hadoop. MapReduce exploits data locality by moving the algorithm to the data instead of bringing the data to the algorithm, hence allowing each node of the distributed cluster to process local data only. To this aim, a proper storage system was developed, the Hadoop distributed file system (HDFS), which is probably the most widespread big data storage solution now, being compliant with almost all analytics software solutions.

More recently, the Apache Spark [2] framework has been developed. Apache Spark is a general purpose in-memory distributed platform supporting many development languages. Although it maintains full compatibility with the MapReduce paradigm, it overcomes MapReduce limitations and has become the favorite platform for large-scale data analytics, by enabling distributed data caching within the nodes main memory, and reducing slow disk access.

Thanks to the availability of such distributed platforms, different libraries provide many open-sourced algorithms for machine learning. Mahout [5], designed for Hadoop, is among the most widespread ones. It contains implementations in the data mining areas such as algorithms to address the cluster analysis, classification, and to support recommendation systems. All the current implementations are based on Hadoop MapReduce and has been exploited to support different data warehousing applications [6,7]. MLlib [8], instead, is the Machine Learning library developed on Spark, and it is rapidly growing both in development and adoption (e.g., network traffic analysis [9], social networks [10]).

3. Contribution of This Work

The aim of this work is to provide to energy scientists a scalable engine to predict fine-grained power consumptions tailored to each specific building under exam. The proposed engine, named SPEC, is customized to efficiently manage energy data, and includes different data mining algorithms to forecast both real and categorical values of power consumption. Furthermore, self-evaluation metrics are included to help the domain experts in assessing the quality of the results obtained. SPEC is built upon state-of-the-art distributed-computing solutions: the current implementation exploits Apache Spark and is able to effectively scale to huge data collections. SPEC is designed to be helpful in various energy-related applications, such as heating and electricity consumptions. As a case study, the proposed engine has been validated to forecast heating consumption every five minutes on 12 buildings.

To analyze the robustness of the proposed engine, it has been tested in two peculiar conditions: (i) from 6:00 a.m. to 10:00 p.m., which includes both transient states with peak values and steady states with more stable consumption values, and (ii) from 5:00 p.m. to 10:00 p.m. with only the steady-state phase. As expected, the forecasting activity in the former case is more challenging.

The energy scientist can successfully exploit this engine without specific algorithm-related knowledge, by setting only the energy-relevant parameters to reach her goals. No development of ad-hoc procedures in a specific programming language is required. For example, energy scientists can choose between fine-grained (every five minutes) or coarse-grained (e.g., hour/day/week) measurement periods, they can select the best metrics to evaluate the results, and the corresponding time frame of interest (e.g., the complete day versus specific hours of the day). In addition to the simple exploitation of the proposed engine, knowing in advance the expected power consumption on a per-building basis can provide interesting knowledge to the energy providers, that can devise proper strategies to efficiently satisfy the energy demand for each building as well as for the overall network. Finally, predicting the power consumption allows both a more accurate energy network sizing, hence providing a more reliable provisioning service, and a more informed energy usage by end-users (customers), who can be encouraged with specific rewards to apply ad-hoc strategies to reduce power consumption during the transient phases, when the energy peak demand is critical to satisfy.

4. Related Work

Recently, energy-related data have gained traction as the focus of many machine-learning-enabled analysis. The wide diffusion of smart and tiny sensor devices has been throughly exploited to monitor indoor and outdoor environmental parameters supporting the collection of a large volume of measures with temporal and spatial information. The knowledge extraction process applied on these data collections discover an interesting subset of actionable knowledge to effectively support the decision making process of facility managers.

Many research contributions on energy-related data have been carried out for: (i) identifying the main factors that increase energy consumption (e.g., floors and room orientation [11], location [12], weather [13]); (ii) characterizing consumption profiles among different users [12,14]; (iii) supporting data visualization and warning notification [15]; (iv) efficient storing and retrieval operations based on NoSQL databases [16];

A parallel research effort has been devoted to designing and developing systems providing innovative and widespread analytics services based on big data technologies. General purpose solutions [17] have been proposed together with specific techniques tailored to a given application domain, such as thermal energy consumption [18], residential energy use [19], renewable energy [20], air pollution levels [21].

In [22], various big data services based on a Hadoop large-scale energy-distribution platform have been analyzed. Authors conclude that most services goal is the energy efficiency improvement and the cost cut in heating maintenance and consumption. Other approaches contribute with algorithmic and technological solutions, such as clustering techniques and association rule mining: in these works, big data mining techniques are exploited both for prediction from historical data and for data exploration.

Similarly, different combinations of such techniques have also been successfully exploited in other domains, e.g., for scaling network data characterization [23], and for social network data exploration [10].

Addressing the wider energy-data management field, in [24,25] two different platforms and infrastructures for managing building-specific data from different smart-city sources are presented. In [26,27] different GIS-enabled frameworks for modelling urban district energy consumption are presented, with the former [26] applied in New York city, and the latter [27] in a North-western Italian city. However, they do not focus on energy consumption prediction but on classifying energy intensity in buildings [26] and estimating space heating [27].

Focusing on the energy forecasting techniques, in [28], a predictor of energy consumption in buildings is proposed, consisting of four building blocks, from data acquisition to performance evaluation. In [29], a thermal-load forecast approach combining multiple data-driven methods is presented, with experimental results on the next-day hourly load forecast.

In [30] a prediction model for district energy consumption in medium (mothly) and long terms (yearly) is presented, based on an ensamble of three different data-mining techniques. In [31] two data-driven thermal-load forecasting models are compared, one based on support vector machines, and another with two nonlinear autoregressive exogenous recurrent neural networks. In [32], two heating distribution network substations in Changchun, China, were analyzed by means of data mining techniques. As a result, six operating states are identified in each heating season. Finally, in [33] the predictions of three energy consumption linear models are compared on two Swedish cities. Forecasts target the next-day prediction and not the specific building prediction model. Overall, current state-of-the-art works do not address specifically the large-scale challenges while keeping focus on the fine-grained data-driven forecasts.

Similarly, research efforts have been devoted to characterizing energy consumption at a large scale [34,35] as well as energy efficiency based on real consumption data [6] or estimated data [36,37]. The study presented in [34] exploits a NoSQL technology to support the collection, storing and analysis of large volumes of energy-related data. In [34], a datawarehouse-style solution targeting KPI computation based on the leading NoSQL database MongoDB [38] has been proposed, exploiting the map-reduce approach. The proposed indicators consider the energy consumption during specific outdoor conditions (temperature range) to characterize the energy consumption of single buildings and groups of buildings in the same neighborhood. As a further step, a more advanced KPI computation approach is presented in [6]. The work in [6] presented the energy signature analysis (ESA) system. It is based on a big data methodology exploiting the map-reduce paradigm. It is able to characterize the building's energy efficiency through the energy signature. The latter estimates the total heat loss coefficient of a building and it is computed by a linear regression of the power used for heating on the difference between the internal temperature and the external temperature. The building signature has been exploited to compute two KPIs: "(i) The intra-building KPI to compare latest observations with past energy demand in the same conditions, for example in a similar outdoor temperature and indoor temperature; and (ii) the inter-building KPI to rank the overall building performance with respect to nearby and similarly characterized buildings by considering spatial co-location, building size, and usage patterns (e.g., residential, office, public building)". In [39] an engine exploiting unsupervised machine learning approaches (clustering) and association rule mining is used to explore energy consumption in buildings.

Differently from the previously-mentioned research papers [6,34,35,39,40], the current work presents an engine based on scalable machine learning approaches to forecast fine-grained power consumption. The previously-cited works focus on diverse targets and proposed different analytics approaches. They also describe a significantly dissimilar architecture, wheres the datawarehouse design is the same. In particular, no state-of-the-art solution provides a large-scale prediction with a fine-grained five minute resolution for each building. Specifically, the target of [34,39] is the characterization of the power consumption and the focus of [6] is the energy efficiency characterization, to define a building ranking, whereas the present paper targets the predicting of sliding-windows power consumption. Furthermore, the methodologies proposed in [6,34] exploit the map-reduce paradigm, while this work exploits the more advanced and high-performing Apache Spark framework; and the approach proposed in [35] does not include the forecasting data mining techniques exploited by the currently proposed methodology.

First attempts towards the prediction of fine-grained energy/power consumption over a sliding window have been proposed in [41,42]. The current paper significantly extends the study in [41] by (i) providing a new algorithm to perform the regression task (i.e., the ridge regression model); (ii) including a new analytics method (i.e., classification approach) to address the prediction of power consumption labels through the random forest classifier; (iii) adapting the prediction models to a longer time frame, including the first hours of the morning when a large number of energy consumption spikes occur; (iv) introducing new and longer prediction horizons; (v) adding an exploitation use case of the proposed approach, based on a real-world district heating network, and (vi) providing a more in-depth and extensive experimental validation (almost three-fold expansion in experimental results).

5. The SPEC Engine

SPEC is a distributed analytics engine aimed at predicting fine-grained power consumption. Its architecture is presented in Figure 1 and consists of different components, each addressing one of the main steps of the knowledge-extraction process. The scope of the current paper is to discuss SPEC performance and usage in the context of thermal energy consumption. To this aim, the dataset under analysis consists of thermal energy consumption data, collected every five minute from a large number of smart meters deployed in 12 buildings. As proposed in [34], energy data are enriched with meteorological information, collected from open-data web services [43]. Added meteorological information includes temperature, relative humidity, precipitation, wind direction, UV index , solar radiation and atmospheric pressure, as defined and provided by [43]. The data collection and integration component is in charge of collecting energy consumption data and integrating meteorological information with the right temporal and geographical correlation. Since the focus is on thermal energy consumption in residential and office buildings, only measurements of the winter season are considered.

The other SPEC components are presented in the next subsections.

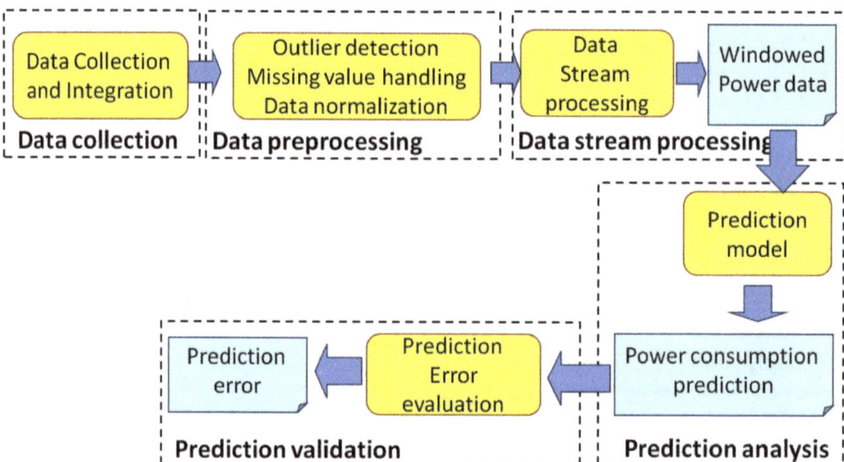

Figure 1. The scalable predictor of power consumption (SPEC) architecture.

5.1. Data Preprocessing

The knowledge extraction process is a multi-step process typically starting with a preprocessing phase, whose aim is to smooth the effect of possibly unreliable measurements. SPEC preprocessing component provides three features which have been proved to be crucial in real-world sensor-provided energy data: (i) outlier detection and removal, (ii) missing value handling, and (iii) data normalization.

Outlier detection and removal. An outlier is a measurement that lies outside the expected range of values. It may occur either when the collected value does not fit the model under study or when faulty sensors provide unacceptable measurements for the phenomenon under analysis.

To detect outliers SPEC integrates the leverage measure. It is a coefficient based on the Mahalanobis distance to define whether a power consumption measurement (X_i) is different from the others. For each observation X_i, SPEC computes the leverage as proposed in [44], as provided in Equation (1):

$$H_i = Mahalanobis^2(X_i) + \frac{1}{N},\tag{1}$$

where N is the number of power consumption measurements, and the Mahalanobis distance, in our study, is computed as provided in Equation (2):

$$Mahalanobis(X_i) = \sqrt{\frac{(X_i - mean(E))^2}{\sum_j (X_j - mean(E))^2}},$$ (2)

where $mean(E)$ is the mean of all energy samples, while $\sum_j (X_j - mean(E))^2$ is the the total square difference between all samples in energy consumption and their mean. Only the observations X_i with a leverage value H_i greater than the $CutOff$ threshold are processed in the next analytics step. The $CutOff$ threshold value is computed as provided in Equation (3):

$$CutOff = \frac{2(K+1)}{N},$$ (3)

where K is the number of variables under analysis.

Missing value handling. Many strategies are available to address missing values. SPEC selects different approaches for different attributes, in particular for contextual data sources, such as unreliable weather data providers: (i) replacement with the daily average value or (ii) replacement with the hourly average value computed in the corresponding time of the previous week. The choice is mainly driven by the physical meaning of each attribute. Specifically, strategy (i) is exploited for rain precipitation and wind direction attributes, while strategy (ii) is applied to solar radiation and UV index attributes.

Data normalization is an important task required when differences in scale and measurement unit exist in the data under analysis. Specifically, the normalization technique allows preserving the original data distribution without affecting the relevance of the analytics results. SPEC integrates two normalization techniques: min-max and z-score. The typical state-of-the-practice approach is to iteratively perform different analysis sessions with different data normalization techniques to identify the strategy yielding better results.

5.2. Data Analysis

The core of the knowledge extraction process in SPEC consists of three building blocks: (i) data stream processing, (ii) prediction analysis, and (iii) prediction validation.

5.2.1. Data Stream Processing

In the buildings under analysis, a large volume of energy data is continuously collected since power consumption is monitored roughly every five. Due to the high volume of collected data, the SPEC engine performs the prediction over a sliding time window. Specifically, when a new power consumption measurement is collected from a building, a sliding window over its historical data stream is considered. This window contains a snapshot of the latest power consumption values of the building, together with correlated meteorological data. The time window size is a parameter (w_{length}) to be chosen depending on the temporal context of interest for the analysis: with short time windows, the evaluation is almost in real-time of the building's consumption is performed based only on very recent measurements; instead, a very large time window includes many historical measurements.

5.2.2. Prediction Analysis

Many data mining algorithms are available for prediction purposes. Depending on the nature of the variable to predict, we can identify two broad classes of approaches: if we wish to predict future power consumption measurements as numerical values, then we focus on regression techniques; otherwise, if the prediction were a consumption level (such as high, mid, low), then a classification approach able to predict categorical values is required. SPEC addresses both requests by providing three regression techniques and a classification algorithm.

The prediction analysis component consists of two steps: (i) model building and (ii) prediction task. In (i) a model for the data under analysis is built by analyzing past power consumption together with meteorological data. Then the model is exploited to forecast the upcoming power consumption in (ii). Among the techniques available to our purposes there are regression-based methods, decision trees, naive Bayes approaches, neural networks, and support vector machines. Each technique employs different learning algorithms to build models from historical data. In SPEC a new data model is created for each time window. To self-assess the performance of the proposed prediction models, data are split into training and test sets. The former is used to build the model, whereas the latter to assess its quality.

For the regression purposes, i.e., the prediction of a real value of power consumption, SPEC provides the following techniques: ridge regression (RR), random forest regression (RFR), and artificial neural networks (ANN). While the latter two techniques have been widely and successfully exploited in many different applications , the former was included to provide better insights to the energy-provider company. The selected techniques are briefly presented in the following, specifically focusing on the Apache Spark implementation.

Ridge regression (RR) builds a model based on the linear dependency among data under analysis. Given a set of input features expressed through a n-dimensional vector $x = [x_1, \dots, x_n] \in \mathbb{R}^n$ and a target variable $y \in \mathbb{R}$ representing the objective of the prediction, the algorithm builds a regression model with L2-regularization using stochastic gradient descent that provides a good estimation of the value of y.

Random forest regressor (RFR) is an ensemble learning method that can be used for regression. Given a training set with known predictions, a group (forest) of decision trees is created as a model. The forest approach reduces the risk of overfitting. In MLlib [8], the RFR algorithm creates different trees during the training phase by generating randomness and minimizing overfitting. The randomness is introduced by (i) performing N sub-sampling of the training set (i.e., bootstrapping), (ii) considering different random subsets of input variables. Thus, a parallel execution of the training step is possible. During the application of the model, the single class labels produced by each tree are aggregated and a global result is computed. Labels are replaced by real values in case of regression problems.

The building of each decision tree is a top-down process: an attribute test condition is chosen at each step so that it best splits the records. To this aim, the Gini index can be exploited. Each node in the tree represents a test on an attribute, being each branch, descending from a node, a range of possible values for that attribute. Leaves represent values of the target attribute. This allows to partition the sample space depending on the test conditions of the different attributes at each node.

To generate a prediction, the tree is visited top-down, following the tests in each node, and branching down to a resulting leaf.

Artificial neural networks (ANN) exploited in SPEC are multi-layer networks without restrictions (the source code has been downloaded from https://github.com/yannart/Scala-Neural-Network). They include an input layer, n hidden layers, and an output layer. Each node in a layer takes as input a weighted sum of the outputs of all the nodes in the previous layer, and it applies a nonlinear activation function to the weighted input. The network is trained with back-propagation and learns by iteratively processing the set of training data records: weights in the network nodes are backward updated to minimize the mean squared prediction error.

Among the available techniques suited to the classification problem (i.e., the prediction of a categorical value such as a range of values of power consumption) SPEC provides the random forest classifier (RFC).

The random forest classifier (RFR) is a technique very similar to the random forest regressor. It is an ensemble learning method based on a pool of trees. Differently from the regressor, the predictions are categorical, such as high, mid, or low energy consumption levels. Each level can be associated to a specific range of real consumption values. There is virtually no limit in the number of different categorical values, however, such techniques typically work well with a low number of classes (i.e., different predicted categories), in the order of tens at most. They are useful because often it is not the precise real value to be of interest, but its level in terms of meaningfulness for the phenomena under study. Hopefully, grouping contiguous values into the same level (i.e., discretization into category), helps in improving accuracy results. Technically, the difference with the regressor is in the the the final predicted value, which cannot be computed as the average of the single tree predictions, but a majority voting approach is used: the category voted by the largest number of trees is selected as the final result of the forest classifier.

5.2.3. Prediction Validation

This component evaluates the ability of the SPEC engine to correctly predict the energy consumption of a building. To this aim, SPEC integrates three metrics to evaluate the quality of regression-based models and one metric for the classification-based models: (i) mean absolute percentage error (MAPE), (ii) weighted absolute percentage error (WAPE), and (iii) symmetric mean absolute percentage error (SMAPE), whose formulas are reported in the following, are the regression metrics, whereas the accuracy is used for the classification model. Accuracy is the ratio of the correct predictions with respect to the overall number of predictions. The metrics are provided in the following Equations (4)–(6).

$$MAPE \ = \ \frac{100\%}{n} \sum_{i=1}^{n} \left| \frac{A_i - P_i}{A_i} \right| \tag{4}$$

$$WAPE \ = \ 100\% \cdot \frac{\sum_{i=1}^{n} |A_i - P_i|}{\sum_{i=1}^{n} A_i} \tag{5}$$

$$SMAPE \ = \ \frac{100\%}{n} \sum_{i=1}^{n} \frac{|A_i - P_i|}{|A_i| + |P_i|} \tag{6}$$

In all formulas, A_i is the actual energy consumption at time t_i while P_i is the corresponding predicted value.

MAPE, or mean absolute percentage deviation (MAPD), can evaluate a predictor quality. However, since MAPE is a percentage, it might be less suitable for energy predictions because of its sensitivity to low absolute values: given the same absolute error, MAPE may be very large in presence of low consumption values, while it might hide errors when the absolute consumption is very high. The WAPE and SMAPE metrics have been proposed to address this issue. WAPE suffers from not having a specific meaning as an error on the single prediction but only on all the forecasts, while SMAPE is able to correctly model the prediction error for each forecast individually. The only drawback of SMAPE is that it is not symmetric. Thus, overestimated forecasts and underestimated forecasts do not have the same impact. Specifically, for the same value of prediction error, the underestimated forecast has a greater impact on the overall SMAPE value. Since each metric has benefits and drawbacks, SPEC provides them all to allow the energy analyst to select the best one for her goals.

6. Experimental Results

We tested the efficiency of SPEC by performing different experimental sessions on a real dataset, including power consumption data collected from 12 residential buildings. Measurements are collected over a full winter period in Italy, from October 15th to April 15th. Energy data have been enriched with meteorological information collected from the weather underground web service (the weather underground web service gathers meteorological data from personal weather stations (PWS) registered by users) [43].

The experiments reported in the paper target a subset of the overall district heating buildings, specifically those buildings for which the full historical datasets of real-world data measurements were available.

The datasets have been stored in a Hadoop cluster available at our University, based on the Cloudera Distribution of Apache Hadoop, version 6.1.0. All experiments have been performed on our cluster, which has 8 worker nodes, and runs Apache Hadoop 3.0.0 and Spark 2.4. The current implementation of SPEC is a project developed in Scala exploiting the Apache Spark framework.

Input variables are: time, date, temperature, humidity, precipitations, pressure, dew point, wind direction, and energy consumption, in the configured time window. The target of the prediction task is the upcoming energy consumption in the near future. For the results reported in this study, the SPEC engine configuration featured the normalization and outlier detection through the min-Max technique and the Leverage approach, respectively; the time window size (w_{length}) has been set to three samples (i.e., roughly 15 min). A parameter grid search has been performed to identify values for algorithm parameters. To configure the ridge regression algorithm in MLlib, the following parameters have been set: intercept = false, numIterations = 100, regParam = 0.01, stepSize = 1.0. For both the Random Forest Regressor and Classifier in MLlib, the parameters were: numTrees = 20, featureSubsetStrategy = all, impurity = variance, maxDepth = 4, maxBins = 100. The ANN regressor has been used with perceptronInputNum = 9 and neuronLayerNum = Array [10, 2, 1].

Experimental results targeted the five minute prediction capability. To this aim, we evaluated the prediction error for all algorithms and all buildings separately. The prediction error has been computed as the averaged error of all predictions in the whole (winter) period. To perform the prediction task we considered two different time frames: (i) the complete day, from 6:00 a.m. to 10:00 p.m.; (ii) the afternoon/evening only, from 5:00 p.m. to 10:00 p.m. The former time frame includes both transient and steady-state phases, thus the prediction is more challenging: values were characterized by large variability and spikes. The second time frame includes typical steady-state phases, with more stable consumption values, hence the prediction task is expected to be easier.

Tables 1–3 report the MAPE, WAPE, SMAPE values for each monitored building obtained through the ridge regression (RR), artificial neural networks (ANN), and the Random Forest Regression (RFR) models respectively. Tables 1–3 focus on the 5–10 p.m. time frame, whereas Tables 4–6 report results of the whole day.

All three regression models yield good results by analyzing consumption values in the 5–10 p.m. time frame. Both RR and ANN models present limited errors: ANN has a MAPE of 6–19%, a WAPE of 6–10%, and a SMAPE of 3–5% as reported in Table 2. Such results are very similar to the ones yielded by RR in Table 1). Also the performances of RFR are quite good, although slightly worse than both RR and ANN models: it has a MAPE of 9–20%, a WAPE of 9–14%, and a SMAPE of 5–7% as reported in Table 3).

As expected, errors yielded by SPEC models worsen when the prediction task is performed for the whole day. As shown in Tables 4–6, on average ANN models reach the best results, with an average prediction error lower than both RR and RFR. Results of RFR are better than RR, although slightly worse than ANN.

Table 1. Prediction error for each building: ridge regression model, time frame: 5–10 p.m.

Building	MAPE	WAPE	SMAPE
B1	10.4%	8.5%	4.0%
B2	6.4%	6.3%	3.1%
B3	10.1%	8.4%	4.1%
B4	18.9%	9.9%	4.9%
B5	9.1%	7.9%	3.7%
B6	7.9%	7.7%	3.6%
B7	9.1%	7.3%	3.6%
B8	6.4%	6.5%	3.1%
B9	15.7%	8.6%	4.2%
B10	12.6%	9.0%	4.3%
B11	15.2%	9.9%	4.9%
B12	12.4%	9.2%	4.4%

Table 2. Prediction error for each building: artificial neural network model, time frame: 5–10 p.m.

Building	MAPE	WAPE	SMAPE
B1	13.5%	11.8%	6.0%
B2	7.1%	7.0%	3.5%
B3	12.8%	10.7%	5.5%
B4	20.4%	12.8%	6.6%
B5	9.9%	9.4%	4.6%
B6	7.4%	7.1%	3.7%
B7	11.0%	9.7%	5.0%
B8	7.7%	7.9%	3.9%
B9	18.6%	12.9%	6.7%
B10	15.9%	12.4%	6.4%
B11	16.9%	12.1%	6.2%
B12	12.7%	10.5%	5.2%

Table 3. Prediction error for each building: random forest regression model, time frame: 5–10 p.m.

Building	MAPE	WAPE	SMAPE
B1	13.3%	11.5%	5.6%
B2	10.4%	10.3%	5.1%
B3	13.3%	11.2%	5.6%
B4	19.8%	11.1%	5.8%
B5	11.2%	10.5%	5.2%
B6	12.2%	12.0%	5.9%
B7	11.5%	9.6%	4.7%
B8	9.3%	9.4%	4.6%
B9	19.0%	11.9%	6.1%
B10	14.6%	11.3%	5.6%
B11	19.3%	13.4%	6.5%
B12	16.9%	13.9%	7.0%

Table 4. Prediction error for each building: ridge regression model, time frame: 6 a.m–10 p.m.

Building	MAPE	WAPE	SMAPE
B1	26.7%	10.1%	9.8%
B2	25.5%	16.8%	8.1%
B3	21.5%	20.6%	8.1%
B4	42.8%	31.1%	13.8%
B5	22.2%	16.3%	8.0%
B6	24.3%	16.9%	9.0%
B7	23.8%	20.4%	9.7%
B8	31.2%	31.8%	14.0%
B9	28.6%	23.8%	10.9%
B10	21.8%	22.2%	10.0%
B11	30.1%	26.2%	11.9%
B12	24.8%	19.9%	9.6%

Table 5. Prediction error for each building: artificial neural network model, time frame: 6 a.m–10 p.m.

Building	MAPE	WAPE	SMAPE
B1	21.4%	11.3%	6.0%
B2	21.0%	9.6%	4.8%
B3	13.7%	10.9%	5.4%
B4	31.9%	15.2%	7.2%
B5	18.9%	10.0%	5.0%
B6	17.9%	9.9%	5.4%
B7	17.4%	11.0%	5.4%
B8	16.4%	14.1%	6.4%
B9	20.6%	12.2%	5.9%
B10	11.9%	10.4%	5.1%
B11	23.7%	15.2%	7.5%
B12	20.6%	11.4%	5.6%

Table 6. Prediction error for each building: random forest regression model, time frame: 6 a.m–10 p.m.

Building	MAPE	WAPE	SMAPE
B1	22.9%	13.5%	7.3%
B2	23.9%	10.9%	5.6%
B3	16.6%	16.5%	7.5%
B4	32.7%	17.0%	7.9%
B5	20.9%	11.5%	5.9%
B6	23.0%	12.3%	6.8%
B7	18.0%	12.0%	6.0%
B8	17.2%	16.5%	7.3%
B9	21.6%	13.0%	6.3%
B10	13.2%	11.7%	5.9%
B11	23.5%	15.1%	7.5%
B12	20.3%	13.5%	6.7%

Since the ANN model yielded the lowest errors and on average achieved better results than both RR and RFR on all time frames (i.e., 6 a.m.–10 p.m. and 5–10 p.m.), we analyzed in more details the error trend yielded by ANN in a given day for a representative building, as reported in Figure 2. Building no. 7 is selected as representatives because its consumption time series include multiple peaks. In all cases, the ANN models are able to predict values following the trend of the actual time series, although the error significantly increases for peaks in energy consumption, such as during the early morning (e.g., 7–8 a.m).

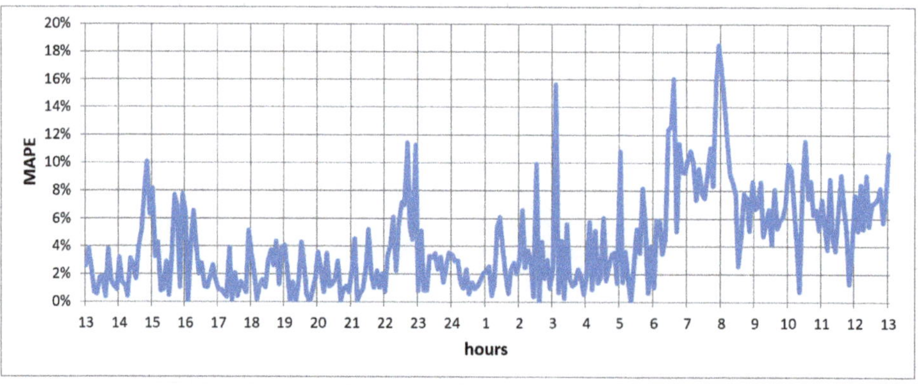

Figure 2. Building 7: detailed results for 24 h of continuous five minute predictions.

We experimentally evaluated the performance of the random forest classifier provided by SPEC. To perform the categorical classification task, the power consumption values per unit of volume have been discretized in six fixed-size bins as in [39]: two bins until 15.5 KW/m^3 (off until 0.05 KW/m^3, low until 15.5 KW/m^3), a bin each 10 KW/m^3 for values until 35.5 (medium consumption until 25.5, high consumption until 35.5) and an additional bin for values exceeding 35.5 KW/m^3. Table 7 reports the accuracy yielded on the 12 buildings under analysis for both the complete day and afternoon/evening time frames. In both cases the classifier achieved good accuracy values: they are in the range from 81–91% when analyzing all collected values (6 a.m.–10 p.m.), and in the range 89–98% (except for building no. 1) on the afternoon/evening time frame.

Table 7. Accuracy for each building: random forest classification model.

Building	Time Frame 6:00 a.m.–10:00 p.m.	Time Frame 5:00 p.m.–10:00 p.m.
B1	81%	77%
B2	91%	98%
B3	91%	93%
B4	85%	95%
B5	88%	94%
B6	88%	94%
B7	84%	93%
B8	85%	98%
B9	87%	93%
B10	89%	97%
B11	85%	89%
B12	88%	89%

Regarding the prediction horizon, we analyze the performance in terms of MAPE, WAPE, and SMAPE of the top-performing approach, i.e., ANN, with longer prediction horizons, specifically 30, 45, and 60 min, as reported in Table 8. The selected 5 buildings are those presenting no missing values, hence not requiring any pre-processing. Such results are compared with two naive techniques: (NAIVE 1) previous-value predictor and (NAIVE 2) same-time previous-day predictor.

Improvements in terms of prediction error reduction of the artificial neural network model over the two selected naive approaches for a 30-min prediction horizon in the time frame 6 a.m.–10 p.m. are reported in Table 9. The results for prediction horizons of 45 and 60 min are reported in Tables 10 and 11 respectively.

The proposed approach always yields to better predictions (i.e., with lower errors) with respect to the NAIVE 2 approach. Similar results are also reported with respect to NAIVE 1 apart from a single exception (SMAPE for building B2). Improvements for building B1 are higher with respect to NAIVE 1 and lower with respect to NAIVE 2, whereas the converse is true for the remaining buildings. This is due to the different usage pattern of residential versus office/public buildings.

Table 8. Artificial neural network model prediction errors for different horizon lenghts: 30, 45, and 60 min; time frame: 6 a.m.–10 p.m.

	MAPE (%)			WAPE (%)			SMAPE (%)		
Horizon (minutes)	30	45	60	30	45	60	30	45	60
B1	8.3	9.1	8.3	8.3	8.8	8.3	4.1	4.3	4.1
B2	18.8	17.6	18.3	19.4	19.4	20.0	8.7	8.5	8.9
B3	10.1	9.6	9.9	9.9	8.9	9.8	4.9	4.6	4.9
B4	8.1	8.0	7.8	8.0	7.8	7.9	3.8	3.7	3.8
B5	11.7	12.5	11.8	11.3	11.4	11.2	5.4	5.6	5.4

Table 9. Improvement in terms of prediction error reduction of the Artificial Neural Network model over naive approaches for a 30-min prediction horizon; time frame: 6 a.m–10 p.m.

Building	NAIVE 1			NAIVE 2		
	MAPE (%)	WAPE (%)	SMAPE (%)	MAPE (%)	WAPE (%)	SMAPE (%)
B1	**12.8**	15.0	5.9	**2.2**	1.8	1.9
B2	3.8	4.3	−0.7	13.4	14.7	4.3
B3	**2.1**	1.1	0.3	**13.8**	13.9	5.5
B4	**1.8**	1.4	1.2	**14.4**	13.7	6.0
B5	1.3	2.5	0.9	11.9	13.1	4.5

Table 10. Improvement in terms of prediction error reduction of the artificial neural network model over naive approaches for a 45-min prediction horizon; time frame: 6 a.m–10 p.m.

Building	NAIVE 1			NAIVE 2		
	MAPE (%)	WAPE (%)	SMAPE (%)	MAPE (%)	WAPE (%)	SMAPE (%)
B1	**7.7**	10.1	3.7	**2.9**	2.0	1.3
B2	**4.9**	8.7	1.2	**13.5**	14.5	4.3
B3	**1.0**	4.1	0.8	**12.0**	12.7	4.9
B4	1.4	3.5	0.8	10.0	9.2	4.1
B5	2.4	5.1	1.3	8.7	7.2	2.5

Table 11. Improvement in terms of prediction error reduction of the artificial neural network model over naive approaches for a 60-min prediction horizon; time frame: 6 a.m–10 p.m.

Building	NAIVE 1			NAIVE 2		
	MAPE (%)	WAPE (%)	SMAPE (%)	MAPE (%)	WAPE (%)	SMAPE (%)
B1	**12.8**	15.0	5.9	**2.2**	1.8	1.7
B2	**6.1**	8.7	1.4	**11.8**	12.8	3.8
B3	**2.7**	6.0	1.6	**13.4**	13.3	5.2
B4	**1.5**	3.0	0.8	**10.1**	9.2	4.4
B5	**2.0**	3.5	0.9	**5.4**	5.6	2.0

Considering the performance of the ANN model on longer prediction horizons, we notice that errors for 30, 45, and 60 min are lower than five minute forecasts. If the energy provider operations can benefit from a 5-min prediction horizon, this is proved to be a more challenging task than 30–60 min horizons. Changing the prediction horizon from 30 to 60 min does not affect significantly the error rates.

Statistical significance of the difference in the values in Tables 9–11 have been computed with the Student t-test for the MAPE metric. Bold values indicate that the t-test has been passed for a p-value of 0.05. We notice that improvements reported for the 60-min horizon are statistically significant for all buildings, and they would be significant also for lower p-values, such as 0.01. On the contrary, the shorter the horizon, the less significant the improvements: for the 45-min horizon, only three buildings out of five pass the t-test with p-value 0.05, whereas all 5 would pass the t-test for p-value 0.10. For the 30-min horizon, again three out of five buildings pass the t-test with p-value 0.05, however the remaining two buildings would require a p-value larger than 0.10.

Finally, we present a sample of how the proposed work could be exploited in a real-world district-heating distribution network. The goal of the solution is to be a tool directly available to energy experts and energy-providing company management to support their decisions in the strategic design and expansion plan of the district heating networks, besides the day-to-day operations. In Figure 3, a district map of building consumption levels with real experimental data is presented for two different days, December 8th in Figure 3a and December 9th in Figure 3b. Each of the five consumption levels is a percentage range of the top consumption among the overall population for that day, and it is associated with a color from green (0–20%) to red (80–100%). The aim is to analyze the total consumption of each neighborhood in the district under exam, hence grouping together the single buildings but keeping a fine-grained spatial resolution. While such a map can be produced for any prediction horizon, the presented results describe a whole day consumption, considering that December 8th is a national holiday. Specific patterns emerge for residential versus office/public buildings; for instance, the two neighborhoods at the top-centre of the map had opposite behaviors: the lower one is green (0–20%) during the national holiday (Figure 3a) and red (80–100%) on the next day (Figure 3b), and indeed it is a neighborhood of office buildings. On the contrary, the upper neighborhood is red (80–100%) during the national holiday (Figure 3a), and light green (20–40%) during the next day, hence showing the behavior of a residential building with most people working out of home.

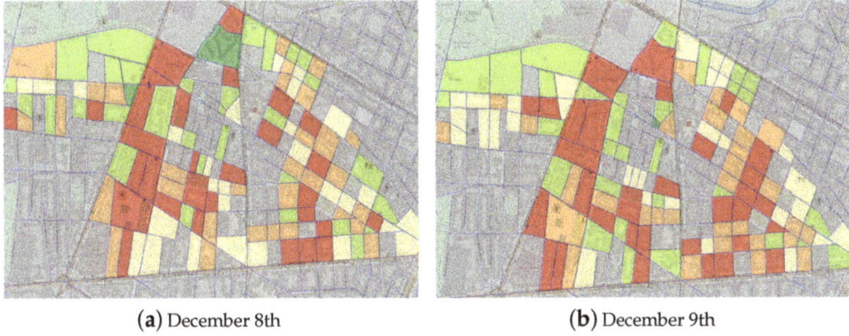

(**a**) December 8th　　　　　　(**b**) December 9th

Figure 3. Total consumption of the buildings in the monitored district, grouped by neighborhood.

The map reported in Figure 4 similarly provides the average consumption per m^3, hence being useful to estimate the energy efficiency of the buildings. It is interesting to note that during the national holiday, even if the previously analyzed office neighborhood was green (0–20%) in total consumption (Figure 3a), it is orange (60–80%) in average consumption (Figure 4a), hence being poorly efficient when using low total energy. On the contrary, it has a very high energy efficiency on December 9th (Figure 4b) with a light-green level (20–40%), when it is a top consuming neighborhood (red level 80–100%, (Figure 3b).

(**a**) December 8th　　　　　　(**b**) December 9th

Figure 4. Average consumption per m^3 of the buildings in the monitored district, grouped by neighborhood.

7. Conclusions and Future Works

In this paper we presented the SPEC (Scalable Predictor of PowEr Consumption) engine to predict power consumption at large scale. SPEC addresses the full analytic stack and exploits a data stream approach over sliding time windows to train a prediction model tailored to each building. It integrates a wide range of algorithms to perform the prediction task both in terms of regression and classification.

Experimental results, achieved on real data, demonstrate the potential of the proposed approach in generating accurate prediction model. On average by considering both time frames (complete day versus only few hours, i.e., transient and steady state phases versus only steady state) the ANN model outperforms the others with a lower prediction error. These results are promising and demonstrate the potential of the proposed methodology in addressing the cumbersome task of predicting power consumption over a sliding window.

Currently, we are extending the current version of the architecture towards a cross-building model to perform more accurate fine grained value predictions. Furthermore, we are working on enriching the prediction models with physical model knowledge, to yield better performance in correspondence of power consumption peaks. Furthermore, we are tailoring the SPEC engine to other energy-related applications, such as electricity applications.

Author Contributions: All authors contributed equally to this work.

Funding: This work has been partially funded by the EU under the H2020 EnABLES project, Grant Agreement n. 730957, and the SmartData@Polito center for Data Science and Big Data technologies, Politecnico di Torino, Italy.

References

1. Borthakur, D. The Hadoop distributed file system: Architecture and design. *Hadoop Proj.* **2007**, *11*, 21.
2. Zaharia, M.; Chowdhury, M.; Das, T.; Dave, A.; Ma, J.; McCauley, M.; Franklin, M.J.; Shenker, S.; Stoica, I. Resilient Distributed Datasets: A Fault-tolerant Abstraction for In-memory Cluster Computing. In Proceedings of the 9th USENIX conference on Networked Systems Design and Implementation, San Jose, CA, USA, 25–27 April 2012.
3. Apiletti, D.; Baralis, E.; Cerquitelli, T.; Chiusano, S.; Grimaudo, L. SeaRum: A Cloud-Based Service for Association Rule mining. In Proceedings of the 12th IEEE International Conference on Trust, Security and Privacy in Computing and Communications, Melbourne, VIC, Australia, 16–18 July 2013.
4. Dean, J.; Ghemawat, S. MapReduce: Simplified data processing on large clusters. OSDI '04 Technical Program OSDI'04; 3 October 2004. Available online: https://static.googleusercontent.com/media/research.google.com/en//archive/mapreduce-osdi04.pdf (accessed on 24 May 2019).
5. Mahout. 2019. Available online: https://mahout.apache.org/ (accessed on 29 May 2019).
6. Acquaviva, A.; Apiletti, D.; Attanasio, A.; Baralis, E.; Bottaccioli, L.; Castagnetti, F.B.; Cerquitelli, T.; Chiusano, S.; Macii, E.; Martellacci, D.; et al. Energy Signature Analysis: Knowledge at Your Fingertips. In Proceedings of the 2015 IEEE International Congress on Big Data, New York, NY, USA, 27 June–2 July 2015; pp. 543–550.
7. Attanasio, A.; Cerquitelli, T.; Chiusano, S. Supporting the analysis of urban data through NoSQL technologies. In Proceedings of the 7th International Conference on Information, Intelligence, Systems & Applications, IISA 2016, Chalkidiki, Greece, 13–15 July 2016; pp. 1–6.
8. The Apache Spark Scalable Machine Learning Library. 2019. Available online: https://spark.apache.org/mllib/ (accessed on 29 May 2019).
9. Apiletti, D.; Baralis, E.; Cerquitelli, T.; Garza, P.; Giordano, D.; Mellia, M.; Venturini, L. SeLINA: A self-learning insightful network analyzer. *IEEE Trans. Netw. Serv. Manag.* **2016**, *13*, 696–710. [CrossRef]
10. Xiao, X.; Attanasio, A.; Chiusano, S.; Cerquitelli, T. Twitter data laid almost bare: An insightful exploratory analyser. *Expert Syst. Appl.* **2017**, *90*, 501–517. [CrossRef]
11. Filippin, C.; Larsen, S.F. Analysis of energy consumption patterns in multi-family housing in a moderate cold climate. *Energy Policy* **2009**, *37*, 3489–3501. [CrossRef]

12. Depuru, S.; Wang, L.; Devabhaktuni, V.; Nelapati, P. A hybrid neural network model and encoding technique for enhanced classification of energy consumption data. In Proceedings of the 2011 IEEE Power and Energy Society General Meeting, San Diego, CA, USA, 24–29 July 2011.

13. Di Corso, E.; Cerquitelli, T.; Apiletti, D. Metatech: Meteorological data analysis for thermal energy characterization by means of self-learning transparent models. *Energies* **2018**, *11*, 1336. [CrossRef]

14. Ardakanian, O.; Koochakzadeh, N.; Singh, R.P.; Golab, L.; Keshav, S. Computing Electricity Consumption Profiles from Household Smart Meter Data. In Proceedings of the Workshops of the EDBT/ICDT 2014 Joint Conference (EDBT/ICDT 2014), Athens, Greece, 28 March2014; pp. 140–147.

15. Wijayasekara, D.; Linda, O.; Manic, M.; Rieger, C. Mining Building Energy Management System Data Using Fuzzy Anomaly Detection and Linguistic Descriptions. *Ind. Inf. IEEE Trans.* **2014**, *10*, 1829–1840. [CrossRef]

16. van der Veen, J.; van der Waaij, B.; Meijer, R. Sensor Data Storage Performance: SQL or NoSQL, Physical or Virtual. In Proceedings of the 2012 IEEE Fifth International Conference on Cloud Computing, Honolulu, HI, USA, 24–29 June 2012; pp. 431–438. [CrossRef]

17. Zulkernine, F.H.; Martin, P.; Zou, Y.; Bauer, M.; Gwadry-Sridhar, F.; Aboulnaga, A. Towards Cloud-Based Analytics-as-a-Service (CLAaaS) for Big Data Analytics in the Cloud. In Proceedings of the 2013 IEEE International Congress on Big Data, Santa Clara, CA, USA, 27 June–2 July 2013.

18. Anjos, D.; Carreira, P.; Francisco, A.P. Real-Time Integration of Building Energy Data. In Proceedings of the 2014 IEEE International Congress on Big Data, Anchorage, AK, USA, 27 June–2 July 2014.

19. Wang, C.; de Groot, M.; Marendy, P. A Service-Oriented System for Optimizing Residential Energy Use. In Proceedings of the 2009 IEEE International Conference on Web Services, Los Angeles, CA, USA, 6–10 July 2009.

20. Lu, S.; Liu, Y.; Meng, D. Towards a Collaborative Simulation Platform for Renewable Energy Systems. In Proceedings of the 2013 IEEE Ninth World Congress on Services, Santa Clara, CA, USA, 28 June– 3 July 2013.

21. Rios, L.G.; Diguez, J.A.I. Big Data Infrastructure for analyzing data generated by Wireless Sensor Networks. In Proceedings of the 2014 IEEE International Congress on Big Data, Anchorage, AK, USA, 27 June– 2 July 2014.

22. Song, M.; Choi, J. Demand-oriented Energy Big Data Services using Hadoop-based Large-scale Distributed System Platform for District Heating. In Proceedings of the 2018 International Conference on Big Data and Computing, Shenzhen, China, 28–30 April 2018; pp. 10–13.

23. Apiletti, D.; Baralis, E.; Cerquitelli, T.; Garza, P.; Venturini, L. SaFe-NeC: A scalable and flexible system for network data characterization. In Proceedings of the NOMS 2016—2016 IEEE/IFIP Network Operations and Management Symposium, Istanbul, Turkey, 25–29 April 2016; pp. 812–816.

24. Ferreira, J.; Afonso, J.; Monteiro, V.; Afonso, J. An Energy Management Platform for Public Buildings. *Electronics* **2018**, *7*, 294. [CrossRef]

25. Brundu, F.G.; Patti, E.; Osello, A.; Del Giudice, M.; Rapetti, N.; Krylovskiy, A.; Jahn, M.; Verda, V.; Guelpa, E.; Rietto, L.; et al. IoT Software Infrastructure for Energy Management and Simulation in Smart Cities. *IEEE Trans. Ind. Inf.* **2017**, *13*, 832–840. [CrossRef]

26. Ma, J.; Cheng, J.C. Estimation of the building energy use intensity in the urban scale by integrating GIS and big data technology. *Appl. Energy* **2016**, *183*, 182–192. [CrossRef]

27. Moghadam, S.T.; Toniolo, J.; Mutani, G.; Lombardi, P. A GIS-statistical approach for assessing built environment energy use at urban scale. *Sustain. Cities Soc.* **2018**, *37*, 70–84. [CrossRef]

28. Fayaz, M.; Kim, D. A Prediction Methodology of Energy Consumption Based on Deep Extreme Learning Machine and Comparative Analysis in Residential Buildings. *Electronics* **2018**, *7*, 222. [CrossRef]

29. Geysen, D.; Somer, O.D.; Johansson, C.; Brage, J.; Vanhoudt, D. Operational thermal load forecasting in district heating networks using machine learning and expert advice. *Energy Build.* **2018**, *162*, 144–153. [CrossRef]

30. Ahmad, T.; Chen, H. Potential of three variant machine-learning models for forecasting district level medium-term and long-term energy demand in smart grid environment. *Energy* **2018**, *160*, 1008–1020. [CrossRef]

31. Koschwitz, D.; Frisch, J.; van Treeck, C. Data-driven heating and cooling load predictions for non-residential buildings based on support vector machine regression and NARX Recurrent Neural Network: A comparative study on district scale. *Energy* **2018**, *165*, 134–142. [CrossRef]

32. Xue, P.; Zhou, Z.; Fang, X.; Chen, X.; Liu, L.; Liu, Y.; Liu, J. Fault detection and operation optimization in district heating substations based on data mining techniques. *Appl. Energy* **2017**, *205*, 926–940. [CrossRef]

33. Suryanarayana, G.; Lago, J.; Geysen, D.; Aleksiejuk, P.; Johansson, C. Thermal load forecasting in district heating networks using deep learning and advanced feature selection methods. *Energy* **2018**, *157*, 141–149. [CrossRef]

34. Acquaviva, A.; Apiletti, D.; Attanasio, A.; Baralis, E.; Castagnetti, F.B.; Cerquitelli, T.; Chiusano, S.; Macii, E.; Martellacci, D.; Patti, E. Enhancing Energy Awareness Through the Analysis of Thermal Energy Consumption. In *Proceedings of the Workshops of the EDBT/ICDT 2015*; Fischer, P.M., Alonso, G., Arenas, M., Geerts, F., Eds.; CEUR-WS.org: Darmstadt, Germany, 2015; Volume 1330, pp. 64–71.

35. Acquaviva, A.; Apiletti, D.; Attanasio, A.; Baralis, E.; Bottaccioli, L.; Cerquitelli, T.; Chiusano, S.; Macii, E.; Patti, E. Forecasting Heating Consumption in Buildings: A Scalable Full-Stack Distributed Engine. *Electronics* **2019**, *8*, 491. [CrossRef]

36. Attanasio, A.; Savino Piscitelli, M.; Chiusano, S.; Capozzoli, A.; Cerquitelli, T. Towards an Automated, Fast and Interpretable Estimation Model of Heating Energy Demand: A Data-Driven Approach Exploiting Building Energy Certificates. *Energies* **2019**, *12*. [CrossRef]

37. Cerquitelli, T.; Corso, E.D.; Proto, S.; Capozzoli, A.; Bellotti, F.; Cassese, M.G.; Baralis, E.; Mellia, M.; Casagrande, S.; Tamburini, M. Exploring energy performance certificates through visualization. In Proceedings of the Workshops of the EDBT/ICDT 2019 Joint Conference EDBT/ICDT 2019, Lisbon, Portugal, 26 March 2019.

38. Chodorow, K.; Dirolf, M. *MongoDB: The Definitive Guide*; O'Reilly Media: Newton, MA, USA, 2010.

39. Cerquitelli, T.; Corso, E.D. Characterizing Thermal Energy Consumption through Exploratory Data mining Algorithms. In Proceedings of the Workshops of the EDBT/ICDT 2016 Joint Conference EDBT/ICDT Workshops 2016, Bordeaux, France, 15 March 2016.

40. Cannistraro, G.; Cannistraro, M.; Cannistraro, A.; Galvagno, A.; Trovato, G. Evaluation on the convenience of a citizen service district heating for residential use. A new scenario introduced by high efficiency energy systems. *Int. J. Heat Technol.* **2015**, *33*. [CrossRef]

41. Cerquitelli, T. Predicting Large Scale Fine Grain Energy Consumption. *Energy Procedia* **2017**, *111*, 1079–1088. [CrossRef]

42. Cannistraro, G.; Cannistraro, M.; Cannistraro, A.; Galvagno, A.; Trovato, G. Technical and economic evaluations about the integration of co-Trigeneration systems in the dairy industry. *Int. J. Heat Technol.* **2016**, *34*, 332–336. [CrossRef]

43. Weather Underground web service. 2019. Available online: https://www.wunderground.com/ (accessed on 29 May 2019).

44. Dagmar Blatná. Outlier in Regression. 2019. Available online: www.laser.uni-erlangen.de (accessed on 29 May 2019).

MDPI
St. Alban-Anlage 66
4052 Basel
Switzerland
Tel. +41 61 683 77 34
Fax +41 61 302 89 18
www.mdpi.com

Energies Editorial Office
E-mail: energies@mdpi.com
www.mdpi.com/journal/energies